Encyclopedia of Electronic Circuits

Encyclopedia of Electronic Circuits

Leo G. Sands
and
Donald R. Mackenroth

Parker Publishing Company, Inc.
West Nyack, New York

Library of Congress Cataloging in Publication Data

Sands, Leo G
  Encyclopedia of electronic circuits.

  1. Electronic circuits--Dictionaries. I. Mackenroth, Donald R., joint author. II. Title.
TK7867.S26   621.3815'3'03   74-26516
ISBN 0-13-275404-5

Printed in the United States of America

HOW TO USE THIS BOOK

Countless electronic circuits have been devised and utilized since the invention of the electron tube by Dr. Lee De Forest in 1906. Many of these circuits were original in concept and many reflected another way of doing the same thing that had previously been done. It's something like music—an arranger can improvise on the original composer's score.

All of these various circuits should be preserved for future reference. Unfortunately, no one has access to all the circuits that have been devised, nor is there room in a volume of this size to show more than a practical assortment of various approaches.

New circuits are devised and old ones revamped by engineers for various reasons, such as cost reduction, improved performance, and patent rights, as well as because of the "NIH" ("Not Invented Here") factor.

Electronic circuits are simple when viewed as individual stages. An apparently complex electronic system consists of a myriad of simple stages. And the obvious complexity of subsystems is similarly simplified when integrated circuit (IC) modules are used.

Many books containing electronic circuits have been published, often confined to special areas. None, including this one, contains all circuits. However, in this book the authors have included as many circuits as space permits with a reasonable amount of explanatory and application information. Within each part, the terms and circuit names are also alphabetically arranged. Cross references are included to simplify location of any circuit in which you have a particular interest.

There was a time when the experimenter and the professional were concerned with every connection of every component. Many still are, but others are now primarily concerned with the connections of a "packaged circuit" to other packaged circuits and external components.

Nevertheless, it is important to understand the functions of packaged circuits and how they can be utilized, as well as how to utilize "discrete components." This book contains a large variety of discrete-components circuits and numerous packaged circuit configurations.

The values of components and specific reference numbers are not given except in the case of a few circuits—for practical reasons. Countless experimenters and even highly-competent engineers have wired and assembled electronic devices utilizing components specified in a book or magazine article and have been disappointed. There are too many "variables" that must be considered, particularly in transistor circuits. Engineers can calculate the required component values, while technicians and experimenters will often remember the approximate values from experience with other circuits for similar applications. Even the seasoned engineer will arrive at the final values by "cut-and-try" experiments. The experimenter can also save much time and gain valuable experience by doing what technicians and engineers do—"breadboard" the circuit before assembling and wiring a device.

(Breadboarding a circuit consists of securing the components by pushing their leads into a piece of styrofoam board or similar material and then connecting them with clip leads or solder-tacked leads. Next, various values of resistors and capacitors are tried in the circuit until measurements indicate that it functions correctly. An additional time-saving technique is to use resistance and capacitance substitution boxes which can be set to any of numerous values of resistance and capacitance. When the correct value has been found, a resistor or capacitor of the indicated value is placed in the circuit.)

Too, once engineers and experimenters wound their own coils and transformers (some still do) because appropriate factory-made units were not available. While most electronics parts stores do not stock a large variety of coils and transformers, they can order the types that may not be available. The experimenter will find a copy of the J. W. Miller Company catalog a valuable reference. This catalog can be obtained through almost any electronic parts distributor. It is one of the few catalogs known to the authors that lists thousands of different RF coils and transformers which can be purchased in single quantities. Other handy references are the Triad and UTC transformer catalogs that list audio and power transformers and which are also available through parts distributors.

(When it is necessary to wind a coil or RF transformer, coil forms with adjustable ferrite cores can be used. Because of the ferrite core, the number of turns is not critical since inductance can be adjusted over a wide range. Invaluable to the engineer and experimenter is a grid dip meter for measuring the resonant frequency of a coil or coil and capacitor combination.)

In this book, in order to simplify the schematic diagrams, connections to power supplies and batteries are often indicated by an arrowhead marked B+, B-, C-, Vcc, etc. In some cases, the voltage is indicated as +20V, etc., denoting the supply voltage with respect to common ground.

Also, tube, transistor, and diode type numbers are not usually specified because there are thousands of types and because so many of them are electrically interchangeable. Essentially similar tube types often differ only in heater voltage and current requirements and base connections. A tube manual (GE, RCA, Sylvania, etc.) is an essential reference for the engineer and experimenter, as is a transistor manual (GE, Motorola, RCA, etc.). IC (integrated circuit) application data sheets and books (GE, Fairchild, Motorola, National, RCA, Signetics, etc.) are also handy. These manuals and data sheets can usually be obtained at or ordered through electronics parts distributors.

Each section of this book offers practical value to technicians, students, experimenters, and engineers. It is a compendium of theoretical, experimental, and practical electronic circuits, as well as some circuits which now have historical value.

The authors express their appreciation to equipment manufacturers and others for furnishing circuits for reproduction in this book, and to the many distinguished members of The Radio Club of America who devised some of the circuits illustrated.

Leo G. Sands

Donald R. Mackenroth

New York City

Contents

1. Amplifier Circuits *(cont.)*

5. Coupling Circuits *(cont.)*

9. Modulation and Demodulation Circuits *(cont.)*

9. Modulation and Demodulation Circuits ***(cont.)***

12. Signal Generation Circuits *(cont.)*

1

Amplifier Circuits

The electronic amplifier came into being when Dr. Lee De Forest added the grid to the thermionic Fleming diode to create the triode electron tube. Of course, all electron tubes depend upon the "Edison effect" which was discovered much earlier. Without knowing about electrons, Edison learned by experimentation that electric current would flow through a vacuum from the heated filament of an electric lamp to metal foil attached to the outside of the lamp bulb. Later, the screen grid, and still later the suppressor grid, were added to the triode tube to form the tetrode and pentode tube which have higher amplification capability.

The solid state amplifier was created with the invention of the germanium transistor by a trio of Bell Laboratories scientists in 1947. Solid state art was advanced a few years later by the invention of the surface-barrier silicon transistor by William H. Forster and his associates at Philco. Since then, new solid state amplifying devices such as the FET (field effect transistor), MOSFET, tunnel diode, and other types of semiconductors have added further advancement. Discrete semiconductor devices are now facing competition from integrated circuits which contain several transistors, diodes, and resistors within a single component.

Therefore, described and illustrated in this chapter are numerous amplifier circuits utilizing tubes and semiconductors.

Broadband RF Amplifier

Figure 1-1 shows an untuned, resistance-coupled, two-stage RF amplifier circuit that uses an FET in the input stage to provide a high input impedance and a high signal-to-noise ratio. The input signal is fed into the high-impedance gate input of the FET Q1,

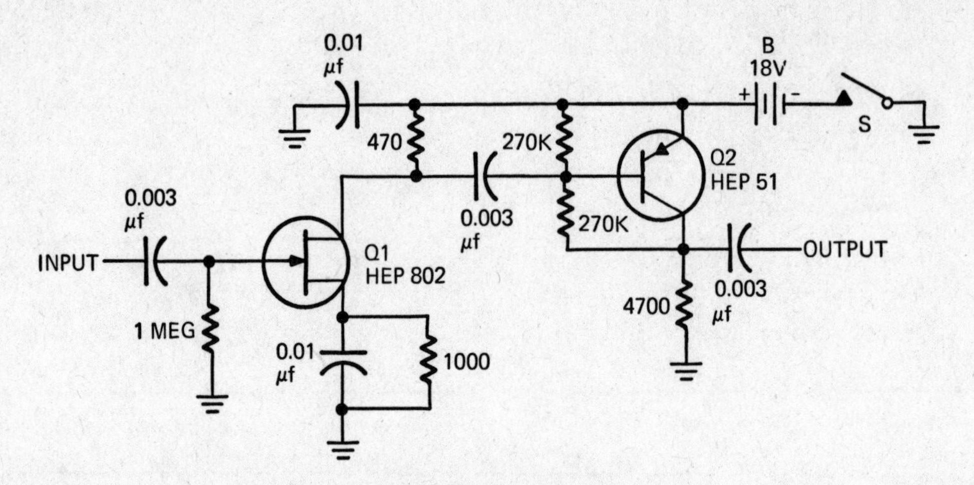

Figure 1-1. Broadband RF amplifier.

which is connected as a common source amplifier. Its amplified output is fed to the base of Q2, which is connected as a common-collector amplifier. The useful frequency range of this RF amplifier extends from below 1 MHz to above 40 MHz. Voltage gain is maximum at 1 MHz, approximately 40 dB (100 times), and drops off to about 26 dB (20 times) at 40 MHz. This RF amplifier is intended to be driven by low-level RF signals. If at 1 MHz input signal level is 10 microvolts, output signal level will be approximately 1000 microvolts. (Circuit courtesy of Motorola Semiconductor Products, Inc.)

Cascode JFET Preamplifier

The circuit of an RF preamplifier, shown in Figure 1-2, illustrates the use of two JFETs in a cascode configuration. The input stage transistor Q1 is connected as a common source amplifier which has a high input impedance and provides about 20 dB of gain. The second transistor Q2, being fed at the source, lowers the impedance seen by the drain of the first transistor. The second transistor operates as a grounded-gate amplifier whose drain load impedance is established by a capacitive voltage divider across output coil L2 which, in combination with the divider C3-C4 and the internal capacitance of the transistor, form a parallel resonant circuit which can be tuned by adjusting the core of L2.

The input signal is transformer coupled to the gate of Q1. The secondary of T1, in combination with the internal capacitance of Q1, forms a parallel resonant circuit which can be tuned by adjusting the ferrite core of T1. The DC drain voltage for the first stage is fed through RF choke coil L1. The signal is coupled to the source of the second stage through capacitor C1, and the source of the second transistor is grounded for DC through L3.

By using resonant circuits only at the input of Q1 and the output of Q2, there is no

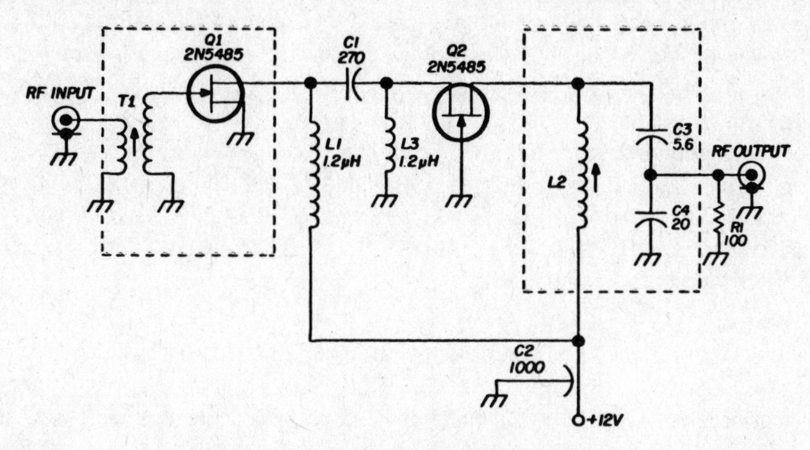

Figure 1-2. Cascode JFET preamplifier.

need for neutralizing the amplifier to prevent oscillation. This particular amplifier was designed for use ahead of a VHF receiver and to match a 50-ohm antenna system at the input and a 50-ohm receiver load at the output. (Illustration courtesy of *Ham Radio*.)

Cascode RF Amplifier

In the tubed cascode RF amplifier circuit shown in Figure 1-3, triode tubes V1 and V2 are essentially connected in series. Current flowing through one tube also flows through the other. The plate voltage of V1 is held constant while the plate current is permitted to vary. This action is like that of a pentode tube, with the advantage that no screen current is required and less noise is introduced.

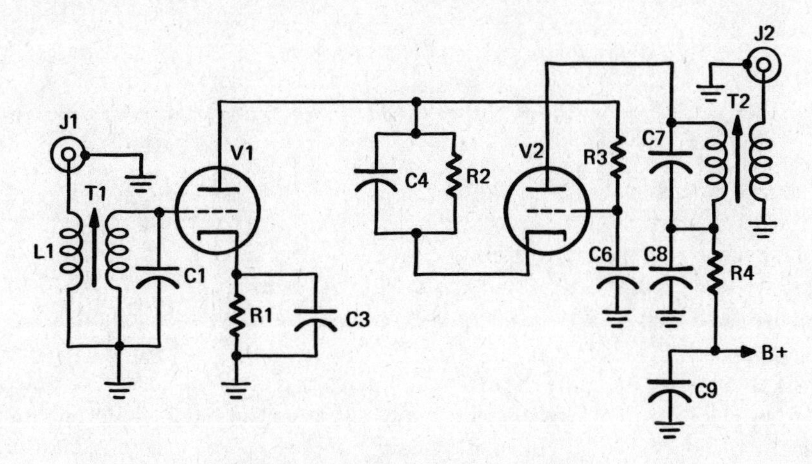

Figure 1-3. Cascode RF amplifier.

Signals from the antenna are fed to the grid of V1 through RF transformer T1 and out from the plate of V2 through T2, both of which are broadly resonant over a band of frequencies. Neutralization is not necessary because the resonant circuits T1-C1 and T2-C7 are not connected to both the input and output of either tube, and because V2 functions as a grounded grid amplifier. The signal voltage across R3 varies the grid-cathode potential of V2. Because the grid is grounded through C6, the signal is effectively applied between the cathode and grid of V2. When a 6ES8 remote cut off dual-triode is used, gain can be varied by applying negative AVC voltage to the grid of V1. Overall gain is typically between 25 dB and 35 dB.

Cathode-Coupled Amplifier

Two triodes are used in the cathode-coupled amplifier whose circuit is shown in Figure 1-4. Changes in cathode current of one tube varies the cathode bias of the other tube. The input signal is fed to the grid of cathode follower V1 and the output is obtained from the plate of grounded grid amplifier V2. Voltage gain is never more than unity, and the phase of the output signal is the same as the input signal (no phase inversion).

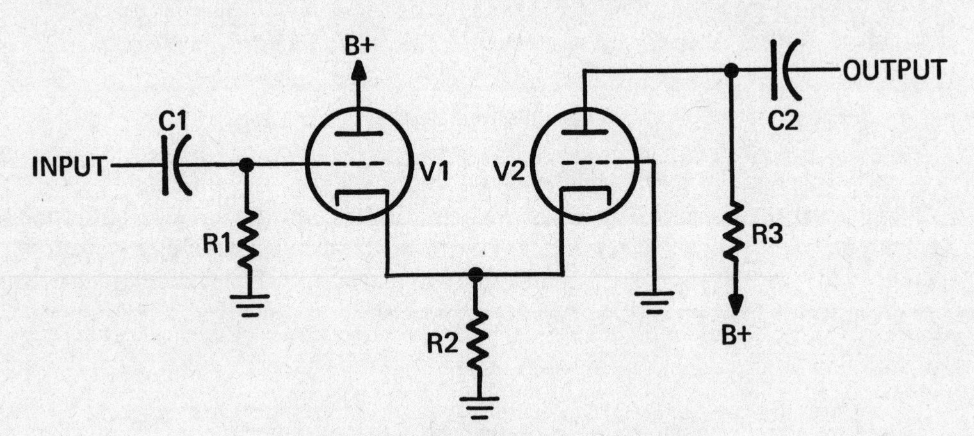

Figure 1-4. Cathode-coupled amplifier.

When the input signal at the grid of V1 is positive-going, the voltage drop across R2 rises, causing the grid of V2 to become more negative with respect to the cathode and, hence, the plate to become more positive. Similarly, when the input signal at the grid of V1 is negative-going, the voltage drop across R2 falls, causing the grid of V2 to become less negative with respect to its cathode and, hence, the plate to become less positive (negative-going).

The frequency response depends upon the values of C1 and C2 and the interelectrode capacitances of V1 and V2. This circuit is often used as an amplitude limiter. Driving V1 into saturation causes V2 to be cut off. Driving V1 to cut off causes V2 to become saturated, depending upon the value of R2 and the characteristics of V1 and V2.

Common-Base Transistor Amplifiers

Figure 1-5A illustrates a PNP transistor connected in a common-base amplifier configuration, and Figure 1-5B shows an NPN transistor connected in the same type of amplifier circuit. The input signal is injected into the emitter-base circuit, and the output signal is taken from the base-collector circuit.

This type of circuit is similar to the electron tube grounded-grid circuit. It has a low input resistance and a high output resistance, and is often used to match a low-impedance circuit to a high-impedance circuit. It can provide a voltage gain of up to 1500, a current gain of less than 1, and a power gain of 20-30 dB. The input and out signals are in phase and the same polarity.

Current gain of a common-base stage is defined as the ratio of the change in collector current to the change in emitter current. The voltage gain is the product of the current gain and the ratio of collector resistance to emitter resistance.

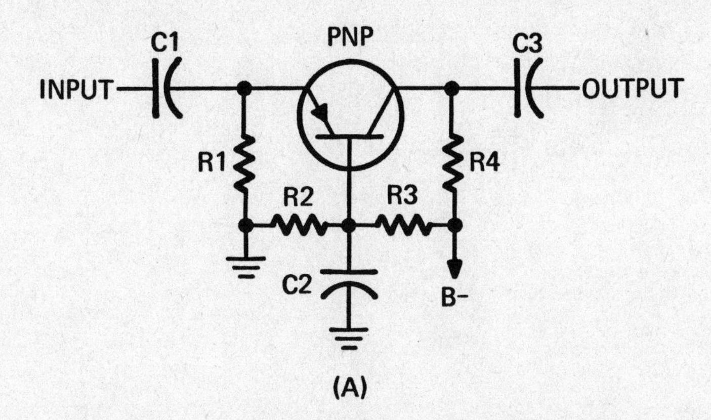

Figure 1-5A. Common base PNP transistor amplifier.

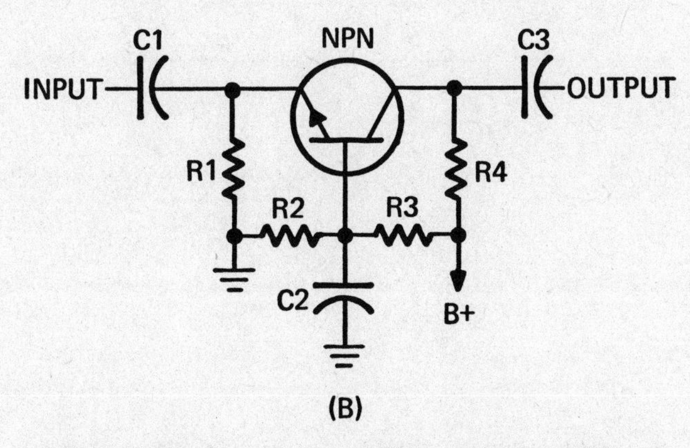

Figure 1-5B. Common base NPN transistor amplifier.

Common-Collector Transistor Amplifier

Figure 1-6 illustrates two configurations for a common-collector bipolar transistor amplifier. A PNP transistor is shown in Figure 1-6A and an NPN transistor is shown in Figure 1-6B.

In the common-collector circuit, the input signal is applied to the base-collector junction, and the output signal is taken from the emitter-collector junction. The common-collector circuit is similar to the electron-tube cathode follower. It has a high input resistance and a low output resistance. It provides good current gain but no voltage gain. The output signal is in phase with the input signal. This type of circuit is used most often for impedance-matching and isolation of stages, and is sometimes called an emitter-follower.

The emitter-to-base current ratio γ (gamma), is a measure of the gain of a common-collector circuit.

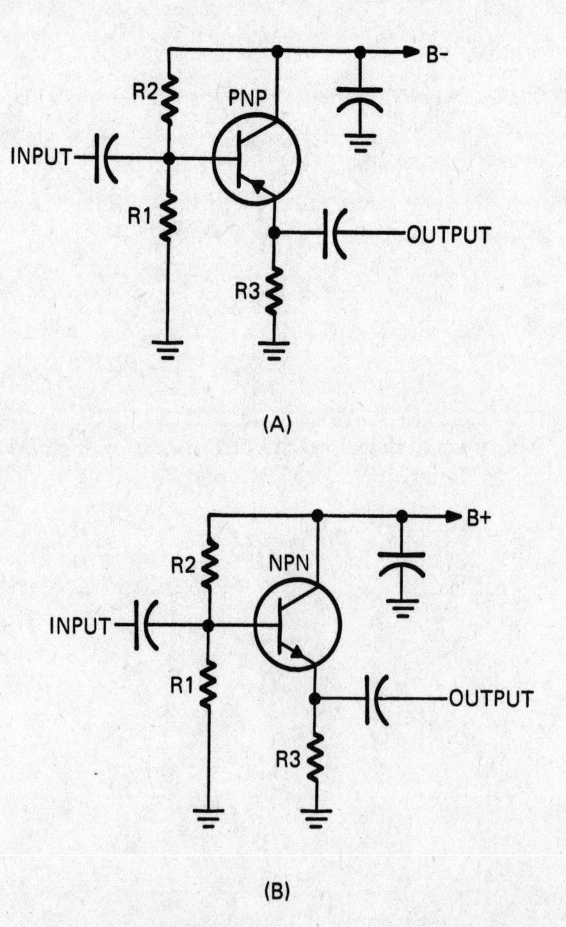

Figure 1-6. Common collector transistor amplifier.

Common-Emitter Transistor Amplifier

Figure 1-7A illustrates a PNP transistor connected in a common-emitter configuration, and Figure 1-7B illustrates a similar circuit using an NPN transistor. In the common-emitter amplifier, the input is fed to the base-emitter junction. This type of circuit is similar to the employment of an electron tube in the grounded-cathode configuration, and is the most commonly used transistor amplifier circuit.

The circuit can provide high power gains, and a current gain of 25-60 is possible with this type of circuit. The output signal is out of phase with the input signal (output polarity is opposite to that of the input).

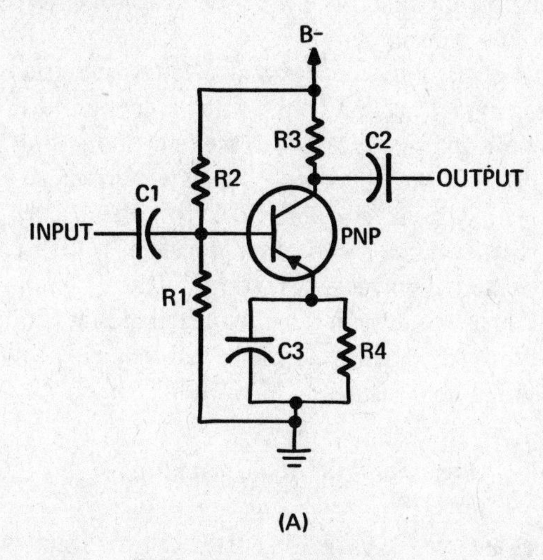

Figure 1-7A. PNP common emitter transistor amplifier.

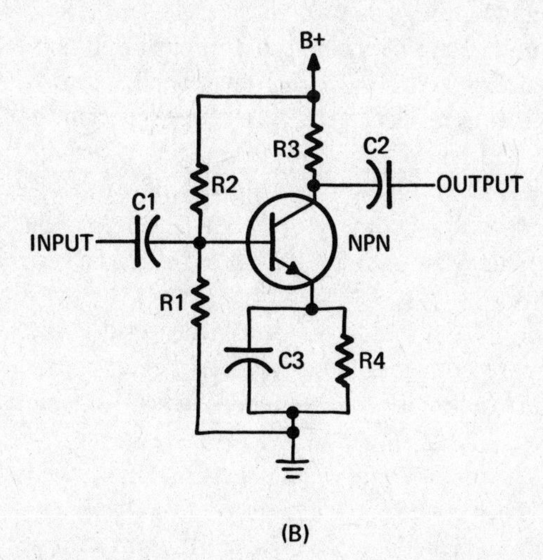

Figure 1-7B. NPN common emitter transistor amplifier.

Compression Amplifier

The function of an audio compression amplifier is to stabilize the level of the AF signal fed to the input of a sound system or radio transmitter as the output level of a microphone varies. When speaking loudly into a nearby microphone, the microphone output level is much greater than when speaking softly into a more distant microphone. The circuit shown in Figure 1-8 employs AGC (automatic gain control) to reduce gain when input level rises, and vice versa.

The microphone is connected to the control grid of V1 (a pentagrid tube) through jack J1. The signal is amplified by V1 and fed to cathode-follower V2A. The output signal level is preset with potentiometer R7 and is fed to J2 which is connected through a cable to the sound system or transmitter.

The amplified signal at the plate of V1 is also fed to the grid of V2B for further amplification and subsequent rectification by the voltage-doubler rectifier CR1-CR2. The resulting DC AGC voltage is fed through R14 to the injection grid of V1. This negative DC voltage controls the gain of V1. As the input signal level rises, AGC voltage becomes more negative and V1 gain is reduced, and vice versa. The attack time (rapidity of compression response) is determined by the time constant of C15 and R14. Operating voltages are obtained from the AC power supply shown at the upper right of the diagram.

When this amplifier is used with a microphone equipped with a PTT (push-to-talk) switch, the PTT line is fed from J1 and J2 and from J2 (through a microphone cable) to the transmit control circuit of a radio transmitter or on-off control circuit of an AF amplifier.

Constant Gain Audio Amplifier

Two push-pull stages are used in the audio amplifier circuit shown in Figure 1-9. The input signal is fed in push-pull to the control grids of pentodes V1 and V2, whose push-pull outputs are coupled through capacitors C3 and C4 to the grids of triodes V3 and V4. The outputs of the triodes are also connected in push-pull and fed out through T2. This amplifier is intended for use in low signal level applications where manual adjustment of signal level is not desired. As can be seen in the diagram, both of the triodes are cathode biased by R15. The gain of the triodes remains constant, but the gain of the pentodes is automatically varied with input signal level. The ground return path from the center tap of the secondary of T1 is through R1, R2, and R4. Shunted across R2 is the plate to cathode path of triode V5 together with R3. V5 acts as a variable resistance in the V1-V2 bias circuit.

Plate voltage for V1, V2, V3, and V4 and screen grid voltage for V1 and V2 is obtained from the half-wave voltage doubler rectifier circuit consisting of C8, C9, and diodes D2 and D3. Ripple filtering is provided by R10 and C10. Note, however, that plate voltage for V5 is not supplied by this rectifier system.

Obviously, triode V5 cannot conduct unless its plate is positive with respect to its cathode. This condition is achieved by obtaining negative DC voltage from the half-wave rectifier circuit which utilizes diode D1. C6, C7, and R8 form a ripple filter. This negative DC voltage is applied to the cathode of V5 through R5. When V5 is conducting, its

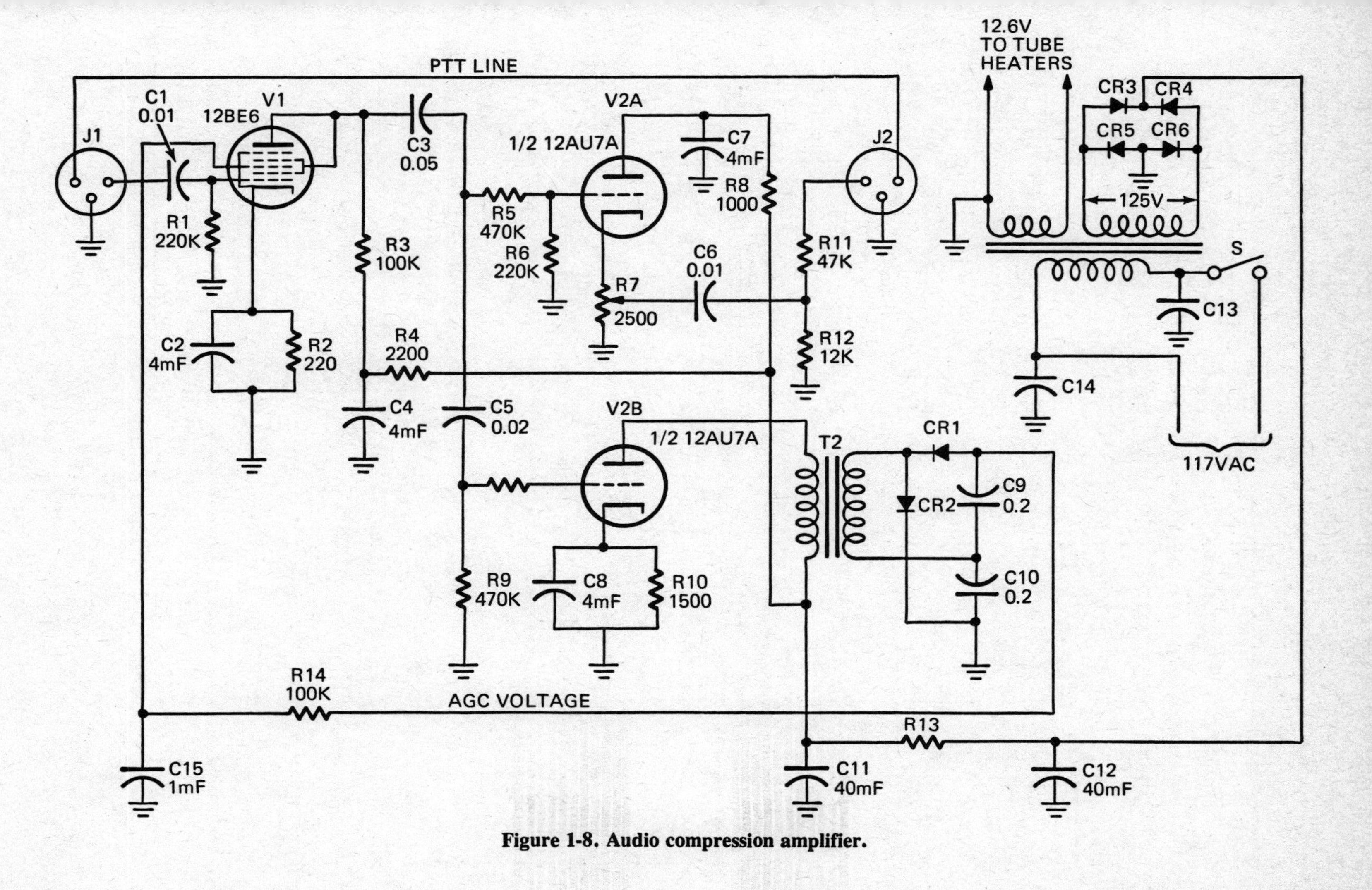

Figure 1-8. Audio compression amplifier.

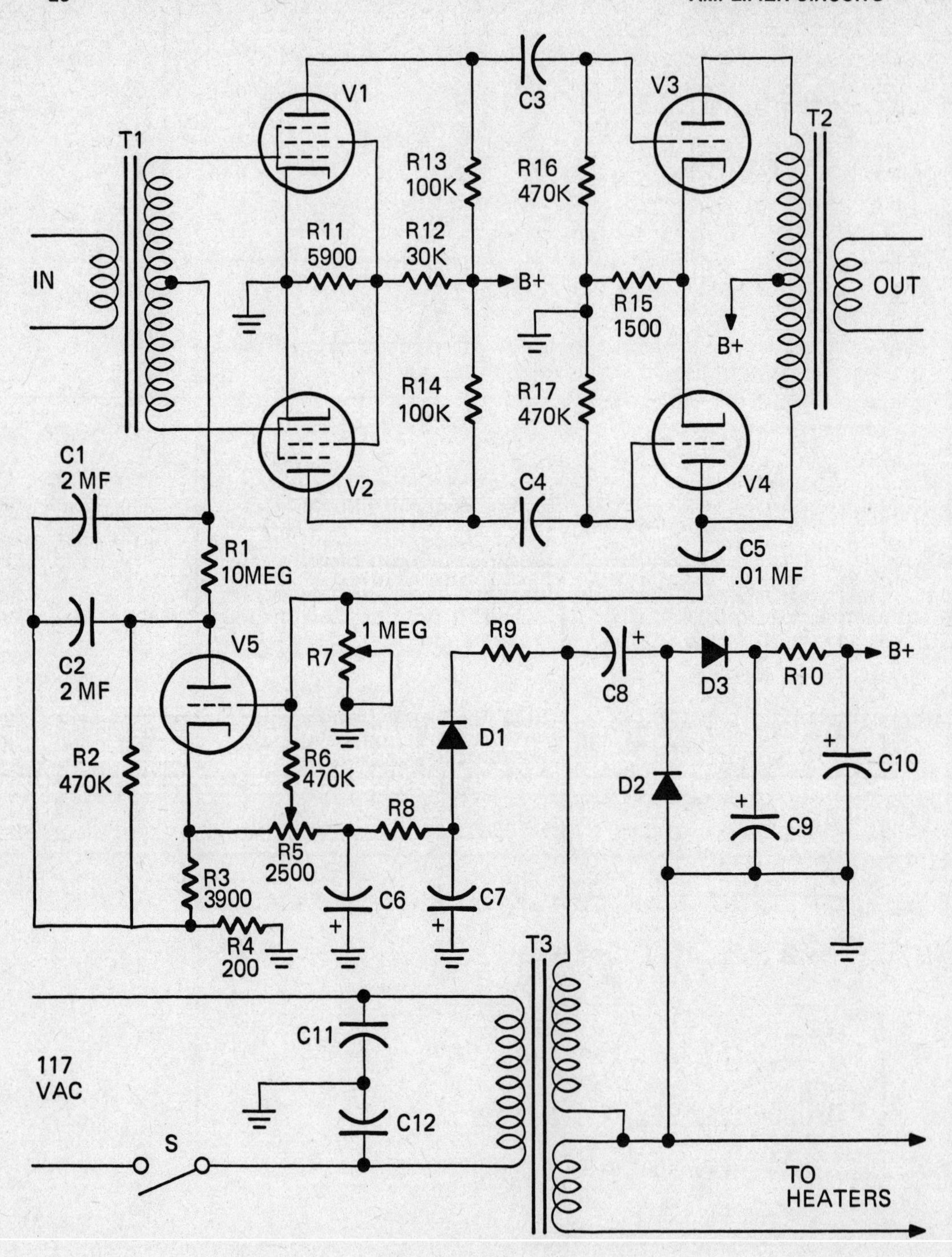

Figure 1-9. Constant gain audio amplifier.

cathode is more negative than its plate, thus making the plate positive with respect to the cathode. The voltage at the junction of R2 and R1 is less negative with respect to ground than the voltage at the junction of R3 and R5.

Triode V5 operates as a grid leak detector. The AF output signal at the plate of V4 is coupled through C5 to the grid of V5. The amount of leak bias is controlled with potentiometer R7. The amount of negative fixed bias is controlled with potentiometer R5. When the signal at the plate of V4 rises, the grid of V5 becomes more negative and the plate-cathode resistance of V5 increases. When the level of the signal at the plate of V4 decreases, the negative bias on the grid of V5 decreases and its plate cathode resistance decreases.

When there is no input signal, there is no signal at the plate of V4, and V5 conduction is at a maximum. If V5 were left out of the circuit, negative bias for V1 and V2 would be obtained from the junction of R3 and R4, which is negative with respect to ground. With V5 in the circuit and conducting, R2 and the plate cathode path of V5, in series with R3, form a voltage divider. V1 and V2 gain is therefore controlled by the conduction of V5.

Current and Voltage Feedback Amplifier

Figure 1-10 shows a circuit that utilizes both voltage and current feedback to compensate for the fact that output impedance is increased by current feedback and decreased by voltage feedback. Therefore, a combination of the two, as shown, may be used to provide constant output impedance when required. Voltage feedback is obtained from the junction of R4 and R5. Current feedback is developed across R2 when it is not bypassed at the signal frequencies.

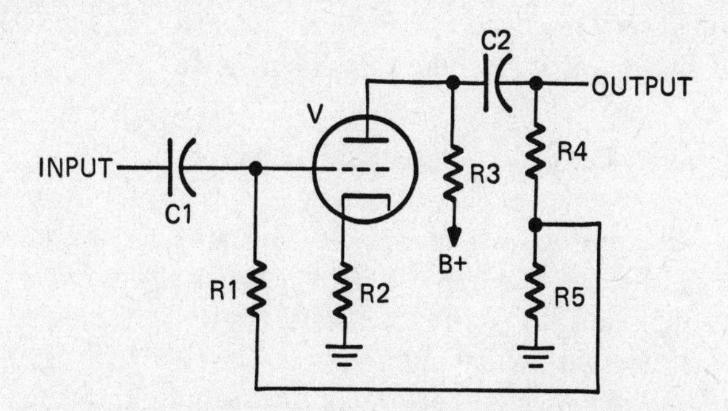

Figure 1-10. Current and voltage feedback amplifier.

Darlington Circuit

The Darlington circuit, shown in Figure 1-11, is a cascaded emitter follower amplifier employing two discrete transistors, or a pair of transistors packaged as a single component. This circuit is widely used because it requires so few components.

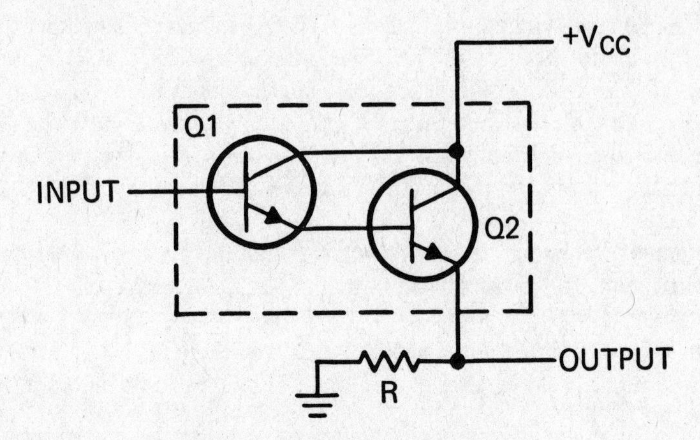

Figure 1-11. The Darlington circuit.

A positive input signal reduces the collector-emitter resistance of Q1 causing Q2 forward bias to rise and the output voltage drop across load resister R to also rise. A less positive input signal increases Q1 collector-emitter resistance, causing Q2 forward bias and output voltage to fall.

Voltage gain is less than unity but current gain is very high and step-down impedance transformation is in the order of 200:1. Since the Darlington circuit is a direct-coupled amplifier with wide frequency response, it is widely used in DC, AF, and RF applications as a power amplifier and/or impedance transformation device.

Darlington Differential Amplifier

Five transistors, all contained within a single integrated circuit, are utilized in the differential amplifier circuit shown in Figure 1-12. The transistors at the left are connected in the Darlington configuration as are the transistors shown at the right. Their gain is controlled by the transistor at the bottom of the circuit. The two input signals are fed in at points X and Y and, as can be seen in the diagram, the output signals are taken from R1 and R6. This type of amplifier circuit has low drift, high common mode rejection and high input impedance.

Degenerative Neutralization

It is well known that regeneration can be used in a detector circuit to greatly increase its sensitivity. It is also well known that degeneration can be employed to stabilize and reduce the gain of an amplifier. In the circuit shown in Figure 1-13, triode V1 is an RF

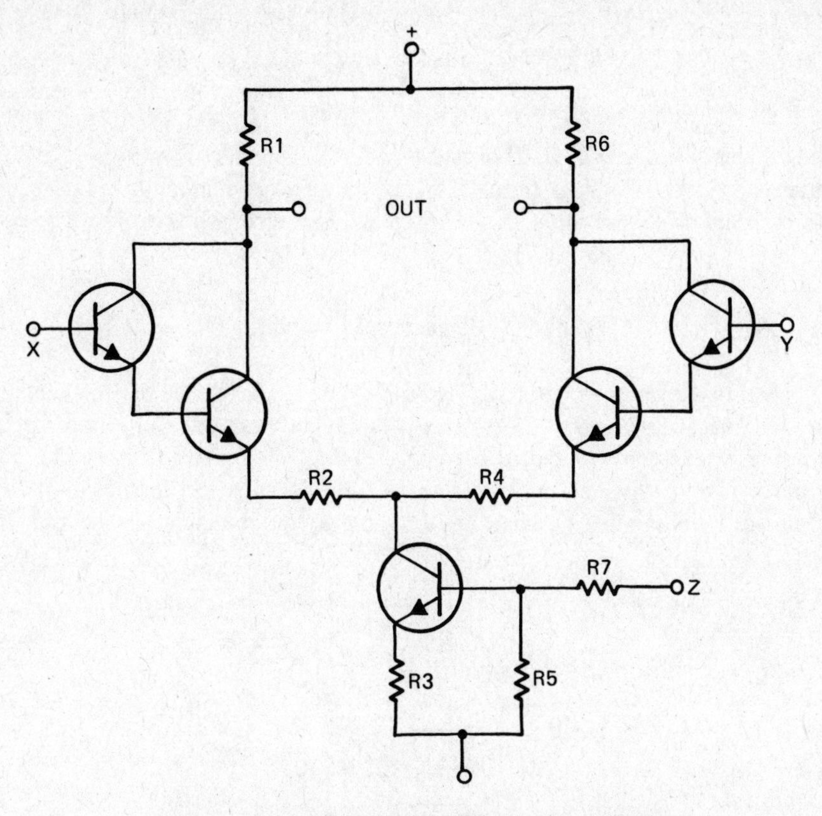

Figure 1-12. The Darlington differential amplifier.

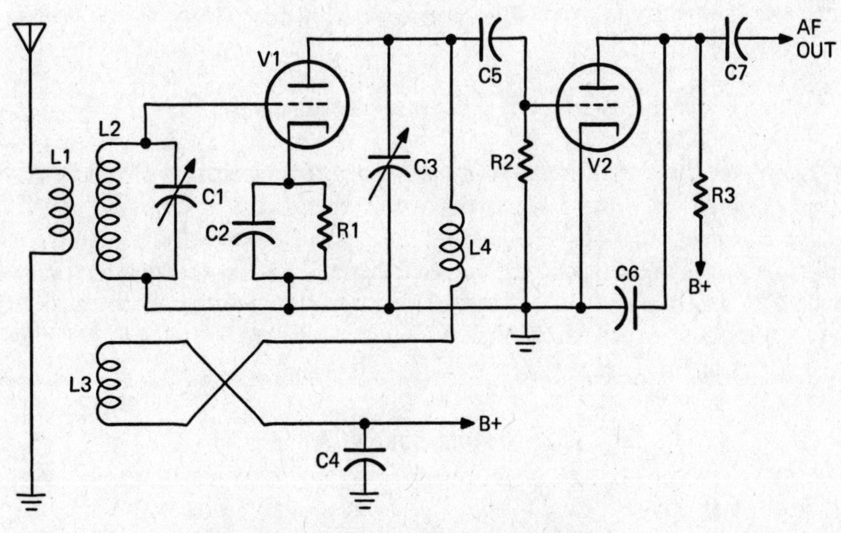

Figure 1-13. Degenerative neutralization.

amplifier and triode V2 is a grid leak detector. So that both the grid circuit and plate circuit of V1 can be tuned to the same frequency without causing the circuit to oscillate, the plate load consists of two coils, L4 and L3, which are connected in series. The combined inductance of these two coils is such that the plate circuit can be tuned over its intended frequency range with variable capacitor C3. L3 is inductively coupled to L2, but is polarized so that the feedback from L3 to L2 will be degenerative instead of regenerative. The amount of degeneration can be controlled by varying the coupling between L3 and L2.

Frequency Compensated Amplifier

Figure 1-14 shows one method for increasing the high frequency gain of an amplifier stage by inserting the inductance L in series with the load resistance RL. The impedance of L increases with frequency and counteracts the effect of shunt capacitance of the tube. This type of circuit is widely used in video amplifiers to offset loss of gain at high frequencies that would otherwise be caused by interelectrode tube capacitances.

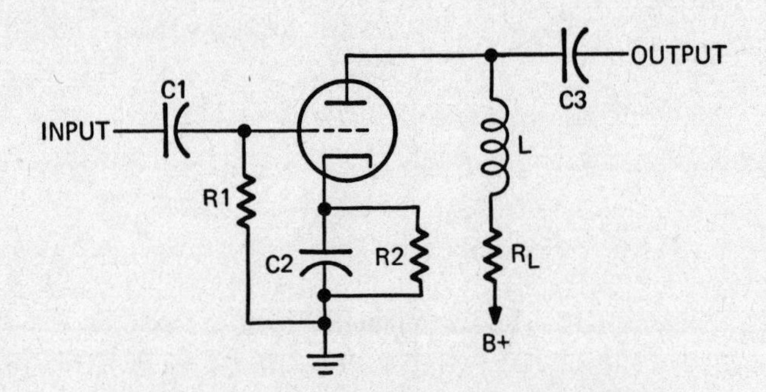

Figure 1-14. Frequency compensated amplifier.

Frequency Compensated Preamplifier

Three PNP transistors are used in the frequency compensated preamplifier circuit, shown in Figure 1-15, which was developed by General Electric Company. By means of the four two-pole, four-position switch, the characteristics of the amplifier can be changed. R9 is the treble equalization potentiometer. Bass boost is controlled with potentiometer R17 which is in series with the 0.7 Henry reactor in the output circuit. When R17 is set for zero resistance, the low frequencies find a low reactance path to the output jack. Conversely, when R17 resistance is maximum, bass response decreases.

Filament-Type Tube AF Amplifier

Figure 1-16 shows a circuit using a filament-type tube as an amplifier. The tube's filament serves as the cathode. Because it is part of the signal circuitry, the filament must

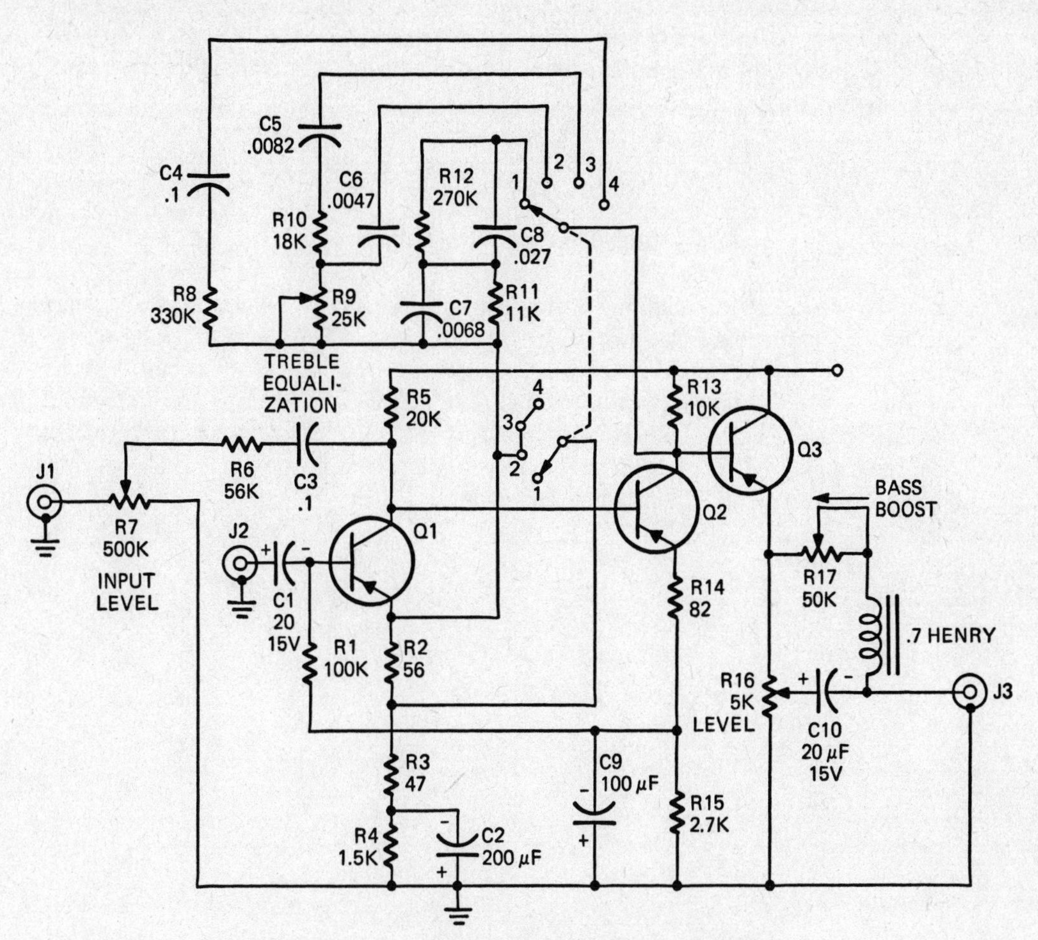

Figure 1-15. Frequency compensated preamplifier.

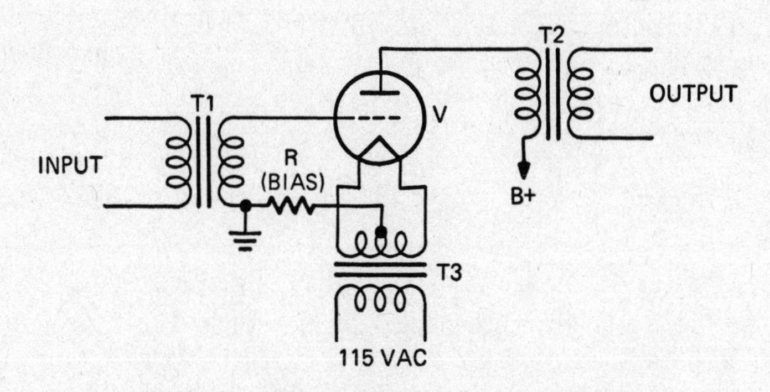

Figure 1-16. Filament-type tube AF amplifier.

be isolated from the source of filament voltage, transformer T3. While not often employed these days, this type of configuration saw wide use in the early days of radio before tubes with indirectly-heated cathodes were common. Use of the center-tapped filament transformer minimizes AC hum. Directly-heated tubes are still used today in high power amplifiers where the hum level is small in comparison to the power level.

Grounded-Grid DC Amplifier

In the grounded-grid amplifier circuit shown in Figure 1-17 the grid is at ground potential, but normally negative with respect to the cathode because of the voltage drop across R1. The grounded grid isolates the input circuit from the output circuit, and the only path for positive feedback is the tiny plate to cathode capacitance. Since the input impedance of this amplifier is low, it is often used as the input stage of a video amplifier whose input signal is fed through a low-impedance transmission line. Impedance match is obtained without the use of a matching transformer.

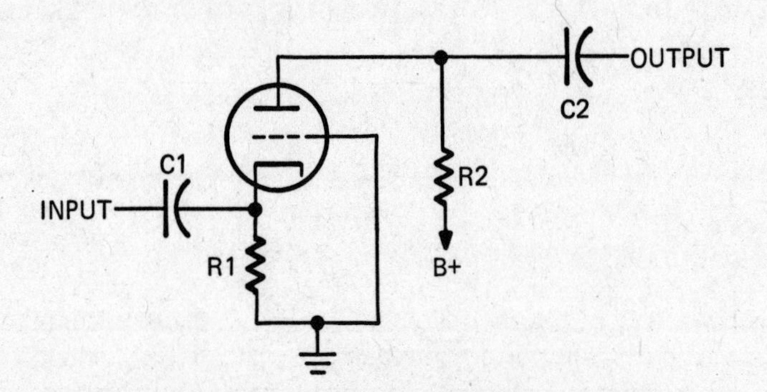

Figure 1-17. Grounded-grid DC amplifier.

The grounded-grid amplifier circuit is also widely used in transmitters as an RF power amplifier which does not require neutralization. The input signal is fed from a driver through a step-down transformer directly or through a low impedance transmission line. The output signal is developed across a parallel resonant circuit in series with the plate (in lieu of R2) or is fed to a pi-network, in which case R2 is replaced by an RF choke coil.

Grounded-Plate Amplifier

Figure 1-18 shows a grounded-plate amplifier circuit. L1 serves as a neutralizing coil, and L2 acts as an interstage coupling autotransformer. This type of circuit is often used in UHF amplifiers to drive a grounded-grid stage (shown as V2) because the lower induced grid noise at high frequencies and the low input capacitance outweigh the disadvantage of poor stability.

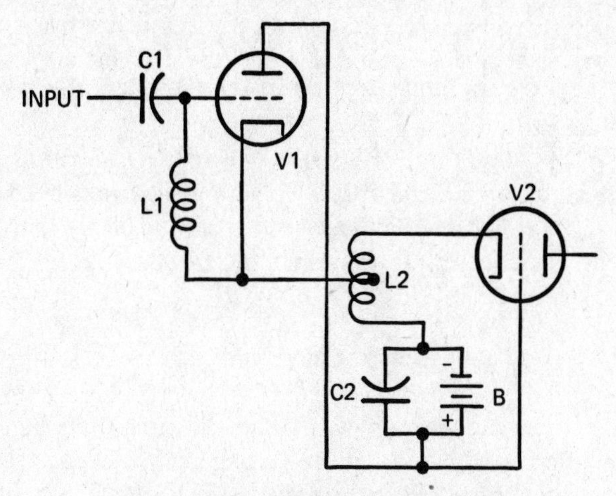

Figure 1-18. Grounded-plate amplifier.

Hazeltine Neutralization

A triode tube RF amplifier circuit employing Hazeltine (or plate) neutralization is shown in Figure 1-19. When both the grid and plate circuits of a triode RF amplifier are tuned to the same frequency, it is highly probable that unwanted oscillation will occur because of the positive feedback (in phase) from the plate to the grid via the Cgp (plate-to-grid interelectrode capacitance) of the tube. (The amplifier functions as a tuned grid-tuned plate oscillator.) Oscillation can be prevented by neutralizing this capacitance. Neutralization is accomplished by feeding back an out-of-phase signal from the plate circuit to the grid through a neutralizing capacitor (Cn).

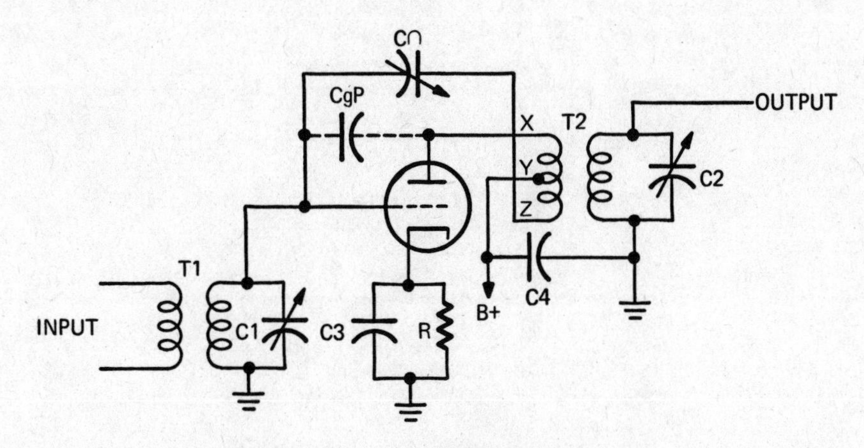

Figure 1-19. Hazeltine neutralization.

As shown in the diagram, the primary winding (plate side) of RF transformer T2 is tapped. There is no RF signal at point Y because of the presence of bypass capacitor C4. However, there is a signal at points X and Z. It is from point Z that the signal is fed back through Cn to the grid. Careful adjustment of Cn (a low-value trimmer capacitor) is required to achieve neutralization.

The invention of this circuit by Professor Alan Hazeltine spurred the growth of the entertainment radio industry in the mid-1920's because it made possible the manufacture of multi-tube TRF (tuned radio frequency) receivers that were highly sensitive and selective. Radio sets using this circuit were known as "neutrodyne" receivers.

IC Power Amplifier

Figure 1-20 is a schematic diagram of an integrated circuit, the Signetics NE 540, used as a power amplifier. Input is to the non-inverting input, while the feedback loop is through R7 to the inverting input. Power capacity is 35 watts. (Illustration courtesy of *Ham Radio*.)

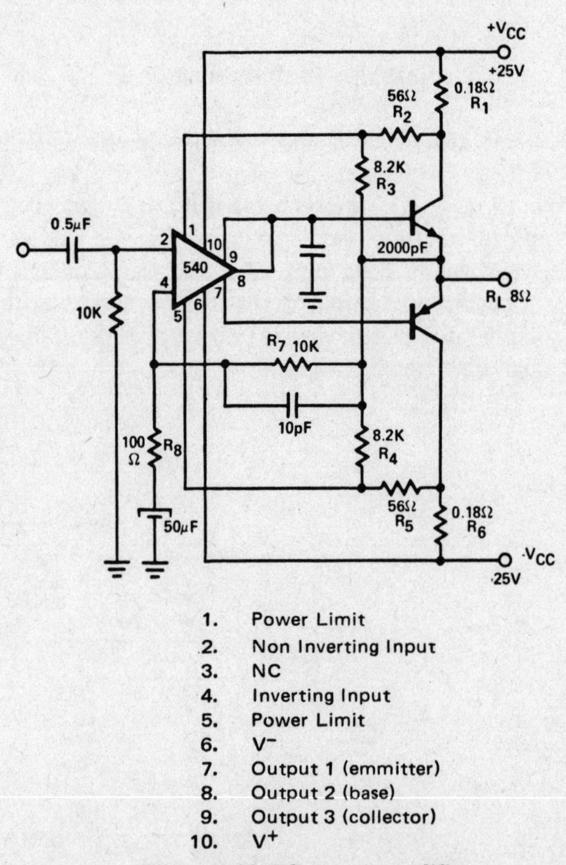

Figure 1-20. IC power amplifier.

IC Stereo Preamplifier

The stereophonic preamplifier shown in Figure 1-21 uses a Fairchild uA 723 dual low-noise operational amplifier in a circuit that gives about 60 dB channel separation (when used with an identical circuit using the other half of the uA 739 chip). This preamplifier is particularly suitable for tape input or other application where a low-noise, high-gain preamplifier is required.

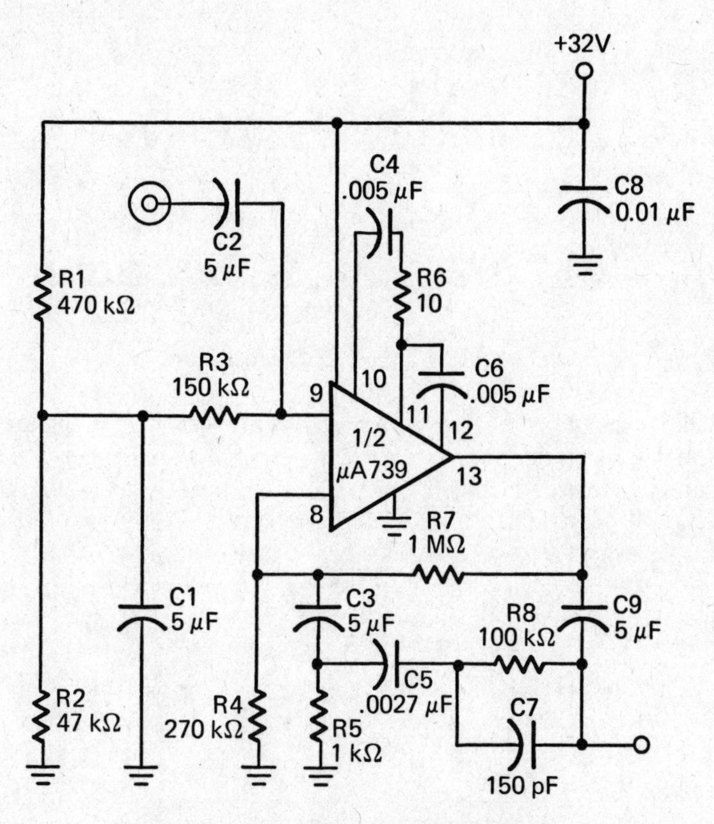

Figure 1-21. IC stereo preamplifier.

Integrated Circuit AF Amplifier

The single IC AF amplifier, whose circuit is shown in Figure 1-22, provides 60 dB of gain over the 10 Hz to 16,000 Hz frequency range. Input impedance is 75,000 ohms and maximum output signal voltage to a 100-ohm load is 7 volts RMS. The Motorola MFC 8040 IC contains seven transistors and two diodes on a single silicon monolithic chip. Since the amplifier can provide 60 dB of gain, a 1-millivolt input signal will produce a 1-volt output signal.

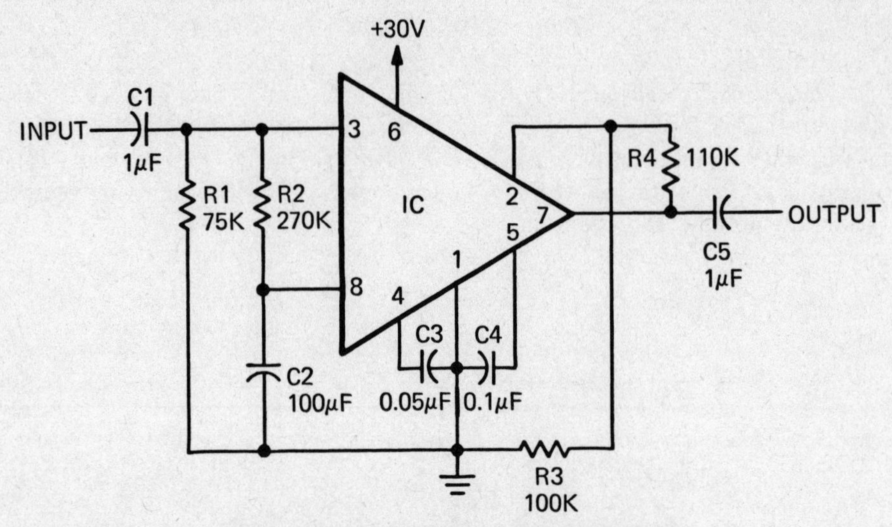

Figure 1-22. Integrated circuit AF amplifier.

Integrated Circuit AF Preamplifier

The audio preamplifier whose circuit is shown in Figure 1-23A provides a voltage gain of 100 times within the frequency range 30-30,000 Hz. A single type PA230 integrated circuit (IC) is used to provide the gain. The input signal is fed through a 1 microfarad capacitor to terminal 12 of the IC. The output signal is taken from terminal 7 of the IC and is developed across the 20,000-ohm volume control.

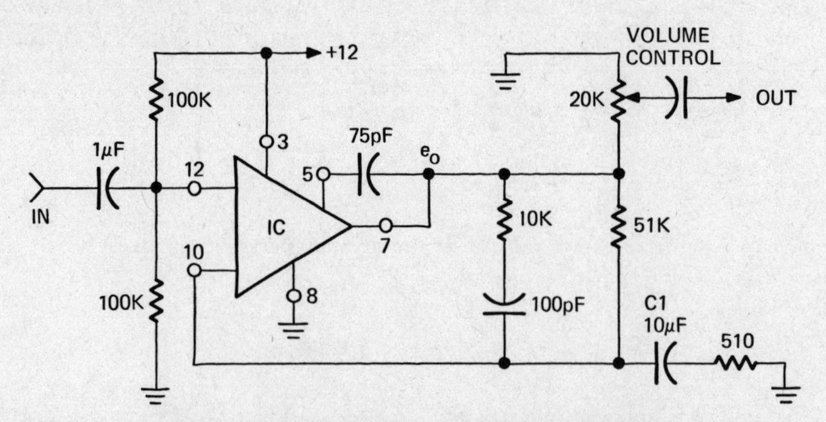

Figure 1-23A. Integrated circuit AF preamplifier.

The internal circuitry of the IC is shown in Figure 1-23B. The IC contains eight NPN transistors, three diodes, and a number of resistors. (Illustration courtesy of General Electric Company.)

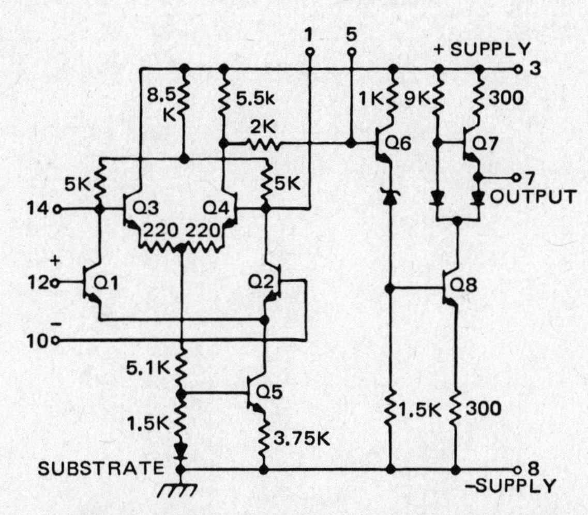

Figure 1-23B. Internal circuitry of the integrated circuit (IC).

Integrated Circuit IF Amplifier

The single IC intermediate frequency amplifier, whose circuit is shown in Figure 1-24, will provide up to 62 dB of gain at 455 kHz. The integrated circuit used in this amplifier contains 16 transistors, four diodes, and 19 resistors. The 455-kHz input signal

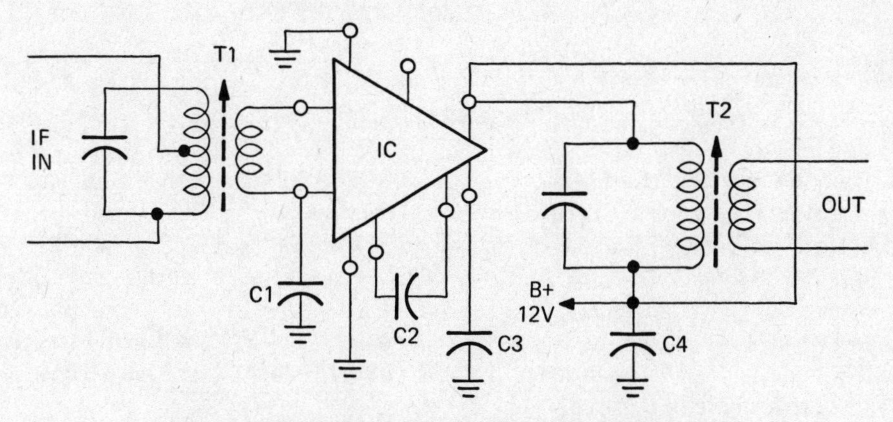

Figure 1-24. Single IC intermediate frequency amplifier.

from the mixer of a super-heterodyne receiver is fed in through IF transformer T1 or a ceramic or crystal selectivity filter (not shown) to the IC. The amplified output signal is fed to the detector or a second IF amplifier through IF transformer T2. Both T1 and T2 are tuned to 455 kHz.

Integrated FM IF Amplifier

Only a single integrated circuit is required to obtain 70 dB of gain at 4.5 MHz or 50 dB gain at 10.7 MHz when using the circuit shown in Figure 1-25. The input signal is coupled to the input of the IC through tunable transformer T1. The amplified signal is coupled to the next stage through tuned transformer T2.

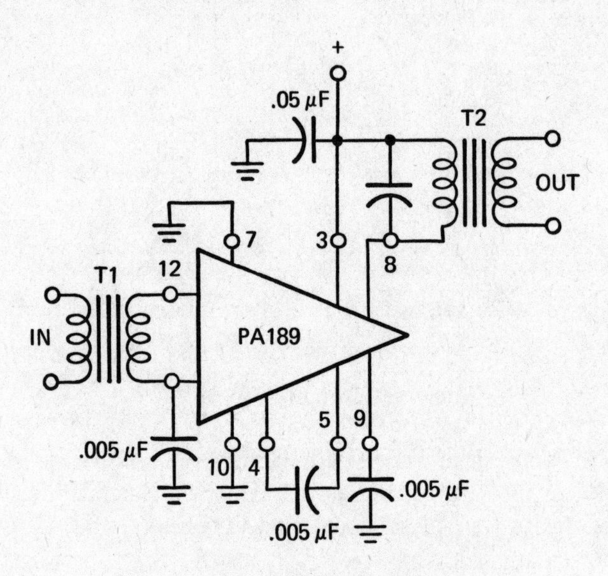

Figure 1-25. Integrated FM IF amplifier.

Integrated IF Strip

Only a single National LM172 integrated circuit and three external capacitors are required to form a complete IF amplifier and AM detector system. The IF input signal is fed to terminal 2 through external capacitor C1 and a band pass filter (not shown). The recovered audio signal is obtained from terminal 6. As can be seen in the diagram, a chain of seven diodes (D1 through D7) is used as a regulated voltage divider. Concurrent with the development of the recovered audio signal at terminal 6 is a DC voltage whose level varies with the level of the input signal. This DC voltage is the AGC voltage which is fed internally to the base of transistor Q4.

Low-Level Audio Amplifier

Figure 1-27 shows an RCA circuit of well-designed low-level audio amplifier. The operating current of this circuit is 1 mA DC and is at a point where the alpha is maximum. This current is held at a constant level by the use of degenerative feedback. Since the

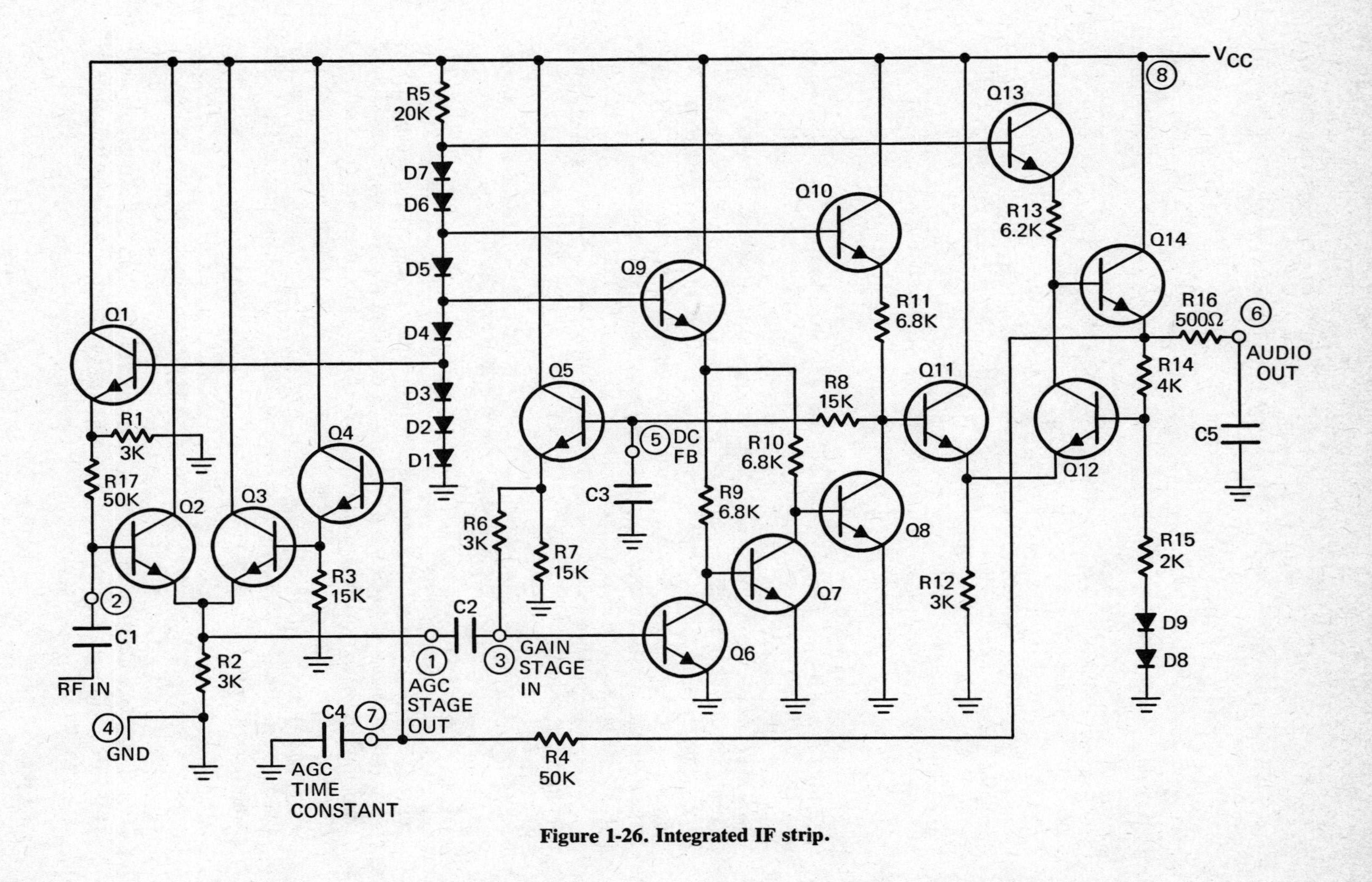

Figure 1-26. Integrated IF strip.

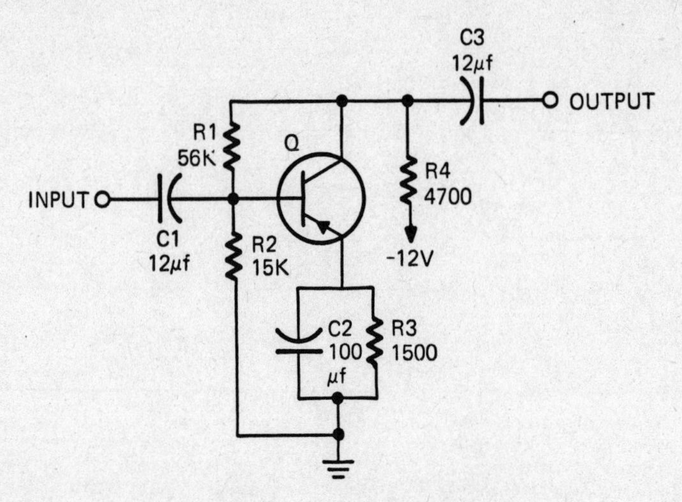

Figure 1-27. Low-level audio amplifier.

amplifier can operate over a wide temperature range, it is suitable as a general purpose amplifier building block.

Emitter resistor R3, which provides DC degeneration, is bypassed by capacitor C1 to avoid AC degeneration. But R1, which is not bypassed, reduces the current gain, linearizes the output, and extends the frequency response. The DC voltage drop across R4 makes the quiescent collector voltage about half the supply voltage. Because of the great, dynamic range of this amplifier, the instantaneous output voltage may swing from zero up to the supply voltage.

Magnetic Phonograph Pickup Preamplifier

The preamplifier, whose circuit is shown in Figure 1-28, is suitable for amplifying the output signal of a magnetic phonograph pickup. The magnetic pickup is connected to jack J and its signal is applied to pin 2, the grid of one of the two sections of the 7025 twin-triode tube. Output from the plate of the first section is coupled through C3 and R6 to the grid of the second triode. R10 and C6 form an RIAA equalization network that compensates for the recording characteristic introduced by the record manufacturer into the recorded disc. The preamplifier provides an overall voltage gain of about 150. (Illustration courtesy of RCA Corporation.)

MOSFET RF Amplifier

Figure 1-29 illustrates the use of a single-gate MOSFET in an RF amplifier stage. The RF signal developed across R_p is fed back from point X to point Y through neutralizing capacitor Cn to prevent unwanted self-oscillation.

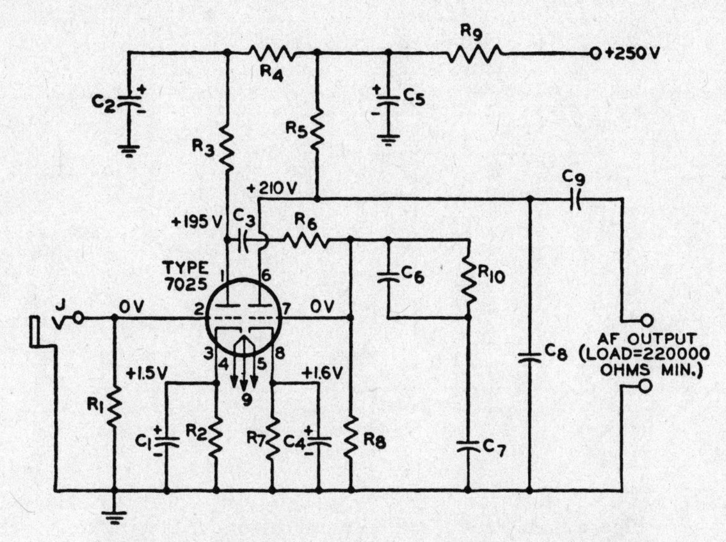

Figure 1-28. Magnetic phonograph pickup preamplifier.

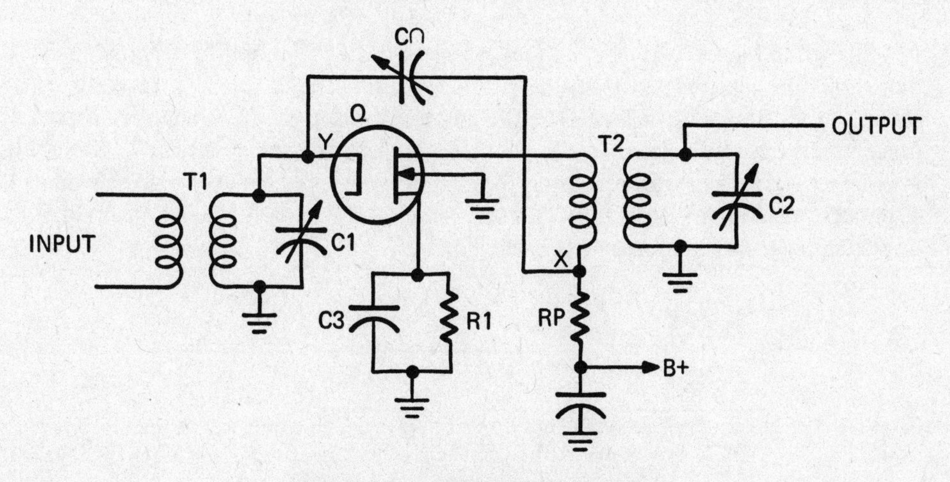

Figure 1-29. MOSFET RF amplifier.

Neutralized Push-Pull Triode RF Amplifier

Figure 1-30 shows the circuit of a push-pull RF power amplifier employing two triode tubes. The capacitance (Cgp) between the plate and grid of triode tubes forms a regeneration feedback path and can cause the amplifier to oscillate. This problem can be overcome by neutralizing the amplifier—feeding back a voltage (through C1 and C2) in phase opposition to the existing feedback.

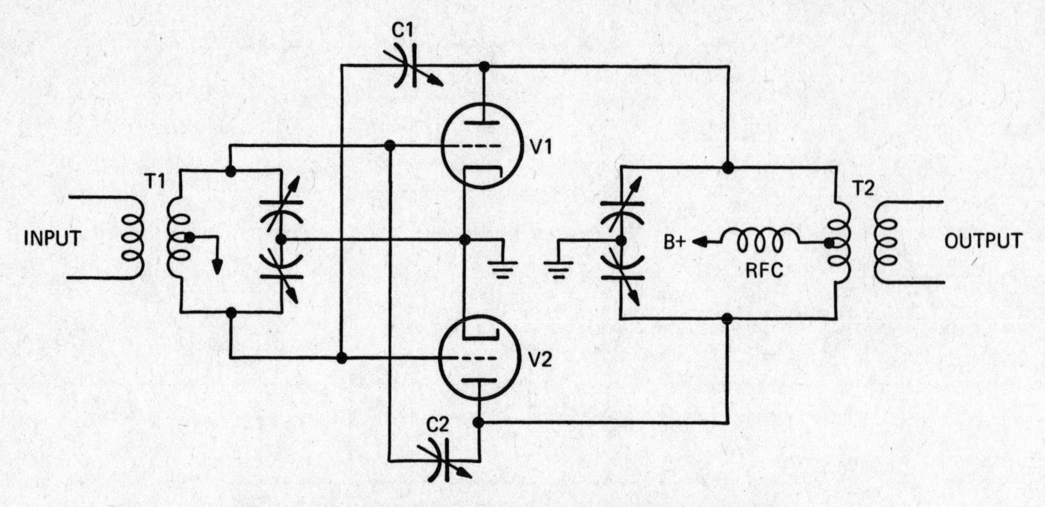

Figure 1-30. Neutralized push-pull triode RF amplifier.

Neutralized Transistor Amplifier

Shown in Figure 1-31 is the circuit of a neutralized IF amplifier employing a PNP transistor. The input signal from the previous stage is fed through a transformer (not shown) to the base of transistor Q1. The amplified output signal is coupled through IF transformer T to the detector diode CR1. To prevent unwanted self-oscillation due to internal incapacitance of the transistor, the output signal is fed back from the secondary of T through neutralizing capacitor C4 to the base of the transistor. This neutralizes the transistor capacitance and prevents oscillation.

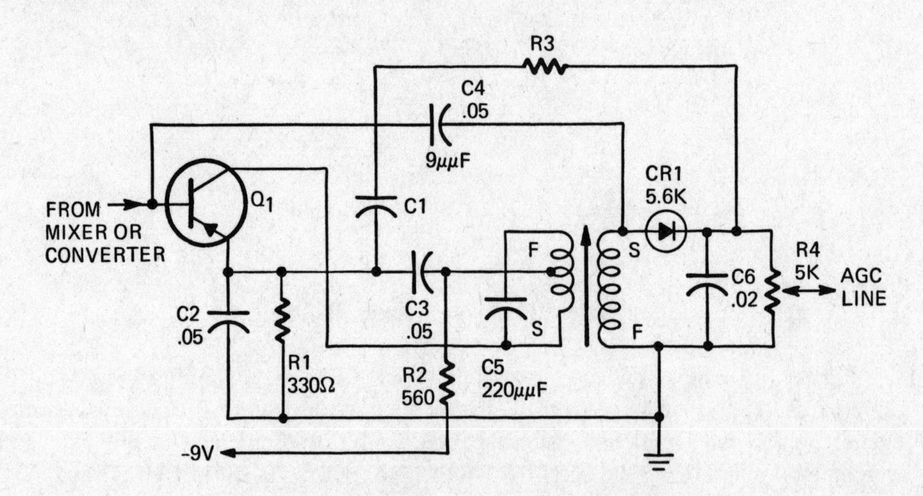

Figure 1-31. Neutralized transistor amplifier.

Neutralized Triode RF Amplifier

Various techniques were used in early tuned radio frequency radio receivers (circa 1928) to prevent RF amplifier stages from oscillating. The most commonly used technique was neutralization, of which there are several variations. In the circuit shown in Figure 1-32, the amplified output signal is fed from a tap on RF transformer T2 back to the grid (input) through neutralizing capacitor CN. Without CN in the circuit, the grid plate path through the tube forms a capacitor which is a positive feedback path. In this circuit, CN neutralizes the interelectrode capacitance of the tube and prevents oscillation.

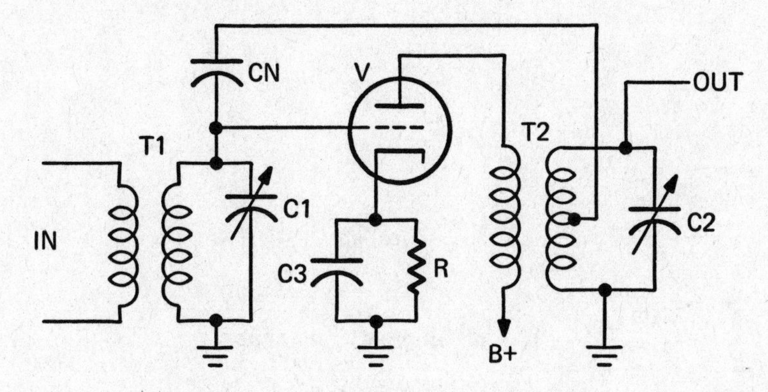

Figure 1-32. Neutralized triode RF amplifier.

To adjust CN (a low capacitance trimmer capacitor), the tube is made inoperative by cutting off either its heater or plate voltage. With an input signal applied, the signal level at the output is measured with an RF voltmeter. When CN is not correctly adjusted, the signal flows through the capacitance of the grid and plate to T2. CN is then adjusted for minimum signal at the output. When heater or plate voltage is restored, maximum amplification without oscillation is achieved.

OP-AMP

Figure 1-33 shows a typical operational amplifier (Op-Amp) circuit. The amplifier consists of a difference amplifier, NPN transistors Q1 and Q2, whose output is further amplified by PNP transistor Q3. The output from the difference amplifier is taken from the collector of Q2. This output reflects the difference between the two inputs. If the inputs are identical, no matter what their frequency or amplitude, there will be no output. To use an Op-Amp as a conventional linear amplifier, one of the inputs is usually tied to ground. In that way, the output of the Op-Amp reflects the amplified difference between one input and ground.

In an operational amplifier, the input indicates that the output signal will be inverted, while the +input indicates that the output signal will be non-inverted.

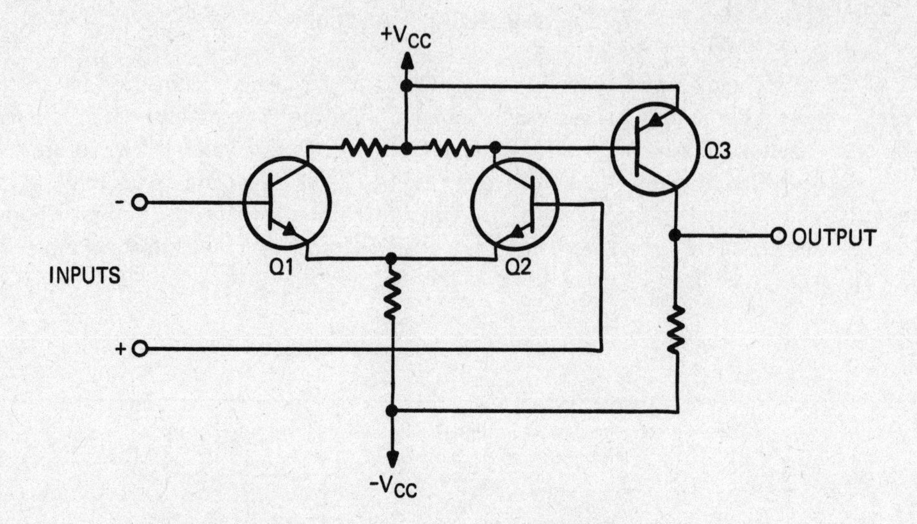

Figure 1-33. Operational amplifier (op-Amp).

OP-AMP Differential Amplifier

Two Fairchild operational amplifier integrated circuits are used to construct the high-performance differential amplifier circuit shown in Figure 1-34. Offset control for the amplifier is provided by R8, the offset null for the uA 741 op-amp. Linearity of the circuit shown is better than 0.1 percent with 50 dB of differential gain from DC to 500 Hz.

Pentode Neutralization

When a triode tube is used as an RF amplifier, with the input and output tuned to the same frequency, unwanted oscillation is apt to occur because of the feedback path through the internal grid-plate capacitance of the tube. Oscillation can be prevented by including a neutralization circuit. When a tetrode or pentode tube is used, neutralization is generally not necessary because the screen grid is bypassed to ground at the signal frequency and isolates the plate from the grid.

However, a pentode tube used as an RF amplifier has a small grid-to-plate capacitance that can cause oscillation. The feedback through Cgp in a pentode (Figure 1-35A) can be neutralized, but the small Cgp value makes the adjustment of the neutralization circuit difficult.

The use of a common bypass capacitor (Cn) for the plate and screen grid circuit can often solve the problem. If its value is properly chosen, this capacitor forms a bridge circuit, such as the one shown in B. It consists of Cgp, Cn, the plate-suppressor grid capacitance Cp, and the grid-screen capacitance Cgs. At terminal 1, in the grid circuit, the voltages at terminals 2 and 3 cancel each other.

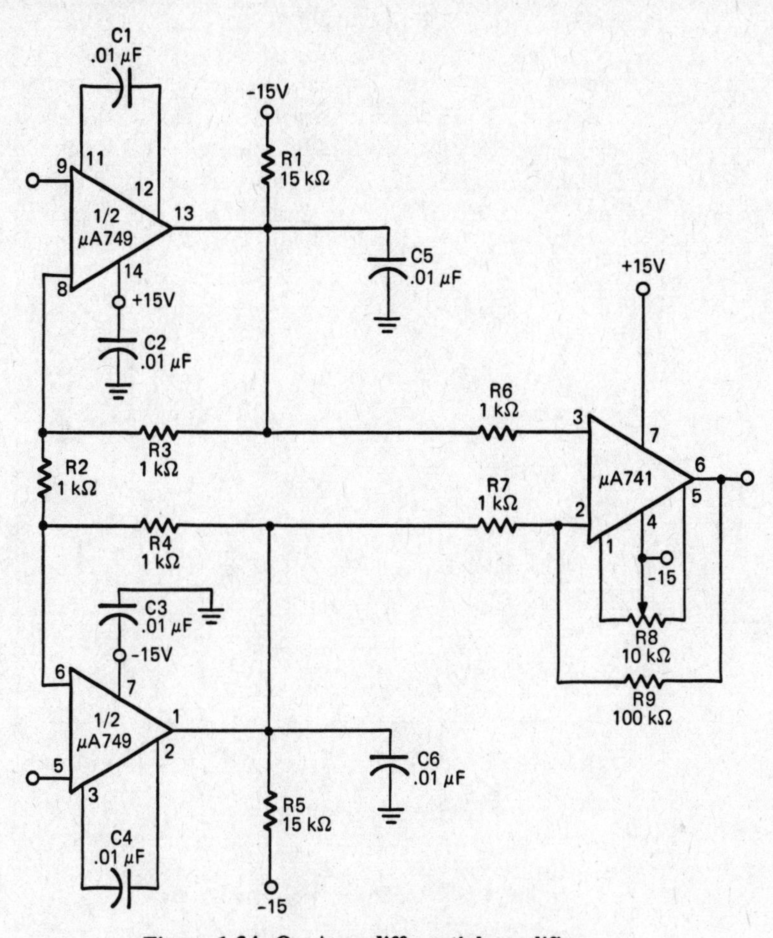

Figure 1-34. Op-Amp differential amplifier.

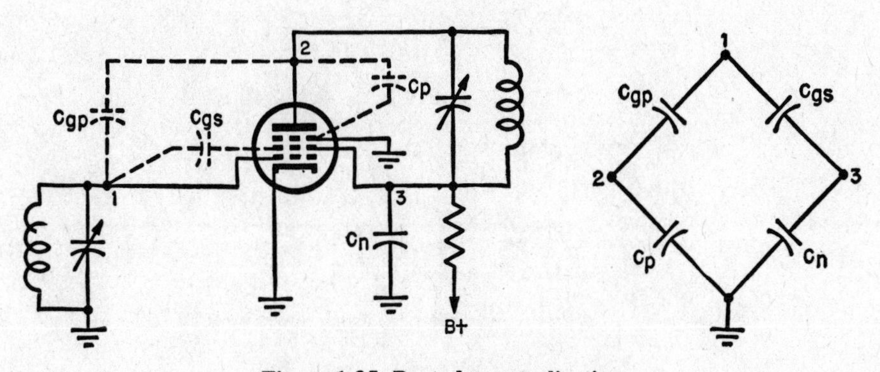

Figure 1-35. Pentode neutralization.

Phonograph Amplifier

Only a single integrated circuit is used in the phonograph amplifier circuit shown in Figure 1-36A. The input signal is obtained from a crystal phonograph pickup whose output level is controlled with potentiometer R1. Potentiometer R2 is the tone control which is used for controlling treble response. The amplified output signal is obtained from terminal 10 of the IC and fed through C8 to the speaker voice coil. Figure 1-36B is a schematic diagram of the circuitry of the IC.

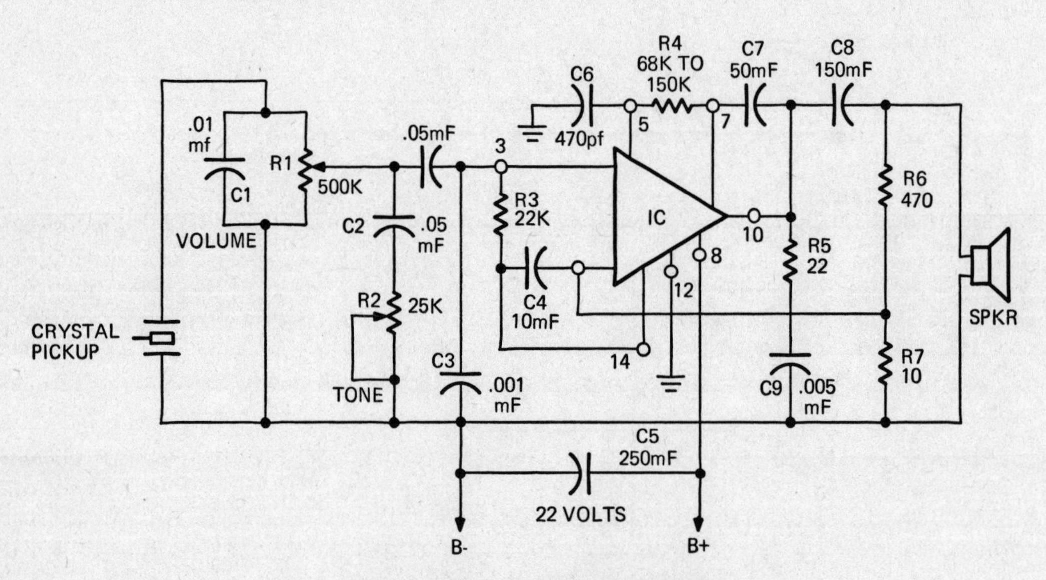

Figure 1-36A. Phonograph amplifier.

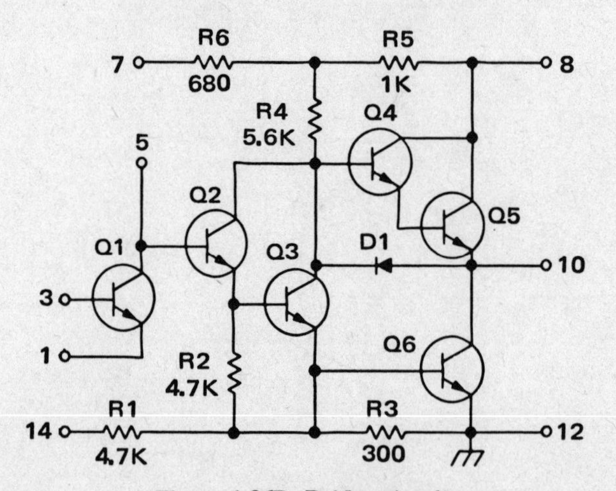

Figure 1-36B. Bridge circuit.

Reflex Amplifier

In the simple reflex receiver circuit shown in Figure 1-37, the same triode tube is used as both an RF amplifier and an AF amplifier. Radio signals coupled from the antenna to the grid of the triode through RF transformer T1 are amplified by the tube and then coupled through RF transformer T2 to the diode detector CR. The recovered audio signal current flows through the primary of AF transformer T3. Since T3 is a step-up transformer, the AF voltage is higher across its secondary than across its primary. This AF voltage is applied to the grid of the tube through the secondary of T1. Plate current is modulated by both the radio signal and the recovered audio signal.

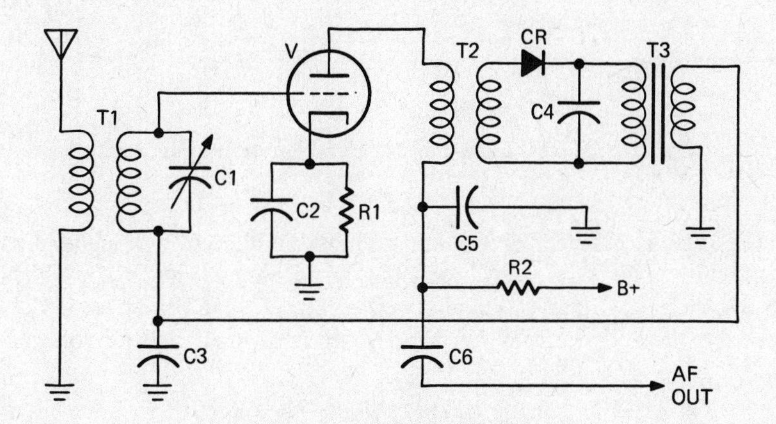

Figure 1-37. Reflex amplifier.

The receiver is tuned by variable capacitor C1. Capacitor C4 across the primary of T3 bypasses RF remaining at the output of the detector, and C3 across the secondary of T3 has low reactance at the radio signal frequencies and allows the RF signal to pass without being attenuated by the high reactance of T3's secondary. C5 serves as an RF bypass capacitor whose reactance at audio frequencies is high. The amplified AF signal is developed across R2 and is fed out to headphones or another AF amplifier stage through C6. The tube is cathode-biased by R1 for Class A operation. Capacitor C2 across R1 bypasses both RF and AF signals and prevents degeneration which would reduce gain.

Resistance-Coupled Transistor Amplifier

The circuits shown in Figure 1-38 are the same except that one uses an NPN transistor (A) and the other uses a PNP transistor (B). The transistors are forward-biased through voltage divider R3-R1 and both are provided emitter bias developed across R2. The AC input signal fed in through C1 alternatively bucks and boosts the net base-emitter current. As base current rises, collector current flowing through R4 rises, causing collector voltage to fall, and vice versa. The resulting variations in voltage drop across R4 are derived at the

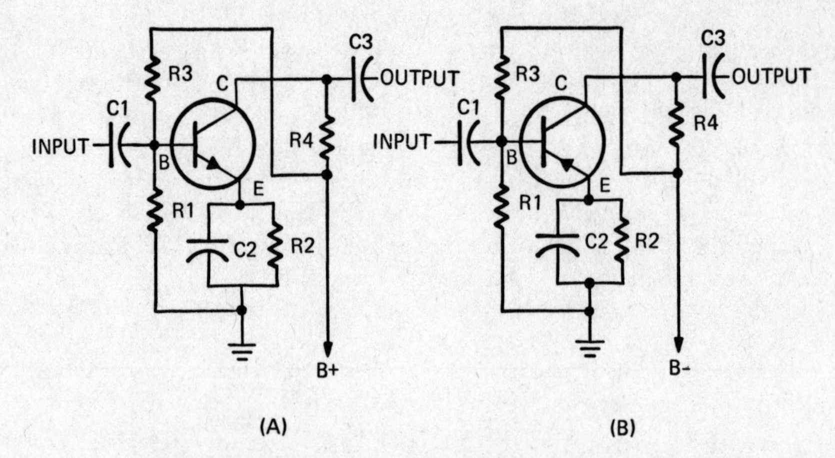

Figure 1-38. Resistance-coupled transistor amplifier circuits.

output (through C3) as an amplified and inverted replica of the AC input signal. Capacitor C2 prevents emitter bias from fluctuating with the AC signal. The chief difference between the two circuits is that the NPN transistor has positive base bias and collector voltage, and that the PNP transistor has negative base bias and collector voltage.

Resistance-Coupled Triode Amplifier

Figure 1-39 shows a triode tube being used in the most common type of AF amplifier circuit, the resistance-coupled amplifier stage. It is called resistance-coupled because the input signal is developed across resistor R1 after being fed from the preceding stage through capacitor C1. The output signal is also developed across a resistor, R3, and is fed on to the next stage through capacitor C3. The tube is cathode-biased by R2, the cathode resistor, capacitor C2, connected across R2, prevents degeneration at signal frequencies.

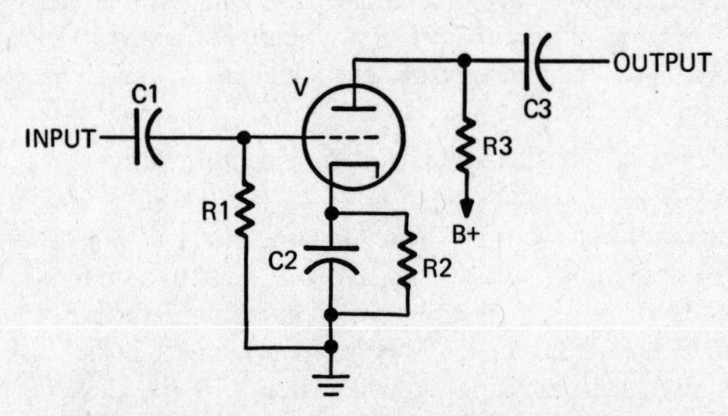

Figure 1-39. Resistance-coupled triode amplifier.

Rice Neutralization

Figure 1-40 illustrates how Rice (or grid) neutralization is used to prevent oscillation in a triode tube RF amplifier caused by in-phase signals being fed back to the grid via the plate-to-grid interelectrode capacitance (Cgp) of the tube. The neutralizing capacitor (Cn) is connected from the plate to the bottom of the grid tank. The signal passing through Cn effectively opposes the signal fed back through Cgp. By balancing the circuit by varying Cn, oscillation can be prevented.

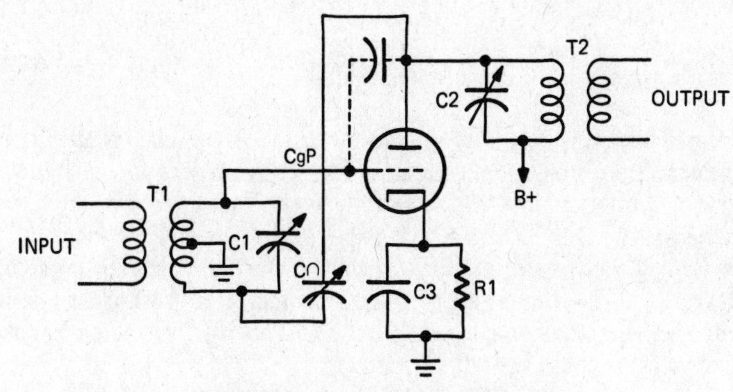

Figure 1-40. Rice neutralization.

Single Transistor Power Amplifier

The CBS circuit shown in Figure 1-41 is of a transformer-coupled Class A audio amplifier and is used where high power gain is required, such as in the output stage of a compact radio receiver. The input transformer (T1) used to match the high output impedance of the driver stage of the audio circuit may have a step down ratio of 6:1 or even higher depending upon the impedance match required, and the output transformer may

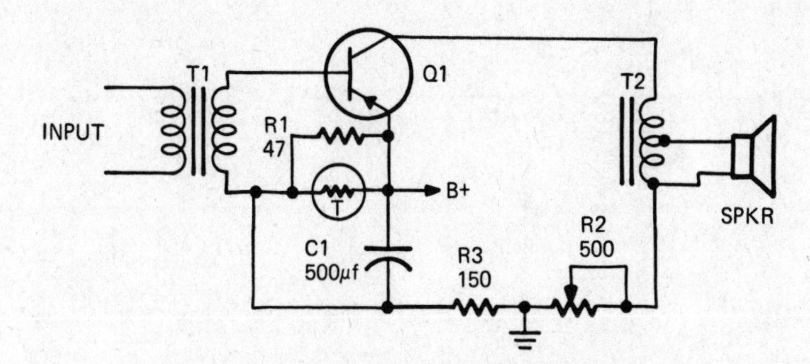

Figure 1-41. Single transistor power amplifier.

very conveniently be of the autotransformer type to match the voice coil to the output of the transistor collector. Temperature stabilization is achieved in this circuit by thermistor RT which provides a compensating bias at the base of Q1. A padding resistor (R1) is connected across the thermistor to obtain the required variation in forward bias with respect to temperature variation. A good heat sink for the transistor is required if the amplifier is to operate safely at high temperatures. A variable resistor (R2) is used in series with R3 to adjust the forward bias of the transistor to the desired operating point since DC characteristics vary among transistors. Typical performance of this circuit as follows:

Power output:	2 watts	Power gain:	36 dB
Input impedance:	30 ohms	Load impedance:	3.2 ohms

Servo Amplifier

Servo systems are widely used in industrial control applications and in some analog computers such as electronic flight trainers. Two NPN power transistors are used in the simple Servo amplifier circuit shown in Figure 1-42. Both transistors are forward-biased and the bias is stabilized by diode D1. The input signals are coupled to the emitters of the transistors through transformer T. The collectors of the transistors are connected in push-pull configuration to the center tapped stator winding of the two-phase AC motor. The other winding of the motor is connected to an AC source through capacitor C2 which causes 90 degrees of phase shift.

The direction of rotation of the motor depends upon the phase of the input signals. When the input signal causes the collector current of Q1 to increase, the motor rotates in one direction. When the input signal causes Q2 collector current to rise higher than Q1 collector current, the motor rotates in the opposite direction. In an actual application, the motor drives a potentiometer or an indicator.

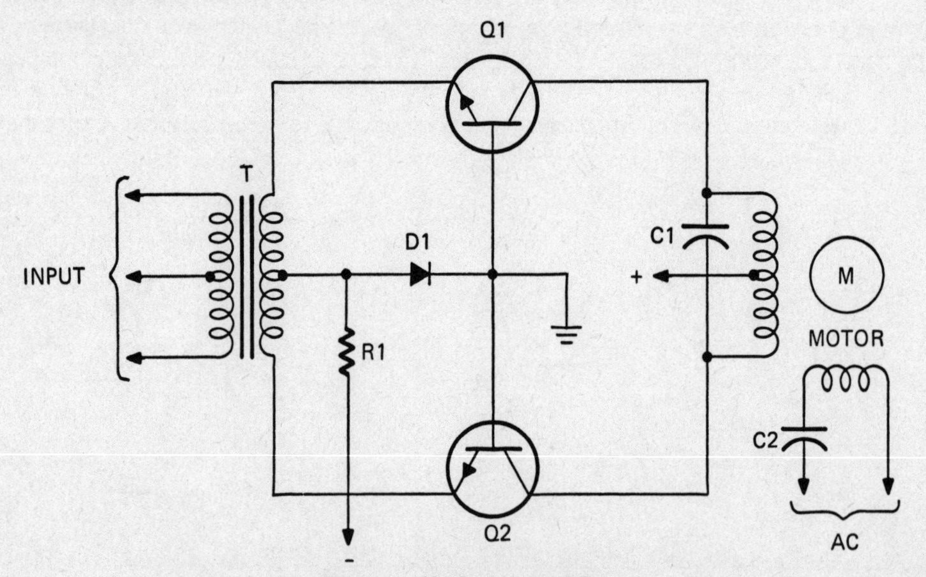

Figure 1-42. Servo amplifier.

Shunt Degeneration

A shunt degeneration (negative feedback) circuit is shown in Figure 1-43. (Figure A is for NPN transistors and Figure B is for PNP transistors.) This type of degeneration is most effective in reducing distortion when the signal source impedance is high compared to the input impedance. Distortion is due mostly to the fall-off of current amplification as emitter current increases. Resistor R1 provides a negative feedback path. An out-of-phase signal is fed back from the collector to the base. The amount of this feedback is approximately equal to:

$$a_{fe} = \frac{R_L}{R_F}$$

where:

a_{fe} is the amplification factor,
R_L is the resistance of the output load in ohms, and
R_F is the feedback resistance in ohms.

(Note: the output current is equal to the current amplification factor times the input current.)

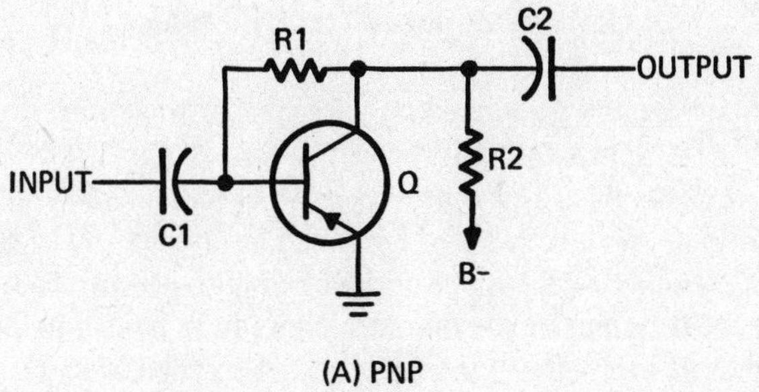

(A) PNP

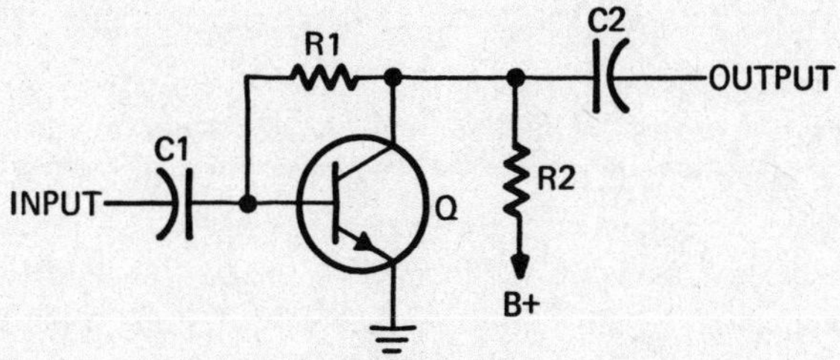

(B) NPN

Figure 1-43. Shunt degeneration.

Shunt-Fed Triode Power Amplifier

Figure 1-44 shows how plate voltage can be supplied to a tube without allowing DC to flow through the load resistance (or resonant tank circuit). If the series reactance of capacitor C is very low and the reactance of inductor L is high compared to the load resistance, direct current flows through L and the output signal is developed across load resistance R. In radio transmitters, L is an RF choke coil and R may be either a parallel-resonant transformer or pi network.

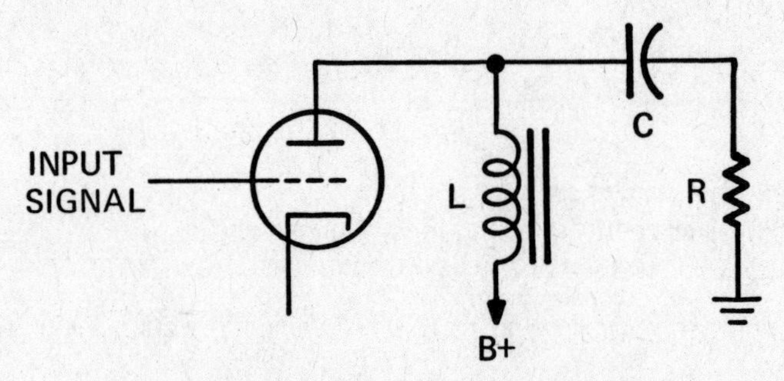

Figure 1-44. Shunt-fed triode power amplifier.

Solid State Audio Power Amplifier

A Fairchild type uA 739 integrated circuit and four transistors are used in the audio power amplifier circuit shown in Figure 1-45. The input signal is fed through C1 to the non-inverting input of the IC. The amplified output of the IC is fed to the two Darlington connected transistors shown at the upper right. The output signal is also fed through diodes D1, and D2, and R6 to the input of the Darlington connected transistors shown at the lower right of the diagram. These four transistors form a push-pull output amplifier in spite of the fact that their input signal source is single ended. Feedback is obtained from the junction of R8 and R9 and is fed to the inverting input of the IC through R5.

Stereophonic Phonograph Amplifier

The stereophonic amplifier whose circuit is shown in Figure 1-46 provides one watt of AF output on each of the two channels. Each input signal, from a stereophonic crystal or ceramic phonograph pickup, is applied through a capacitor and a volume control to the control grid of a power pentode tube. The output of each pentode is developed across the 3000-ohm primary of an output transformer. No power transformer is required. The AC line voltage is rectified by a diode and filtered by the combination of C6, C8, and R10. Screen grid voltage filtering and stabilizing is obtained through the use of C7, a 50 uF electrolytic capacitor. The heaters of the two 60XF5 tubes are connected in series across the AC power line. (Illustration courtesy of RCA Corporation.)

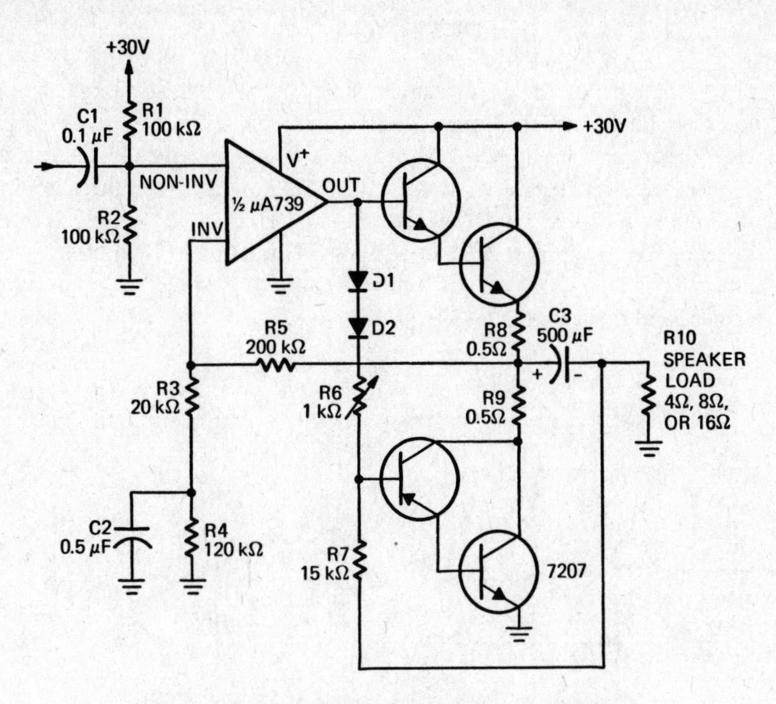

Figure 1-45. Solid state audio power amplifier.

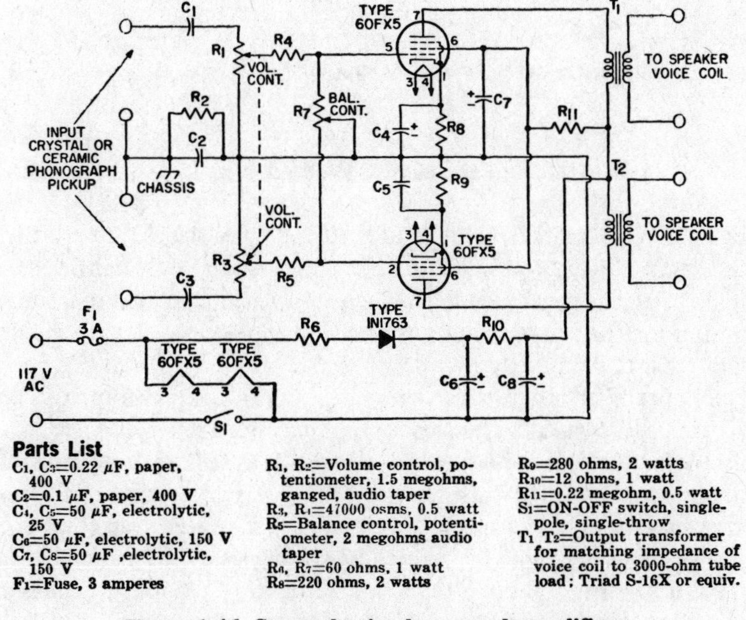

Parts List

C_1, C_3=0.22 μF, paper, 400 V
C_2=0.1 μF, paper, 400 V
C_4, C_5=50 μF, electrolytic, 25 V
C_6=50 μF, electrolytic, 150 V
C_7, C_8=50 μF ,electrolytic, 150 V
F_1=Fuse, 3 amperes
R_1, R_2=Volume control, potentiometer, 1.5 megohms, ganged, audio taper
R_3, R_4=47000 osms, 0.5 watt
R_5=Balance control, potentiometer, 2 megohms audio taper
R_6, R_7=60 ohms, 1 watt
R_8=220 ohms, 2 watts
R_9=280 ohms, 2 watts
R_{10}=12 ohms, 1 watt
R_{11}=0.22 megohm, 0.5 watt
S_1=ON-OFF switch, single-pole, single-throw
T_1 T_2=Output transformer for matching impedance of voice coil to 3000-ohm tube load ; Triad S-16X or equiv.

Figure 1-46. Stereophonic phonograph amplifier.

Transducer Amplifier

Transducers can be devices that convert mechanical or heat energy to electrical energy, finding application in such areas as environmental and strain analysis, pressure indicators, and thermocouples. Since a transducer's output is typically only in the millivolt range, these devices require an extremely stable, low-drift amplifier, a circuit for which is illustrated in Figure 1-47. The circuit uses a Fairchild MA727 temperature-controlled differential preamplifier IC to drive a MA741 operational amplifier. The circuit shown features .01 percent accuracy over the military temperature range, 300 Megohm input impedance, and open-loop gain of 140 dB.

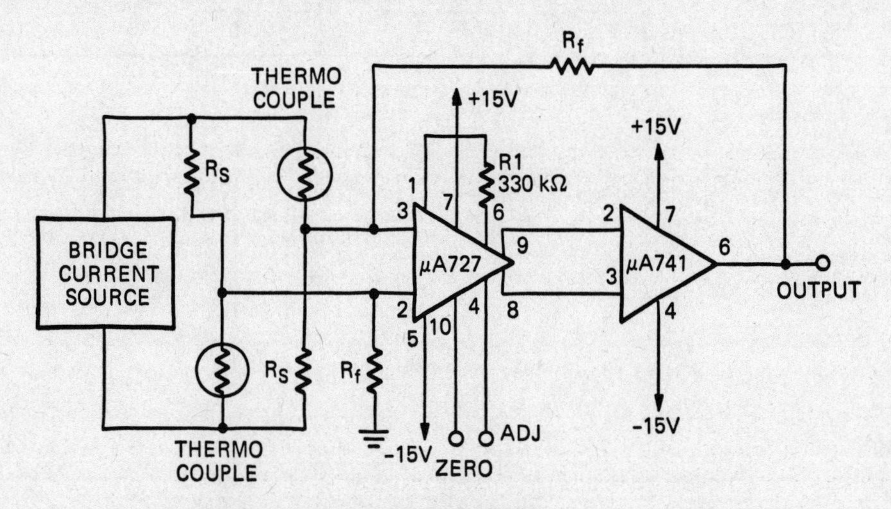

Figure 1-47. Tranducer amplifier.

Transformer-Coupled Triode Amplifier

In early radio receivers built in the 1920's, AF amplifier stages were usually coupled through transformers. Figure 1-48 shows the circuit of a typical early transformer-coupled voltage amplifier stage using a triode tube. The AF input signal is fed through transformer T1 to the grid of the filament type tube. The output signal is coupled to the next stage through another transformer (T2).

This type of coupling has the advantage that the primary of transformer T2 provides a relatively high impedance at audio frequencies, but provides a low-resistance path for DC. Thus the plate voltage source (shown as the battery B3, a "B" battery) does not have to provide as high a voltage as when resistance coupling is used. Since the transformers usually have a step-up ratio of about 3:1, they add to the overall voltage gain.

Negative grid bias is supplied by battery B1 (known as a "C" battery) and filament current is supplied by battery B2 (known as an "A" battery). Amplifier gain is varied by adjusting filament current with rheostat R.

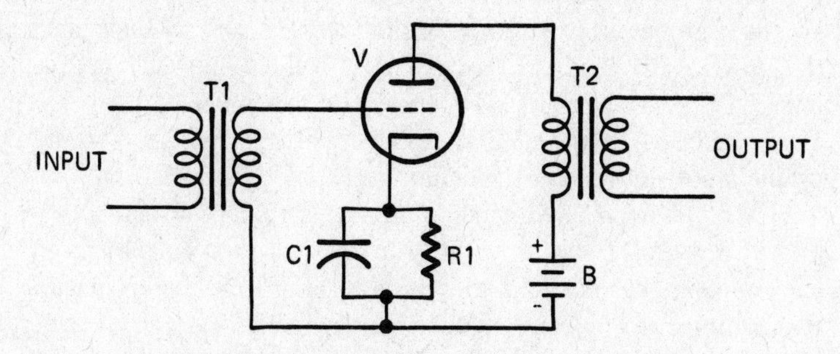

Figure 1-48. Transformer-coupled triode amplifier.

The use of interstage AF transformers was essentially phased out in the early 1930's because of their relatively narrow frequency response and high cost compared to resistance-capacitance coupling components. Now, however, AF interstage transformers which have broad frequency response are available. When used with modern tubes, much higher voltage gain can be obtained than can be provided by resistance-capacitance coupled amplifiers.

Transformer-Coupled Push-Pull Power Amplifier

The circuit shown in Figure 1-49 is used in audio applications and can be biased to operate as either a Class A, Class AB, or Class B power amplifier. When biased for Class A operation, the input signal level must never be great enough to drive the control grids positive. If this occurs, the output is not an amplified replica of the input signal. The input AC signal voltage across the primary winding of the input transformer T1 is stepped up and fed to the grids of the two tubes which may be beam power tubes, pentodes, or triodes. The signals at the grids are 180 degrees out of phase with each other. The

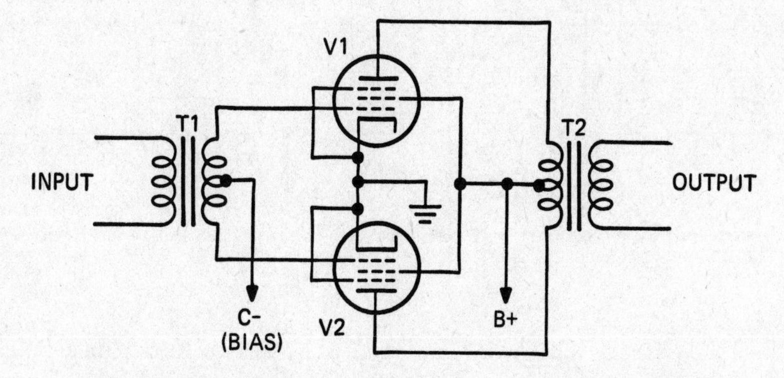

Figure 1-49. Transformer-coupled push-pull power amplifier.

amplified signal is fed to the speaker voice coil through step-down output transformer T2 which matches the high output impedance of the tubes to the low impedance speaker.

Higher AF power output and efficiency can be obtained, but with more distortion, by biasing the tubes for Class AB or Class B operation by making the negative bias voltage at C- higher. When so biased, V1 and V2 plate current is very low when no input signal is present. When the input signal makes the control grid of V1 positive and the control grid of V2 negative, V1 plate current rises sharply and V2 plate current falls to near zero. When the control grid of V1 sees a negative signal and the control grid of V2 sees a positive signal, V1 plate current falls to near zero and V2 plate current rises sharply. Distortion can be minimized by using matched tubes and by the careful selection of bias voltage.

Transistor Amplifier

A single transistor is used as both an RF (or IF) and AF amplifier in the circuit shown in Figure 1-50. The RF input signal is fed through T1 to the base of the transistor. The amplified RF signal is coupled through T2 to detector diode CR. The recovered audio signal developed across volume control R5 is fed through C4 and the secondary of T1 to the base of the transistor. The amplified AF signal is developed across R4 and is coupled to an AF amplifier through C6.

The values of C1 and C5 are such that they effectively bypass the RF signal but do not significantly attenuate the AF signal. C2, on the other hand, has sufficient capacitance to bypass the AF and RF signals so that degeneration will not result.

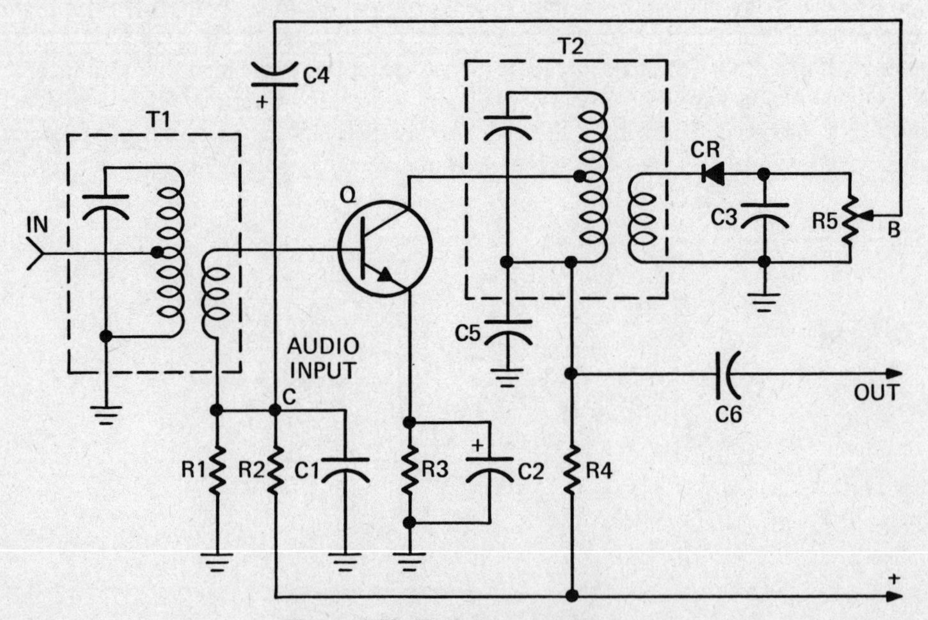

Figure 1-50. Transistor amplifier.

Transistor IF Amplifier

A single NPN transistor is used in the IF amplifier stage circuit shown in Figure 1-51. When a PNP transistor is used in this kind of circuit when the input and output transformers are both tuned to the same frequency, neutralization is usually required in order to avoid unwanted oscillation. When an NPN transistor is used, neutralization may or may not be required, depending upon the gain and band pass of the amplifier. In this circuit, no neutralization is provided.

The input signal is coupled to the base of the transistor through T1 whose primary is parallel resonant and whose impedance is high. To avoid loading down the resonant circuit by the input circuit resistance, the input signal is fed to a tap on the primary of T1, a low impedance point. T1 is a stepdown transformer since it interfaces the relatively low input resistance of the transistor. The amplified output signal is coupled to the next stage through T2. As in the case of T1, the input signal is fed to a tap on the primary of the transformer.

The transistor is forward biased by the voltage divider consisting of R1 and R2. C1 is an RF bypass capacitor. Emitter bias, which prevents thermal runaway of the transistor, is provided by R3 which is bypassed by C2 to prevent degeneration.

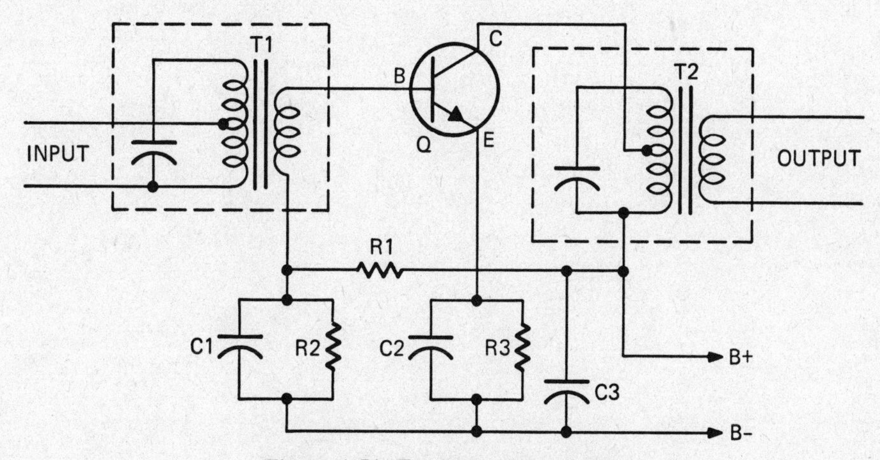

Figure 1-51. Transistor IF amplifier.

Voltage Scaling Inverter Amplifier

Figure 1-52 illustrates how an operational amplifier (Op-Amp) is used as an inverter in which the input is multiplied by a "scale factor," and yielded as an inverted output. Gain of the stage is given by the ratio of the feedback resistor, Rf, to the input resistor, Ri, according to the formula:

$$V_{out} = V_{in} \frac{Rf}{Ri}$$

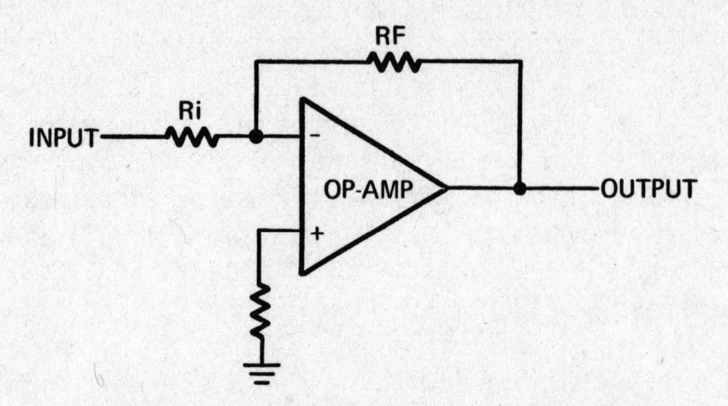

Figure 1-52. Voltage scaling inverted amplifier.

Thus, the gain of the amplifier stage may be changed by adjusting the value of the feedback resistor, Rf, or the input resistor, Ri.

2

Audio Control Circuits

Audio frequency (AF) signals can be controlled in various ways. The level of an AF signal can be controlled with a variable attenuator. In some cases, a simple potentiometer will suffice. In other cases where impedance matching is important, more sophisticated attenuators and/or transformers are required. This chapter describes and illustrates basic circuits for controlling and interfacing AF signals.

Audio Output Power Splitter

One method for feeding three loudspeakers from one source of audio is shown in Figure 2-1. The sound levels of the three loudspeakers (A, B, and C) can be controlled independently with variable T-pads R1, R2, and R3. Transformer T (such as Alco Mix-N-Match) has three 8-ohm windings. It splits the audio power: 50 percent is available to speaker A and 50 percent to speakers B and C combined, when A is an 8-ohm speaker, R1 is an 8-ohm T-pad, B and C are 16-ohm speakers, and R2 and R3 are 16-ohm T-pads.

Audio Take-Off from Radio Receiver

It has been customary for many years to tap the audio signal of a radio receiver at the output of the detector or volume control when using the receiver as a program source for a sound system. This has become increasingly difficult to do when the receiver is very compact and employs a printed circuit board. The alternative, shown in Figure 2-2, is to derive the audio signal at the receiver output. Switch S is added to enable use of the built in loudspeaker and to silence the speaker when feeding the audio to an external amplifier or recorder.

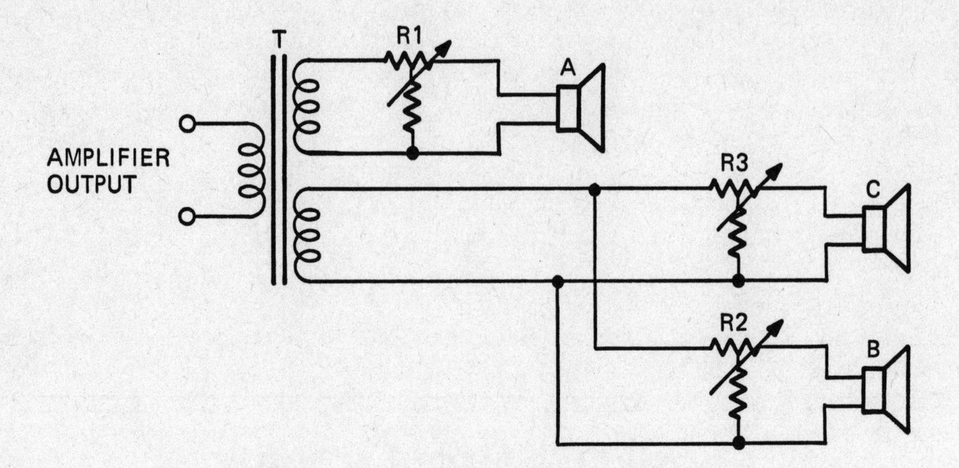

Figure 2-1. Audio output power splitter.

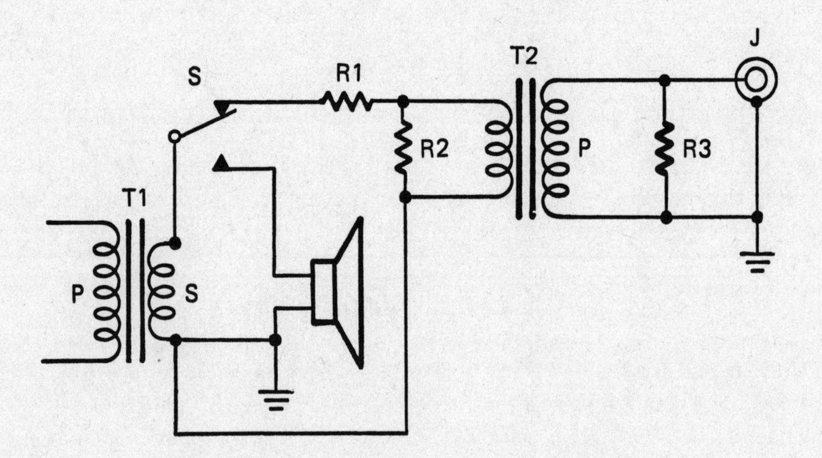

Figure 2-2. Audio take-off from radio receiver.

In this circuit, R1 and R2 form a fixed L-pad which provides a load across the secondary of the receiver's output transformer T1 when the speaker is disconnected by S. The L-pad also reduces the level of the signal fed to transformer T2. This transformer may be a 4-ohm/5000-ohm output transformer with the signal fed to its low-impedance primary. It steps up the signal voltage and provides a higher source impedance to the amplifier. Resistor R3 may or may not be required. J is a phono jack which makes it convenient to feed the signal through a shielded audio cable to the external amplifier or recorder. The shell can be grounded since T2 provides isolation in addition to impedance and voltage transformations.

Constant-Voltage Speaker Feed

The loudspeakers of a public address, sound paging, or music distribution system can be fed through a so-called constant-voltage speaker feedline. The amplifier feeds a 25-, 70-, or 115-volt audio signal to the line, as shown in Figure 2-3. The audio voltage obviously varies, but is held relatively constant at the specified level by an AGC (automatic gain control) circuit in the amplifier, or by the use of regenerative feedback within the amplifier, or by both.

The sound level of each speaker can be set, without affecting the level of other speakers, by selecting taps on a transformer through which the speaker is fed. For example, to feed 12.5 watts of audio to an 8-ohm speaker, the transformer is set to deliver 10 volts to the speaker. If set to deliver 5 volts to the speaker, the power input to the speaker will be slightly greater than 3 watts since $W = E^2/R$ where W is watts, E is volts, and R is speaker impedance.

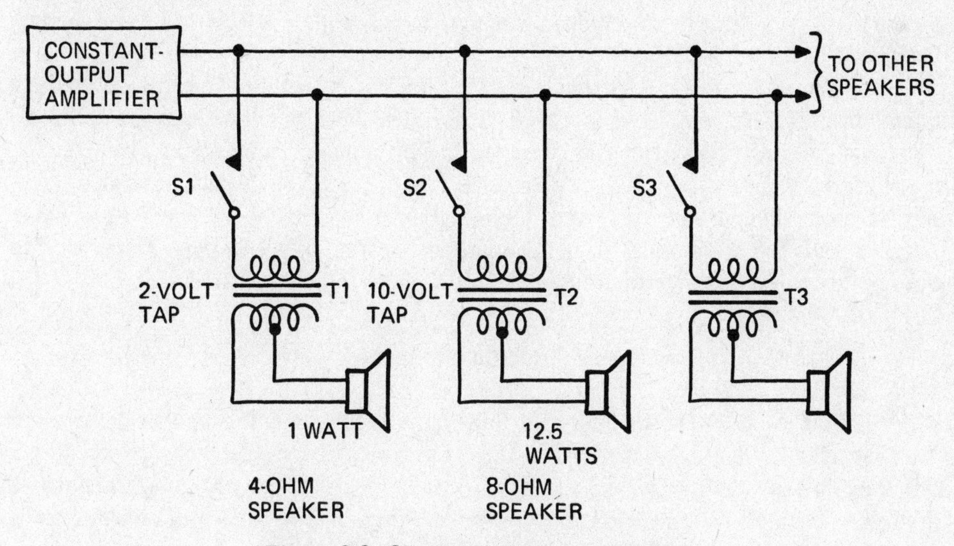

Figure 2-3. Constant-voltage speaker feed.

Extension Speaker Feed System

Figure 2-4 shows one way to feed a local speaker and three distant remote speakers from the low impedance output of an AF amplifier. Assuming that the load on the amplifier output can be as low as 4 ohms, the impedance of all four speakers can be 8 ohms.

Speaker A is connected across the audio amplifier output terminals. Transformer T1 is also connected across the amplifier output terminals. It is a 70-volt speaker feedline transformer used backwards. Its secondary winding is tapped. The output of the amplifier is connected to the 8-ohm tap and common terminal of the secondary of T1. The primary

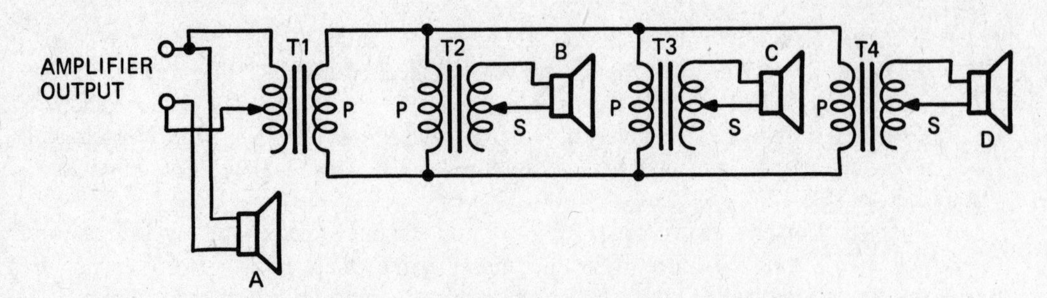

Figure 2-4. Extension speaker feed system.

winding is connected to the speaker line. When the amplifier is set to deliver 10 watts to a 4-ohm load, the voltage at the amplifier output terminals will be 6.3 volts. T1 steps up this voltage to 70 volts.

Speakers B, C, and D are connected to the feedline through 70-volt speaker feedline transformers T2, T3, and T4 respectively which step down the voltage to the speakers. The volume level of these extension speakers can be adjusted independently by selecting transformer secondary taps. If each speaker is supplied 3.6 volts by its associated transformer, each will consume 1.62 watts. Of the 10 watts supplied by the amplifier, speaker A gets 5 watts and speakers B, C, and D jointly get approximately 5 watts. The preceding does not take into consideration the insertion losses of the transformers.

IC Intercom Circuit

Figure 2-5 shows how simply an integrated circuit amplifier can be used in an intercom circuit. The talk/listen switch merely connects the master unit speaker to the amplifier input so it will serve as a microphone when it is in the talk position. When it is in the listen position, the remote speaker is used as a microphone. A step-up transformer is used to increase the signal voltage generated by the 4-ohm speakers when used as microphones. No output transformer is required to drive the speakers because of the low output impedance of the IC. (Illustration courtesy of *Ham Radio*.)

Level Controls

A level control or volume control is often erroneously called a "gain" control. Unless it actually controls the gain of an amplifying device, it is a "level" control. The most common type of AF level control is a potentiometer, shown in Figure 2-6A. It is simply a variable voltage divider. As it is adjusted both its input and output impedances vary.

An L-pad attenuator, shown in B, consists of two rheostats driven by the same shaft. As the resistance of R1 is increased, the resistance of R2 is decreased. The load imped-

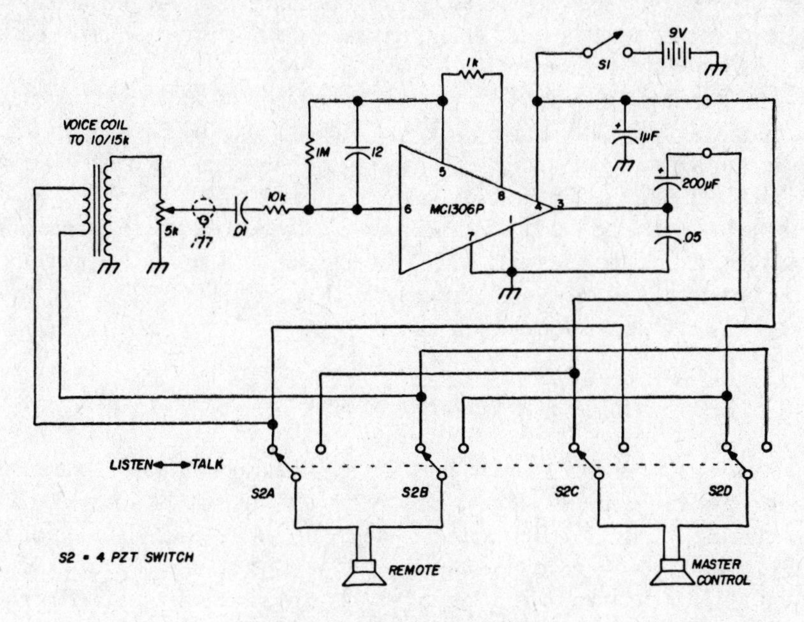

Figure 2-5. IC intercom circuit.

A. Potentiometer.

B. L-pad attenuator.

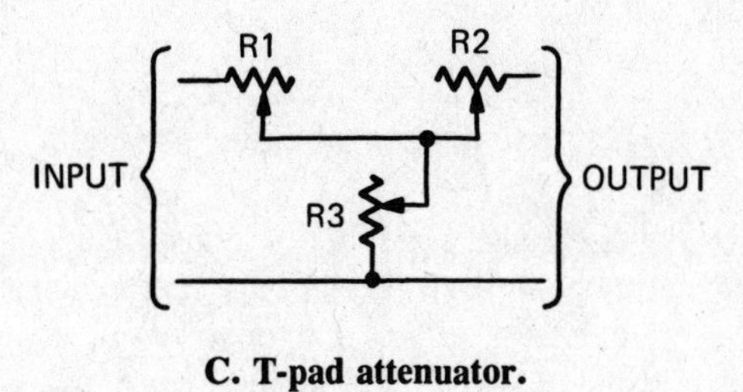

C. T-pad attenuator.

Figure 2-6. Level controls.

ance seen by the input source remains constant, regardless of the amount of signal attenuation introduced by the L-pad. The impedance seen by the load, looking back into the L-pad, varies with the setting of the L-pad.

A T-pad attenuator, shown in C, consists of three rheostats driven by the same shaft. As the resistances of R1 and R2 are increased, the resistance of R3 is decreased. The impedance seen by the input source remains constant. And the impedance seen by the load, looking back into the T-pad, also remains constant.

Both the L-pad and the T-pad are variable voltage dividers. As the resistance of R1 of the L-pad is increased, the resistance of R2 is decreased, and so is the voltage divider ratio. The same is true of R1 and R3 of the T-pad. But, in addition, R2 is one arm of a voltage divider consisting of R2 and the load.

Low-Frequency Attenuator

The reproduction of low frequency sounds by a loudspeaker can be attenuated by connecting a pair of electrolytic capacitors in series with one of the speaker voice coil leads, as shown in Figure 2-7. The capacitors are connected in series-opposing, in regard to polarity, with respect to each other so they can be used in an AC circuit. When the voltage fed to C1 is positive, it is polarized so that its capacitance is in series with the speaker. C2, however, is polarized so that it acts as a resistance. When the signal voltage reverses in polarity, the opposite is true. If both C1 and C2 are 40-mf capacitors they will have the effect of inserting approximately 18 ohms of reactance in series with the voice coil at 240 Hz, but only about 1 ohm at 4000 Hz. The low-frequency response can be varied by using different amounts of series capacitance.

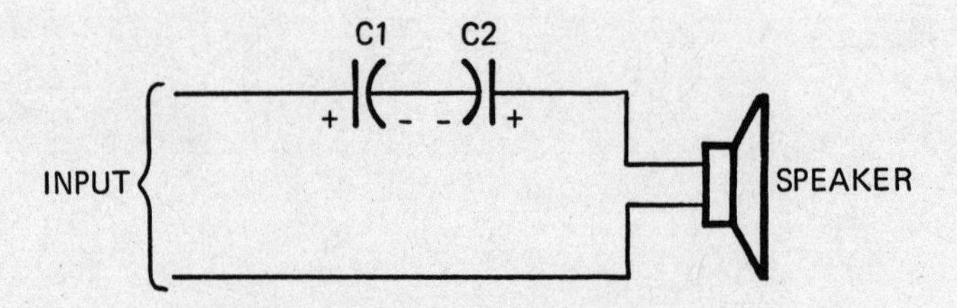

Figure 2-7. Low-frequency attenuator.

Matched Impedance Speaker Feed

Where it is necessary to connect a number of loudspeakers to the 500-ohm output of an amplifier, impedance matching transformers are required at each speaker location. Figure 2-8 shows three methods for varying the reflected impedance of each loudspeaker so that the total load impedance will be 500 ohms. Transformer T1 has a tapped low-impedance secondary. The impedance seen by the feedline depends upon the speaker impedance and the impedance ratio of the transformer. T2 has a tapped primary and an

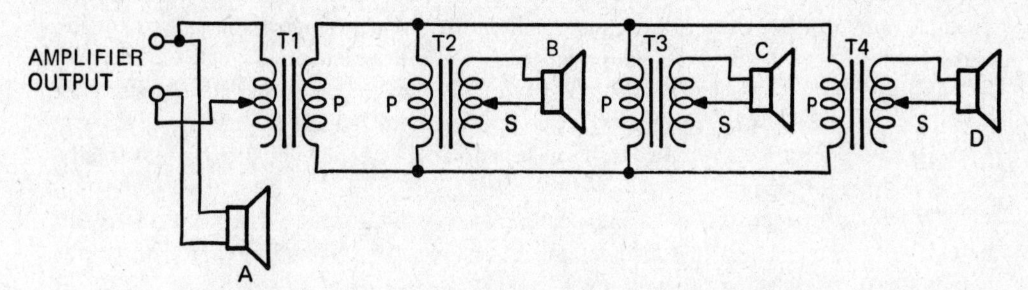

Figure 2-8. Three methods for varying reflected impedance.

untapped low-impedance primary. T3 is a tapped auto transformer. The taps of each of the three transformers should be selected so that the reflected primary impedance is 1500 ohms. When four speakers are used, the reflected primary impedance of each of four transformers should be 2000 ohms, and so on.

Passive AF Signal Mixer

Two audio signals can be mixed by using two potentiometers as shown in Figure 2-9. The levels of inputs 1 and 2 are controlled independently with potentiometers R1 and R2. Resistors R3 and R4 reduce, but do not eliminate, interaction between the potentiometers. The resistances of R1 and R2 depend on the output impedances of the signal sources and of R3 and R4 on the input impedance of the load (usually an amplifier).

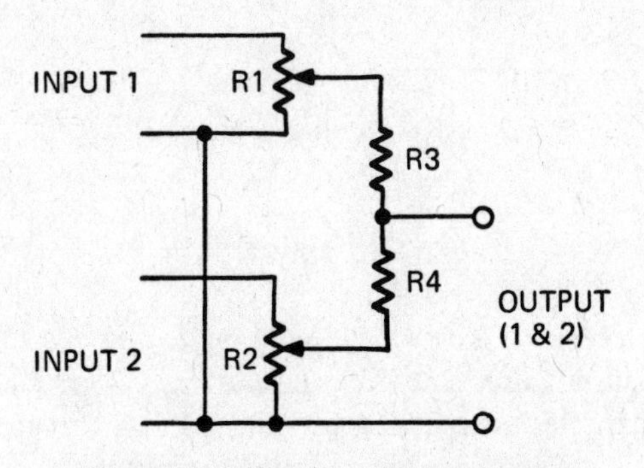

Figure 2-9. Passive AF signal mixer.

Assume that R1 and R2 are each 250,000 ohms and R3 and R4 are each 100,000 ohms and that the input impedance of the load is 50,000 ohms. With both R1 and R2 set at their zero output positions, the load will be terminated in 50,000 ohms (R3 and R4 in parallel). With R1 and R2 set to any position, the load will never be terminated in less than

50,000 ohms. When R1 is set to deliver a signal to R3 at a level of 100 millivolts, for example, and R2 is set to its zero output position, the output voltage at the junction of R3 and R4 will be approximately 33 millivolts because of the voltage divider action of R3 and R4 in parallel with the load impedance. When both R1 and R2 are set to deliver signals (to R3 and R4) at a level of 100 millivolts, the composite signal level at the junction of R3 and R4 will be higher than 33 millivolts because neither R1 nor R2 shunt R3 or R4 totally across the mixer output.

While not perfect, this is an inexpensive mixer which can be used to combine the outputs of two ceramic phonograph pickups, radio tuners or preamplifiers for feeding into a single auxiliary or other medium-level amplifier input.

Speaker Combiner

A single loudspeaker can be fed from two signal sources through a three-winding transformer, such as the Mix-n-Match (Alco Electronic Products Co.). A typical application is illustrated in Figure 2-10. Here, a single speaker is used as an extension speaker of a two-channel stereophonic sound reproducing system. The left channel signal is fed into winding L1 and the right channel signal into winding L2. The combined left and right channel signals developed across L3 are fed to the speaker which reproduces the sounds monaurally.

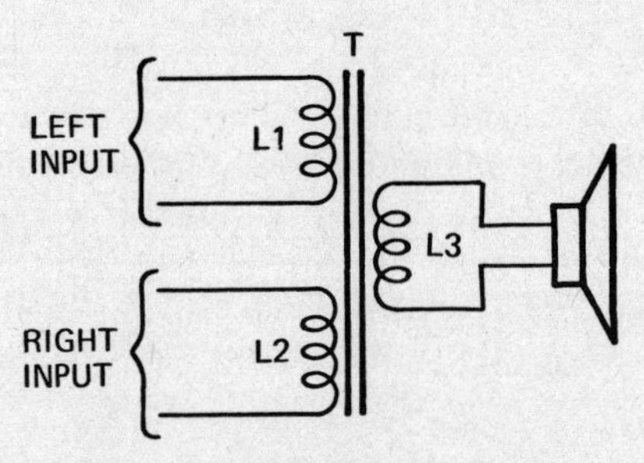

Figure 2-10. Speaker combiner.

Speaker Splitter

Two loudspeakers can be fed from the same signal source directly or through a Mix-n-Match (Alco Electronic Products Co.) transformer (T), as shown in Figure 2-11. The signal source is connected to primary winding L1 and the speakers to secondary windings L2 and L3. All three windings are designed to interface an 8-ohm load or source. The available power is divided equally between both speakers.

Three-Channel Stereo Synthesizer

The effect of three-channel stereophonic sound can be obtained with a two-channel stereophonic sound reproducing system by adding a third loudspeaker and a mixing

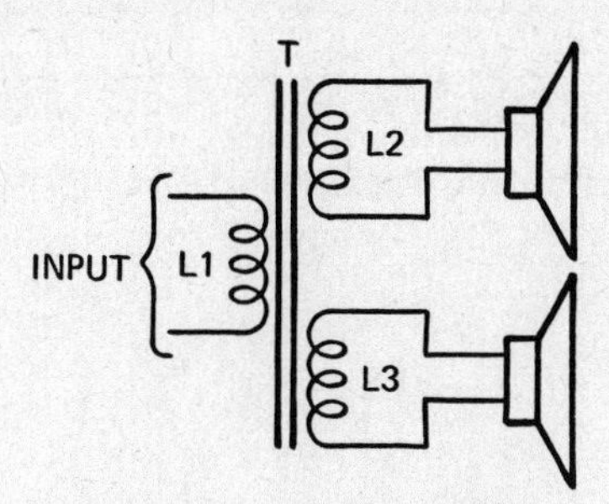

Figure 2-11. Speaker splitter.

transformer, connected as shown in Figure 2-12. Transformer T is a "Mix-n-Match" transformer (Alco Electronic Products Co.) which has three windings of equal impedance. The third loudspeaker is connected to one of the windings. The left channel signal is fed through a variable T-pad (R1) to one of the other windings, and the right channel signal is fed through a variable T-pad (R2) to the remaining winding. R1 and R2 are adjusted to obtain the desired blend of left and right signals through the third loudspeaker.

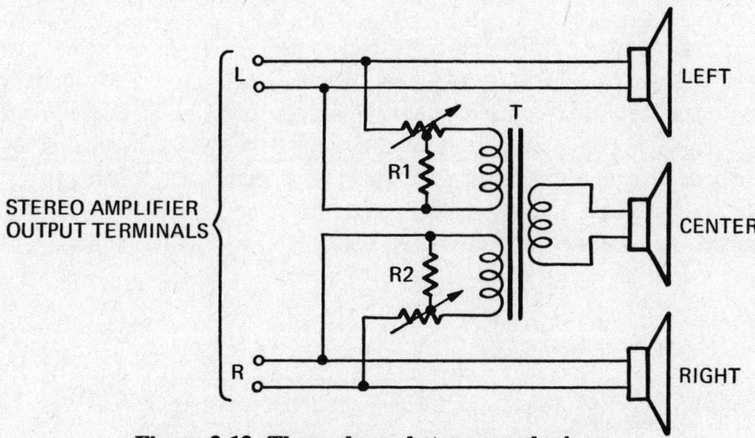

Figure 2-12. Three-channel stereo synthesizer.

Tone Controls

Two methods of tone control used in early radios are illustrated in Figure 2-13. Actually, both the R1-C2 network and the R3-C4 network perform their task in a similar manner. In the case of the R1-C2 network, adjustment of potentiometer R1 toward C1 causes more of the incoming signal to be fed to capacitor C2. However, because the reactance of C2 decreases as frequency increases, it will attenuate higher frequencies more than lower frequencies. The R3-C4 network operates in the same manner. In this case,

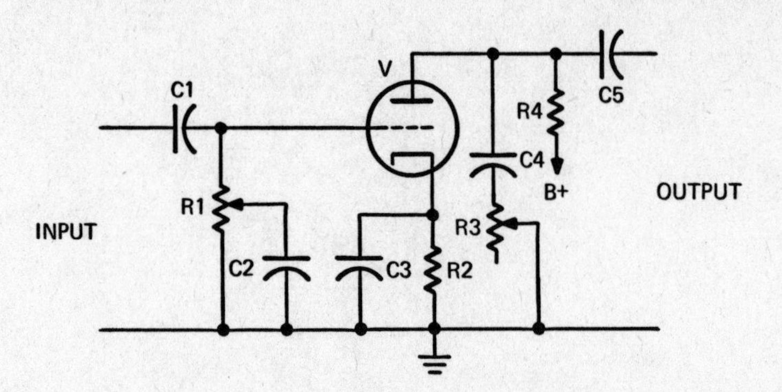

Figure 2-13. Two methods of tone control.

increasing the resistance of R3 reduces the amount of high frequency attenuation, and vice versa. Both circuits produce the illusion of bass boost when the high frequencies are attenuated and volume is increased.

Two-Channel Audio Mixer

Figure 2-14 illustrates the circuit of a two-channel audio mixer using a twin-triode tube. Input signals are applied to the grids of the triodes through level-control potentiometers R3 and R7. The amplified output signals are combined and fed out through C2. If one input signal is larger than the other, the signal developed across its respective load resistor will be commensurately greater. Interaction between the two plate circuits is minimized by R1 and R5. Both triodes share the same cathode bias resistor (4) and cathode bypass capacitor (C1). (Illustration courtesy of RCA Corporation.)

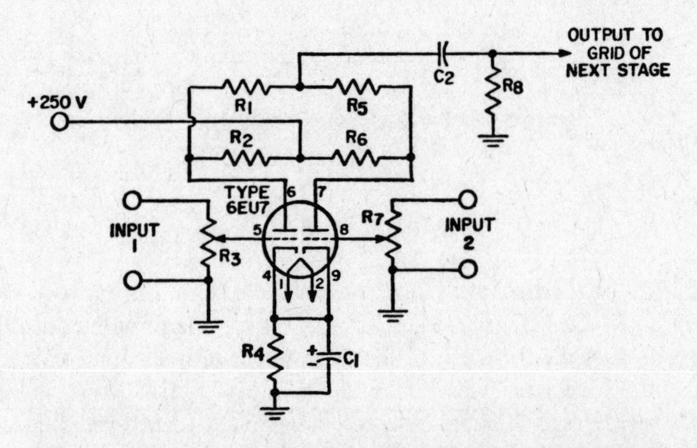

Figure 2-14. Two-channel audio mixer using twin-triode tube.

Wireless Intercom

Wireless intercom systems operate on the "carrier current" principle in that they utilize the powerline as the transmission medium. Figure 2-15 is a schematic of a Lafayette wireless intercom unit which is actually a low-frequency transceiver. It has a talk-listen switch that converts it from a receiver into a transmitter when set in the "talk" position. When set in the "listen" position, the incoming RF signal is fed to the antenna winding of T1 and is coupled to the tuned tank winding. From there, it is fed the grid of the 12AU6(1) tube which functions as a grid leak detector. The recovered audio at the plate of this tube is capacitively coupled to the control grid of the 50EH5 AF power amplifier tube which, in turn, feeds the AF signal to the speaker through the output transformer T3.

The 12AU6(2) tube functions as a squelch. When there is no incoming RF signal at its control grid, its plate current is high and its plate voltage is low. Since the 12AU6(1) tube gets its screen grid voltage from the plate of the 12AU6(2) tube, the 12AU6(1) tube is cut off. But, when an RF signal is present, it is rectified by the grid-cathode path of the 12AU6(2) tube, making the grid negative and causing plate current to fall and plate

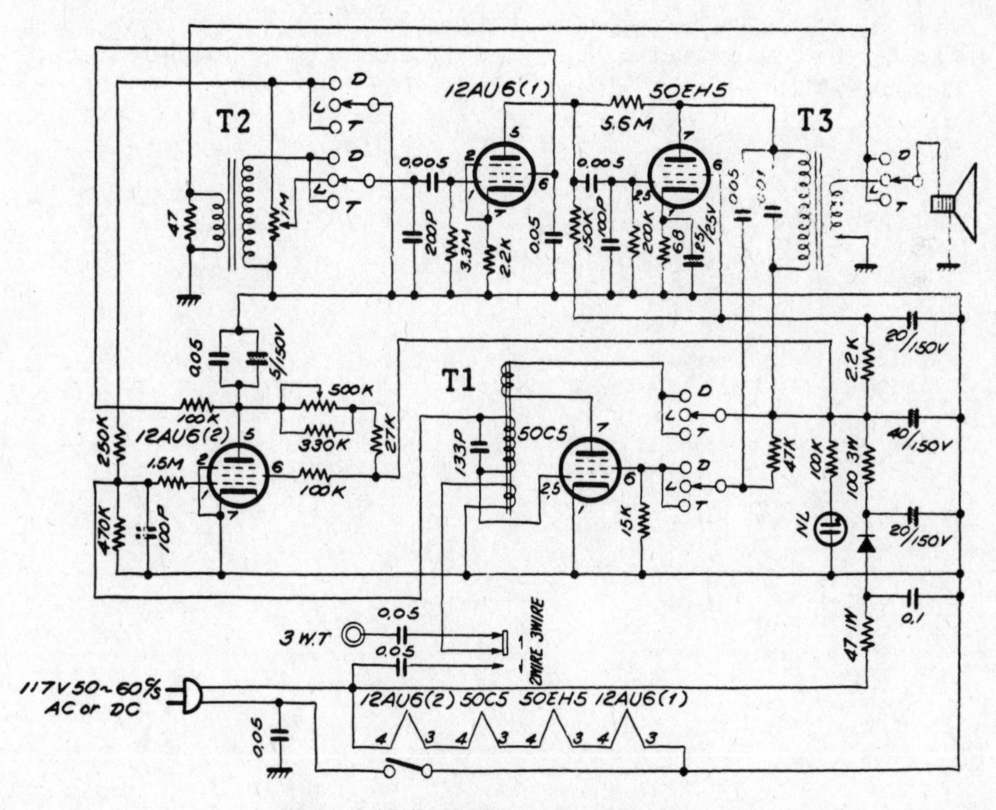

Figure 2-15. Lafayette wireless intercom.

voltage to rise—thus making the screen grid of the 12AU6(1) tube positive enough to allow it to operate and pass signals to the AF power amplifier.

When set to transmit, DC plate voltage is applied to the plate of the 50C5 tube through the tickler coil of T1. The screen grid of the 50C5 tube is connected to the plate of the 50EH5 tube which now functions as the modulator. The speaker, which now acts as a microphone, is connected to the primary of T2 and its output signal is amplified by the 12AU6(1) tube which drives the 50EH5 modulator. Since the voltage across the primary of T3 varies with the audio signal, screen grid modulation of the 50C5 RF oscillator tube is achieved. The amplitude modulated RF output of the 50C5 tube is fed through T1 and a capacitor to the AC power line. Both the receiving and transmitting frequencies are the same and are determined by the tank circuit of T1.

Two or more of these intercom units may be used for intercommunication if they are connected to the same powerline. Normally, all units are set to receive so they will be ready to intercept a signal from any of the other units. Since they all operate on the same frequency, only one unit can transmit at any one time.

As the diagram shows, the unit has an AC/DC power supply. The tube heaters are connected in series across the powerline. When connected to an AC powerline, plate and screen grid voltage is obtained through a half-wave diode rectifier. When connected to a DC powerline, the diode simply acts as a closed switch if the powerplug is inserted into the power outlet so that the anode of the diode will be positive. If plugged in the wrong way, the unit will not operate because the diode will be reverse biased and will therefore act as an open switch.

3

Biasing Circuits

Bias is important for proper operation of electronic components. The control grid of an electron tube is reverse biased to limit plate current. The gate of an FET is reverse biased to limit drain current. But the base of a bipolar transistor is forward biased to induce collector current flow. Described and illustrated in this part of the book are various biasing circuits for electron tubes and transistors.

Cathode Bias

Negative grid bias for an electron tube can be obtained by connecting a resistor (R2) in series with the cathode, as shown in Figure 3-1. The voltage drop across R2 is equal to the product of the cathode current and the resistance of R2. For example, if R2 is 1000 ohms and cathode current is 2 mA, the voltage across R2 will be 2 volts since 1000 x 0.002 = 2. Point Y (cathode) is 2 volts positive with respect to point Z (common ground). Since no current flows through R1, the voltage drop across it is zero. Therefore, point X (grid) is 2 volts negative with respect to Y. Capacitor C2 charges to the Y-Z voltage and prevents changes in this voltage that would otherwise be caused as cathode current is varied by the signal.

For a triode tube, the value of R2 can be calculated by:

$$R2 = \frac{E}{I}$$

where:

R2 = cathode resistor value in ohms,
E = required grid bias voltage, and
I = plate current in amperes.

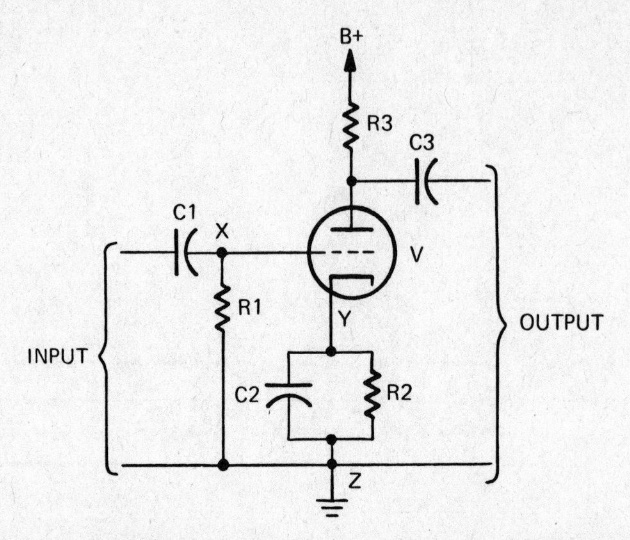

Figure 3-1. Connecting a resistor in series with the cathode for negative grid bias.

For a tetrode or pentode tube, the value of R2 can be calculated by:

$$R2 = \frac{E}{IP + ISG}$$

where:

R2 = cathode resistor value in ohms,
E = required control grid bias voltage,
IP = plate current in amperes, and
ISG = screen grid current in amperes.

FET Drain Bias

The characteristics of a field effect transistor (FET) are similar to those of a triode tube. Its gate can be reverse-biased in the same manner as an electron tube (see CATHODE BIAS). In the circuit shown in Figure 3-2, the gate of an FET is made negative with respect to its source by the voltage drop across R2. Since the source is positive with respect to ground, and since the gate is grounded through R1, the gate is negative with respect to the source. As does happen in an electron tube circuit, C2 minimizes degeneration.

Fixed Bias

In the circuit illustrated in Figure 3-3 the grid is negatively biased by a DC voltage connected to C-. The bias voltage source may be a battery, a rectifier power supply, or a negative point of a voltage divider network.

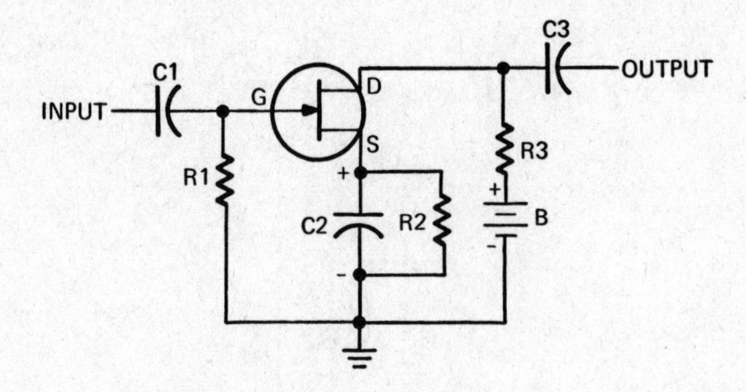

Figure 3-2. Gate of an FET made negative with respect to its source.

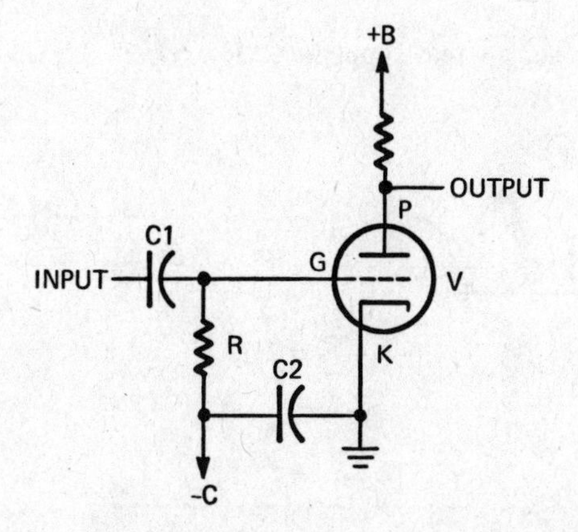

Figure 3-3. Grid negatively biased by a DC voltage.

Grid-Leak and Battery Bias

Figure 3-4 shows how a separate bias supply is used to prevent excessive plate current in the event that the input signal is removed from a tube employing grid-leak bias. With normal grid-leak bias, removing the driving voltage or lowering it below the value necessary to drive the grid positive, and result in grid current, will result in loss of bias, and excessive plate current flow and damage to the tube can result. Using this circuit, there will be a negative potential applied to the grid by battery B1 (or a DC power supply) will prevent excessive plate current.

MOSFET Drain Bias

The gate of a single-gate MOSFET can be made negative with respect to its source by inserting a resistor (R2) in series with the source, as shown in Figure 3-5. (See

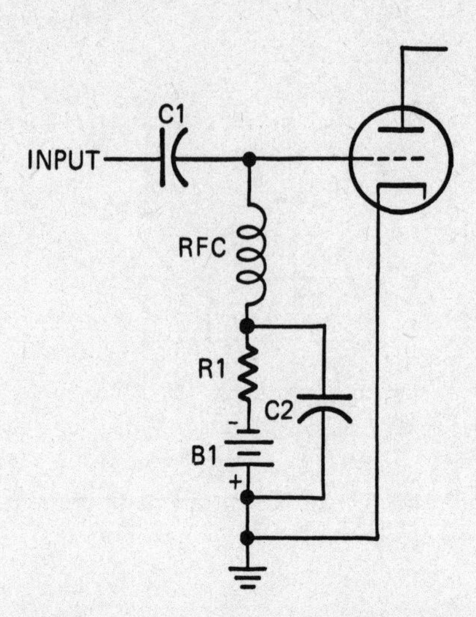

Figure 3-4. Separate bias supply used to prevent excessive plate current.

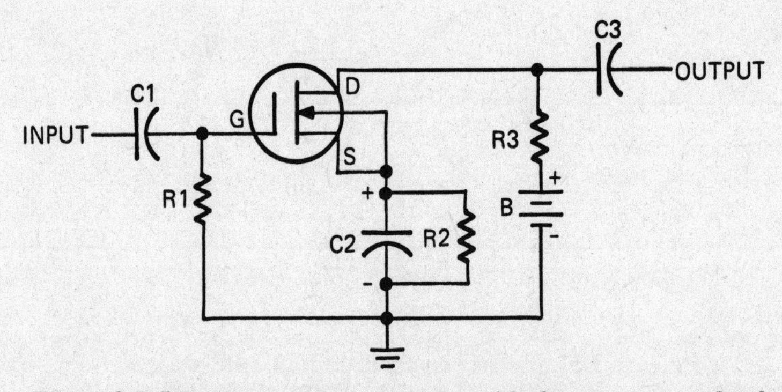

Figure 3-5. Inserting a resistor in series with the source to make the gate of a single-gate MOSFET negative.

CATHODE BIAS and FET SELF BIAS.) Because of R2, the source of the MOSFET is positive with respect to ground and, because the gate is grounded through R1, the gate is negative with respect to the source. Capacitor C2, shunted across R2, minimizes gain loss that would be caused by degeneration if R2 was not bypassed by R2.

Oscillator Generated Bias

Instead of utilizing cathode bias or fixed bias from a DC voltage source, an AF power amplifier pentode (or beam power tube) can be biased by the DC voltage generated by the local oscillator of the radio receiver. As shown in Figure 3-6A, the control grid of the AF

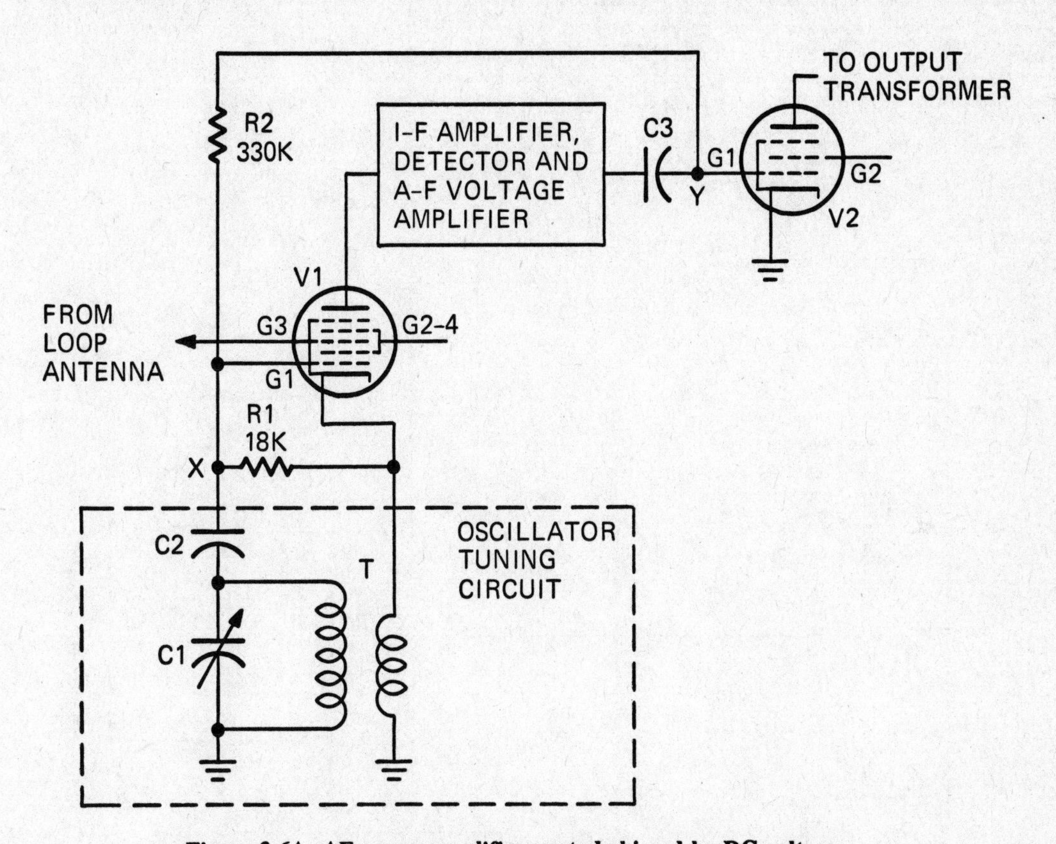

Figure 3-6A. AF power amplifier pentode biased by DC voltage.

power amplifier tube is connected through R2 to the control grid (G1) of a pentagrid converter tube. When the oscillator portion of the pentagrid converter tube circuit is functioning, a negative DC voltage is developed across its grid leak R1 making point X negative with respect to ground. The value of the bias required for the AF power amplifier tube can be varied by changing the value of R1. The advantage of this circuit is that it requires fewer components.

In the complete superheterodyne receiver, Figure 3-6B, this technique is also shown. The 12BE6 is the pentagrid converter tube and the 50C5 is the AF pentode or beam power output tube. Note that the cathode of the 50C5 is grounded directly, not through a resistor. The control grid of the 50C5 is connected through a 330K ohm resistor to the oscillator grid (1) of the 12BE6. When the oscillator is functioning, a DC voltage (bias) is developed across the 18K oscillator grid leak resistor, with pin 1 negative with respect to ground. Since the control grid of the 50C5 does not draw current, there is no DC voltage drop across the 330K resistor and the DC voltage at the control grid of the 50C5 is the same as at the oscillator grid of the 12BE6. Thus the AF amplifier, the 50C5, is biased by the grid leak bias voltage developed by the oscillator.

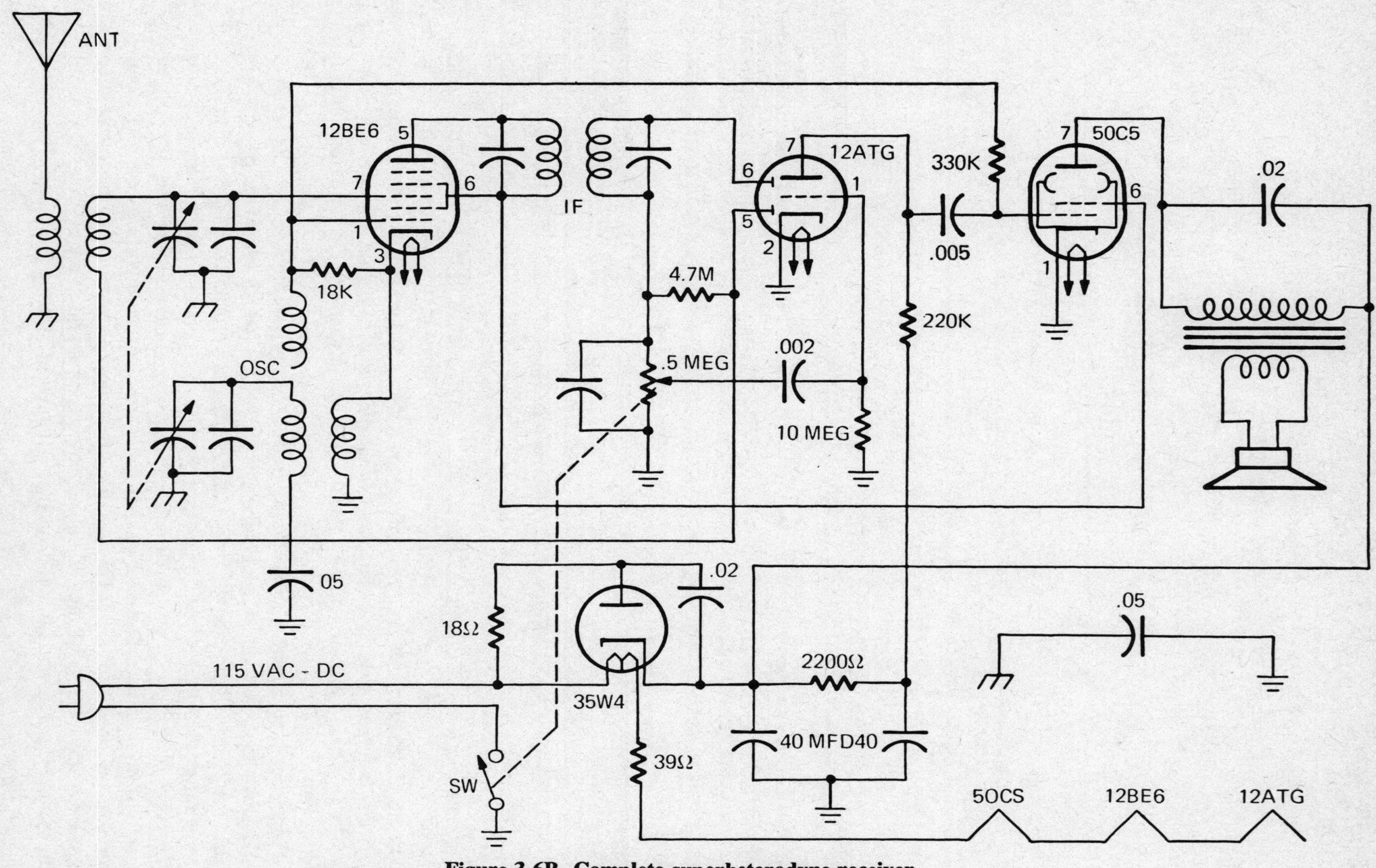

Figure 3-6B. Complete superheterodyne receiver.

Transistor Fixed Bias

Figure 3-7 shows a transistor circuit employing fixed bias. In this case, a PNP transistor is connected in a common-emitter configuration. R1 and R2 form a voltage divider across battery B1. Since the base is connected to the junction of R1 and R2, the base will be negative with respect to the emitter. A separate battery (B2) furnishes collector-emitter voltage.

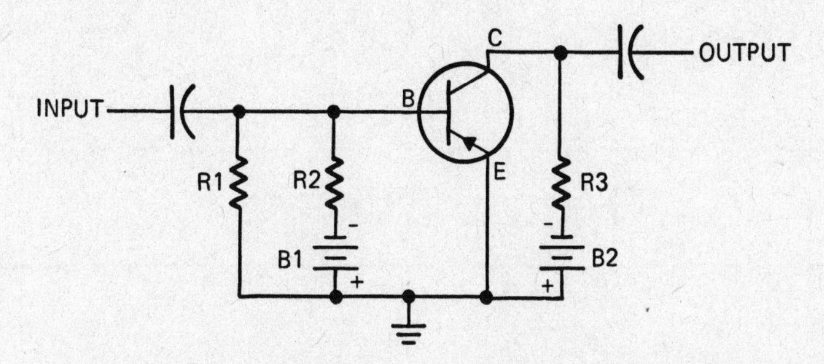

Figure 3-7. Transistor circuit employing fixed bias.

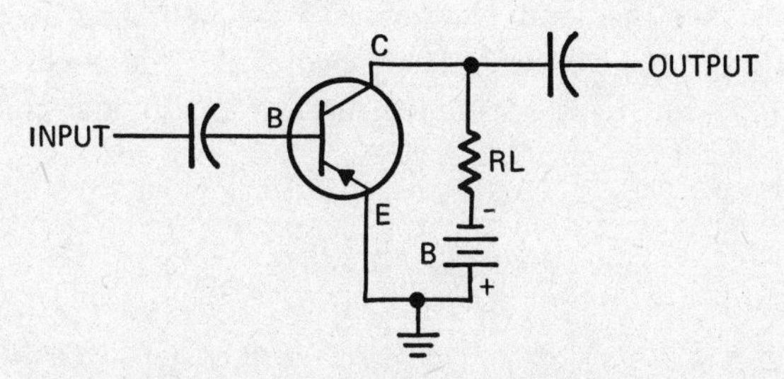

Figure 3-8. Transistor circuit in which self-bias is used.

Transistor Self-Bias

Figure 3-8 shows a transistor circuit in which self-bias is used in a common-emitter PNP amplifier stage. The collector is at the most negative potential, and the emitter is at the most positive potential, making their relationship correct. Since the base is sandwiched between these elements and floating, it is at some intermediate potential, depending upon the internal resistance parameters and the internal current flow. Thus, the base will be positive with respect to the collector, but negative with respect to the emitter.

The potential between the emitter and the base must be kept to a small value compared to that between the collector and the base. This is achieved internally by the high-resistance of the collector-base junction and the low-resistance of the emitter junction, which provide the desired relationships. The supply voltage polarity is reversed for NPN transistors.

4

Control Circuits

The flow of electric current is most often controlled with a manually operated switch, directly or through a relay. Thyratron tubes are used for controlling electric current flow electronically. So are the more recently developed SCR (silicon controlled rectifier), Zener diode, diac, triac, and other semiconductor devices. Various control circuits are described and illustrated in this chapter.

Attenuator Amplifier

An attenuator may be a fixed or variable passive device which reduces signal level, such as a voltage divider, potentiometer, L-pad, T-pad, etc., and which provides gain. The electronic attenuator, whose circuit is shown in Figure 4-1, provides 13 dB of voltage gain when its control is set for zero attenuation. When attenuation control R, a 50,000-ohm potentiometer, is set to insert approximately 7500 ohms of resistance between pin 2 of the MFC 6040 IC and ground, gain is reduced to 0 dB. As the resistance of R is increased, attenuation is increased. When R is set to 33,000 ohms, attenuation is 90 dB.

If input signal level is 100 millivolts RMS, for example, output signal level can be adjusted with R as listed as follows.

R (ohms)	Gain or loss	Output level
0	+13 dB	450 mV
7,000	+ 3 dB	140 mV
10,000	−17 dB	14.13 mV
20,000	−57 dB	141.3 uV
33,000	−77 dB	14.1 uV

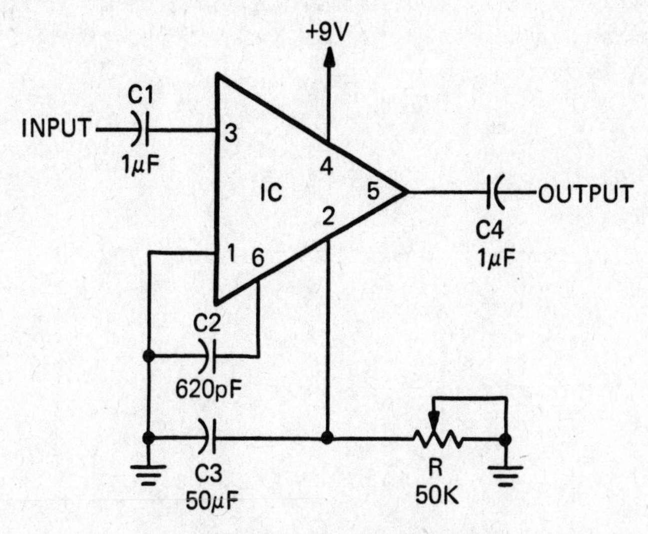

Figure 4-1. Electronic attenuator.

When maximum rated input voltage of 500 millivolts is applied, output voltage can be varied from 706 microvolts to 2.25 volts RMS.

The Motorola MFC 6040 electronic attenuator contains ten transistors and two diodes within a single silicon monolithic chip. Frequency response is flat from 100 Hz to 300 kHz and is down only 3 dB at 5 MHz.

Automatic Gain Control

Figure 4-2 shows an IF stage of a receiver in which two methods of automatic gain control are employed simultaneously. Negative AGC voltage from the detector is applied through R5 to the base of transistor amplifier Q2. Although Q2 is forward-biased by the positive potential applied to it through R2, a stronger signal causes a larger negative potential to be applied, reducing the forward bias on Q2 and reducing its gain. By acting as a shunt across the primary of transformer T1, diode D1 also provides some automatic gain control.

Automatic Level Control

The automatic level control can be used to automatically regulate the level of a signal. It employs a Raytheon CK1103 Raysistor which consists of a light sensitive resistance and a lamp. As shown in Figure 4-3, the input signal is fed in at phonojack J1 and out from phonojack J2. R1 and R2 form a fixed voltage divider. The light sensitive resistance of the Raysistor is shunted across R2. The Raysistor's lamp is connected to the audio output of the amplifier through phonojack J3. When amplifier output level tends to rise, the lamp within the Raysistor glows more brightly and reduces the resistance of the light sensitive resistor. When output level tends to rise, the lamp glows more dimly and the resistance of the light sensitive resistor becomes higher. Therefore, the Raysistor acts as a variable resistor across R2 whose value varies with the output voltage fed to the Raysistor filament.

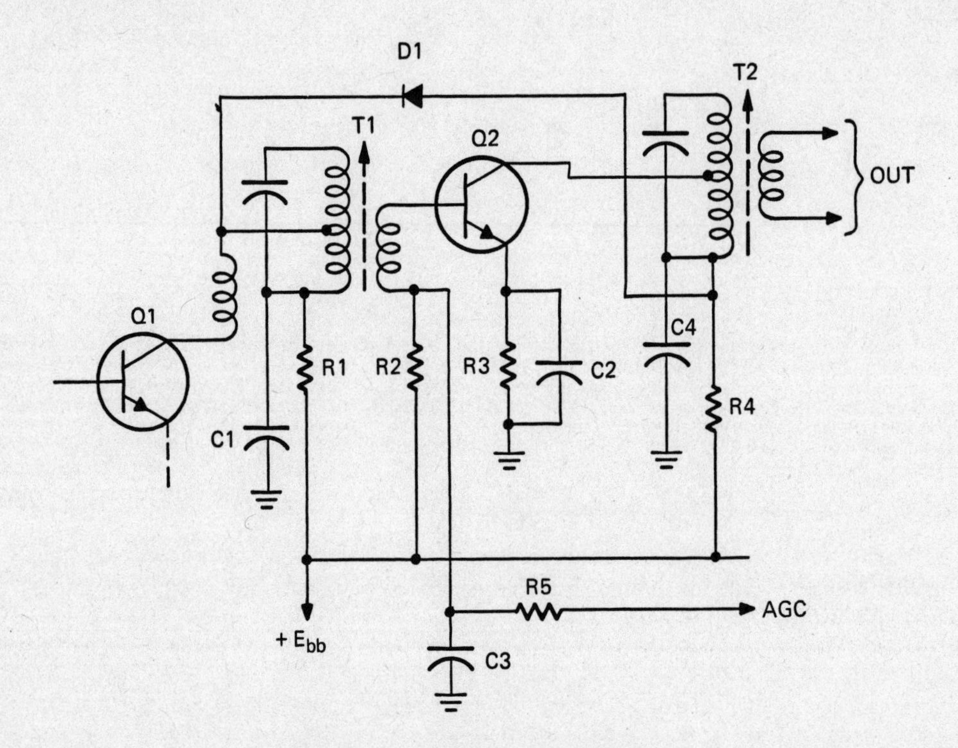

Figure 4-2. Automatic gain control.

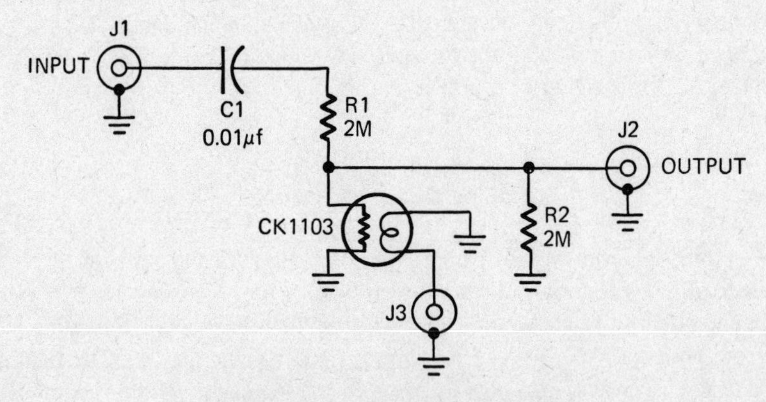

Figure 4-3. Automatic level control.

Capacitive-Discharge Ignition

In the capacitive-discharge ignition system, whose circuit is shown in Figure 4-4, capacitor C1 is charged to a high-voltage level—in this case, 480 volts DC. When the distributor points close, the SCR (silicon controlled rectifier) is triggered, and the capacitor discharges through the SCR and the primary of the ignition coil. The inductive kick causes a very high voltage to be induced into the secondary of the ignition coil. This voltage is fed to the spark plugs through the distributor.

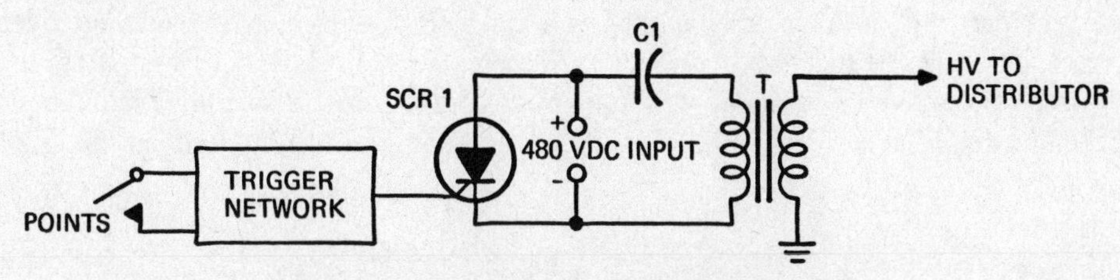

Figure 4-4. Capacitive-discharge ignition circuit.

Discriminator-Derived Squelch

Figure 4-5 illustrates the principle of discriminator-derived squelch. The audio amplifier triode is normally biased to cutoff by applying a positive voltage to its cathode through R14 and R9. Normally, the squelch tube is biased so that it conducts heavily, causing the positive voltage at the junction of R7 and R8 to be low enough to keep the audio amplifier cut off.

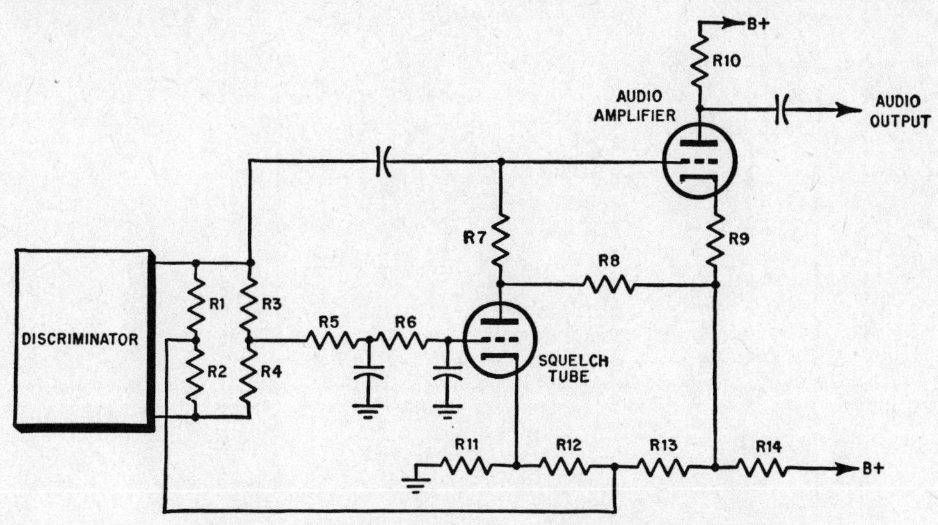

Figure 4-5. The principle of discriminator-derived squelch.

A negative DC voltage is developed at the junction of load resistors R3 and R4 when a signal is applied. This negative voltage is applied through an R-C filter to the grid of the squelch tube, causing it to conduct less heavily. Its plate voltage rises and the voltage at the junction of R7 and R8 becomes positive enough to reduce the bias on the audio amplifier tube bias on the audio amplifier tube so it will conduct and amplify the audio signal.

Electronic Oven Control

In Figure 4-6, a diode, D1, is used as a temperature sensor in an electronic oven control. D1 is in the feedback loop of the first amplifier. If the temperature rises, the resistance of diode D1 decreases, permitting more feedback to the inverting input, pin 8, of the first amplifier and reducing its gain. The input to the non-inverting input, pin 5, of

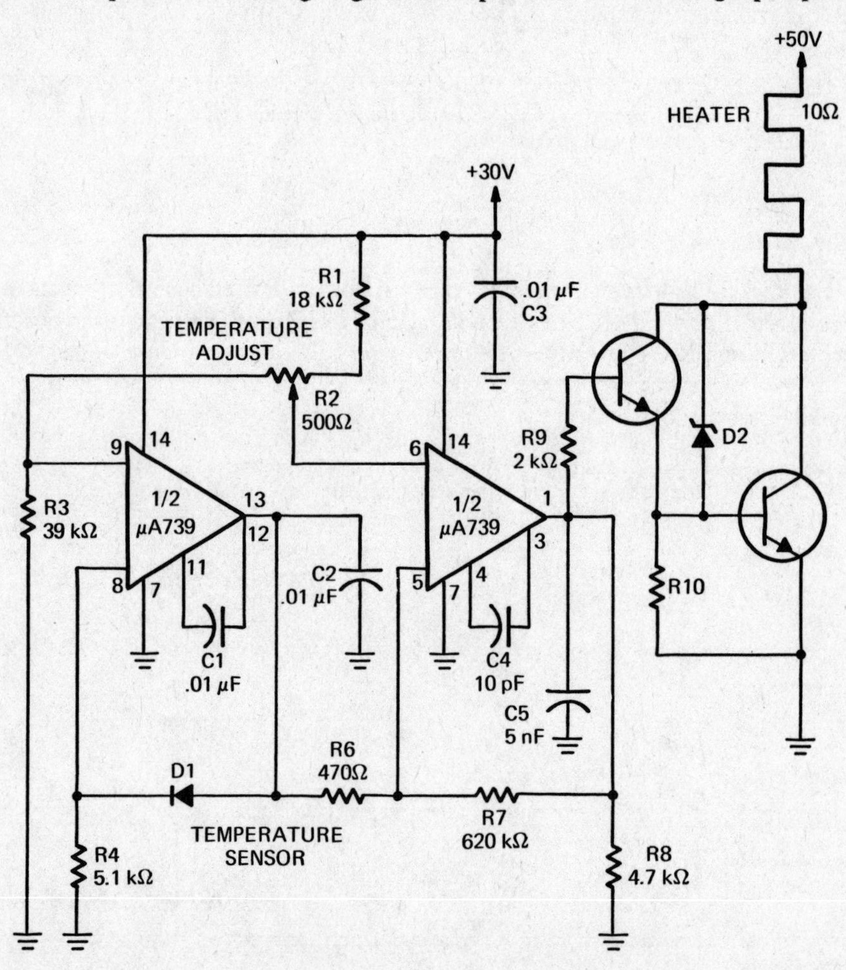

Figure 4-6. Electronic oven control.

the second amplifier stage is also reduced, and the output from the second amplifier stage and the current through the heater element becomes less. By adjusting R2, temperature can be set by controlling the bias level to the inverting input, pin 6, of the second amplifier stage.

Gain Control

A gain control, as opposed to a level control, is used to vary the gain of an amplifier such as a pentode tube. The gain of a remote cutoff pentode (variable mu) tube can be changed by varying its cathode bias as shown in Figure 4-7A, with a potentiometer. In this circuit, the input signal is fed to grid G1. Gain is controlled with potentiometer R2. Fixed resistor R1, in series with R2, prevents minimum cathode bias from being reduced to zero when R1 is set for minimum resistance to obtain maximum gain. Capacitor C2 places the cathode at ground potential at the signal frequency and prevents degeneration at that frequency.

The gain of a sharp cutoff pentode can be changed by varying its screen grid (G2) voltage, as shown in Figure 4-7B, with potentiometer R2. The gain increases as G2 voltage is increased by adjusting R2.

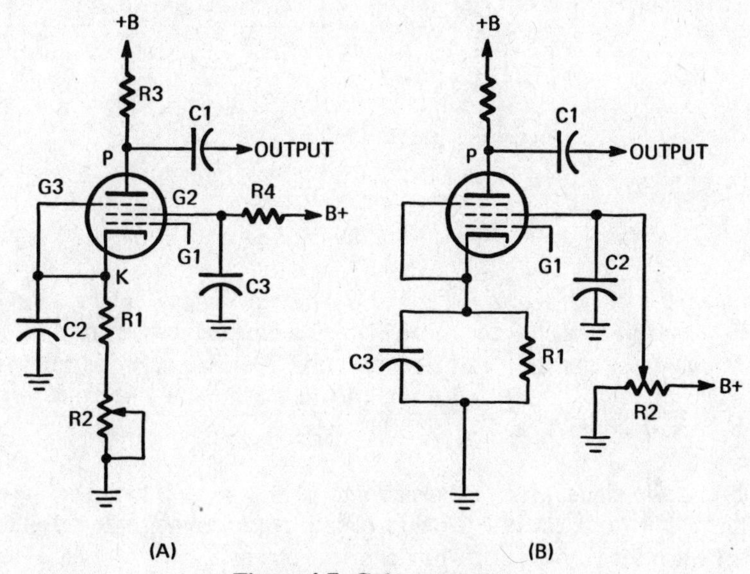

Figure 4-7. Gain control.

Instant-On Radio Control

To eliminate having to wait for a tube-type radio to start operating until the tube heaters reach operating temperature, the tube heaters can be left turned on continuously. To silence the receiver, only the plate supply voltage need be cut off. In an AC-DC type radio, in which the tube heaters are connected in series, the "instant-on" feature can be added by connecting a diode across the on-off switch, as shown in Figure 4-8. (V1, V2, V3, V4, and V5 represent the tube heaters.)

When switch S is open, rectified AC heater current flows through diode CR during the power source half-cycles so that the cathode of the diode is made negative with respect to its anode. With the diode polarized as shown in the diagram, the plate of the receiver's rectifier tube (V5) receives negative voltage pulses and cannot conduct. When S is closed, it shorts the diode. Now, unrectified AC flows through the tube heaters and the AC line voltage is applied to the plate of V5. Since the plate of V5 is made positive during half of each AC cycle, it functions normally and the receiver operates. Thus, when S is closed, the receiver starts to operate immediately since the tube heaters are already warmed up.

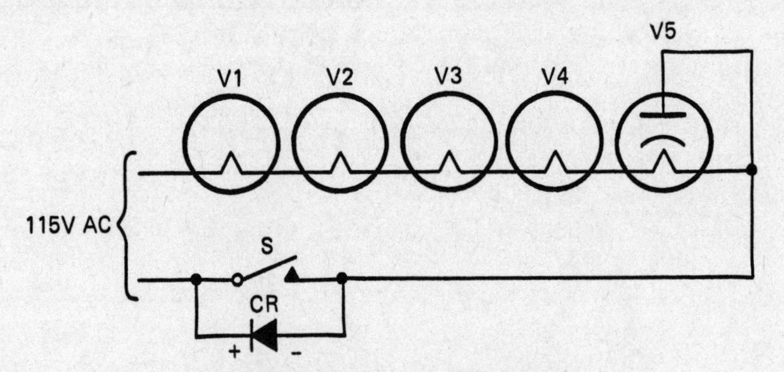

Figure 4-8. Instant-on radio control.

Noise-Actuated Squelch

In the squelch circuit shown in Figure 4-9, the audio amplifier tube is biased on and off by the squelch tube which is controlled by the quieting of receiver noise and by the rise in limiter voltage when a radio signal is intercepted. The no-signal noise output of the FM discriminator is amplified by the noise amplifier tube and fed to one diode section of the dual-diode noise rectifier tube. The rectified noise voltage is applied to the grid of the squelch tube through R6.

With no radio signal present, squelch tube plate current is high and the audio amplifier tube is biased to cut off. When a signal quiets the noise at the output of the discriminator and increases the first limiter grid voltage, squelch tube plate current falls and the audio amplifier tube bias is reduced so it can function.

The output of the noise amplifier is tuned by L1-C3 so that only high-frequency noise will be fed to the noise rectifier. The positive half of the no-signal noise causes C5 to charge to a positive value. The negative half causes C6 to charge to a negative value. Since C5 and C6 are joined through R4, and because of the negative limiter voltage fed to them through R5, the voltage across C7 is equal to the algebraic sum of the voltages across C5 and C6.

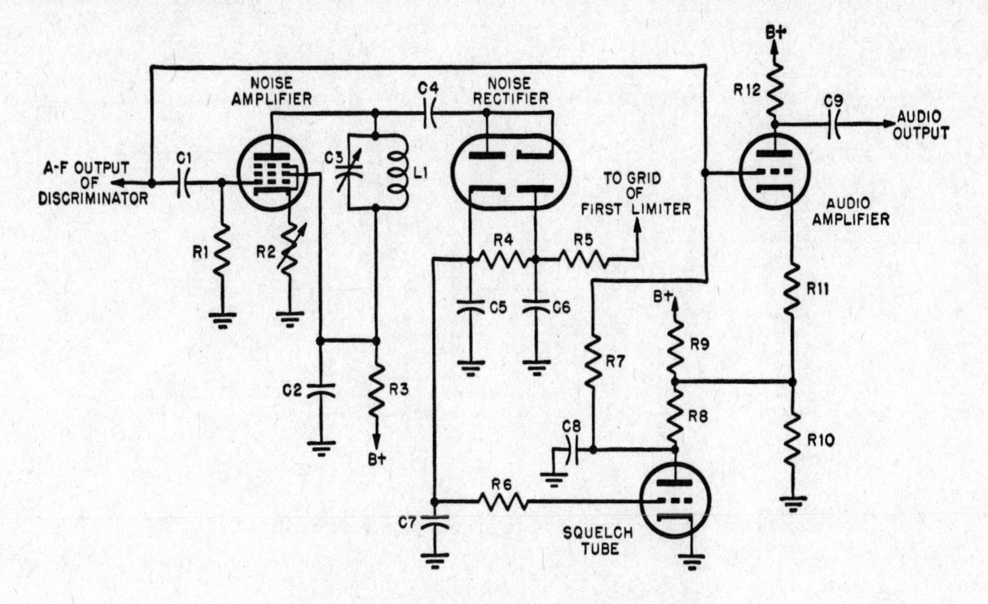

Figure 4-9. Noise-actuated squelch circuit.

Simplexed-to-Ground Control Circuit

The circuit shown in Figure 4-10 is useful for transmitting both audio signal information and DC control information over a single pair of wires. Transformers T1 and T2 are telephone repeating coils with twin primaries and secondaries, connected as shown. L1 and L2 of T1 are connected in series, as are L3 and L4 of T2. L3 and L4 of T1 and L1 and L2 of T2 are also connected in series so that they effectively form center-tapped windings.

Audio information is fed through the primaries of T1, transmitted from the secondary of T1 to the primary of T2, and transferred to the secondary of T2 and fed on to an amplifier or transmitter. When DC control is required, the switch S in the secondary circuit of T1 is closed, causing DC to be transmitted to the primaries of T2. Direct current flows through both wires in the same direction and from the junction of L1 and L2 of T2 through the coil of relay K and back to the battery and S through earth ground. The relay is pulled in when current flows through its coil. The audio signal being transmitted over the lines will not be affected by the DC across the line.

Step Attenuator

Figure 4-11 shows the circuit of a step-type attenuator. The attenuator shown has two sections, a 6 dB and a 12 dB section, but commercially available attenuators may have many sections. In the case of each section, with the switch in the top position, the signal is fed through without attenuation, while with the switch in the bottom position, a portion of

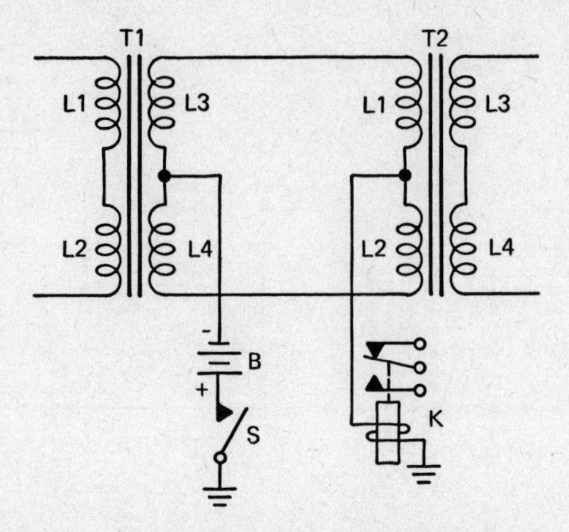

Figure 4-10. Simplexed-to-ground control circuit.

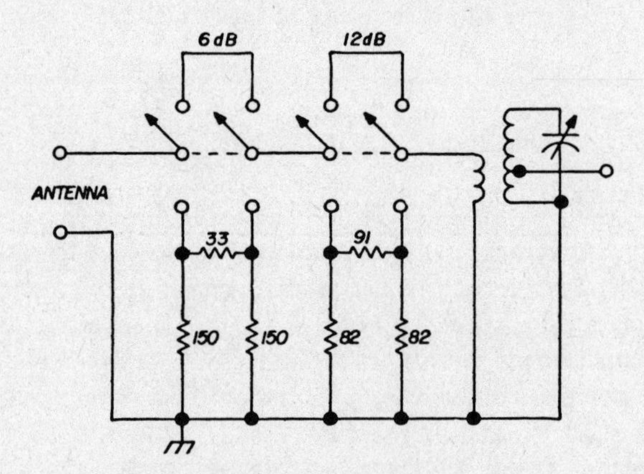

Figure 4-11. Step attenuator circuit.

the signal is bypassed to ground. This particular circuit shows the attenuator used at the input of a radio receiver to reduce signal level. (Illustration courtesy of *Ham Radio*.)

Stepping Switch Driver

A stepping relay, rotary solenoid actuated switch, and other similar devices can be driven at automatically timed intervals by the circuit shown in Figure 4-12. The AC power line voltage is rectified by diode CR and filtered by C1. R1 limits diode current when C1 is being charged initially. The DC voltage across X and Y is fed to a relaxation oscillator

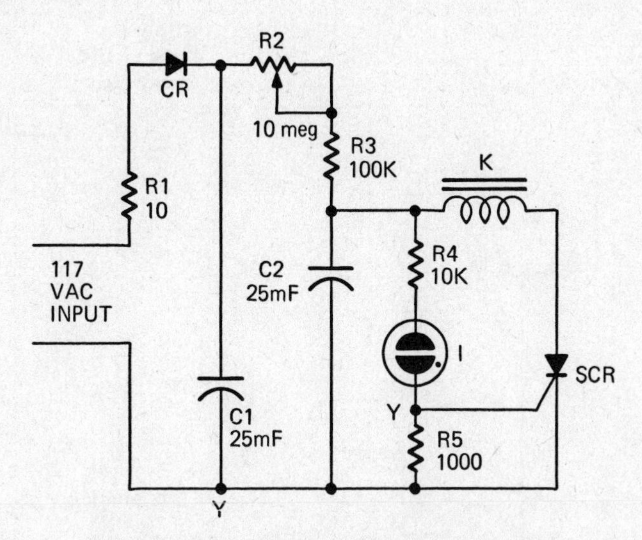

Figure 4-12. Stepping switch driver circuit.

circuit consisting of R2, R3, R4, R5, C2, and neon lamp I. The load (relay coil, etc.) is represented by K.

No current flows through K until C25 is charged through R2 and R3 to a level of about 65 volts. Then neon lamp I fires and allows a momentary positive voltage to appear at Y. This positive voltage triggers SCR, a silicon controlled rectifier, causing it to conduct and allow current to flow through the load L. Load current flows only momentarily until C2 is discharged sufficiently through I to reduce the voltage across SCR so it will cease conducting.

The cycle resumes and repeats itself at intervals determined by the setting of R2. The repetition rate can also be varied by using a higher or lower value capacitor for C2.

Switch Failure Monitoring Device

The switch failure monitoring circuit shown in Figure 4-13 can be used to insure that interlock switches are operating properly. The device is constructed so that the two separate switches, the monitoring and the monitored, must be mechanically actuated at approximately the same time. Normally, even a slight lag from one switch to the other is not sufficient to cause the time-delay relay to actuate because of its inherent delay.

If, however, either the monitoring switch or the monitored switch fails to change with the other, a circuit is completed from one side of the powerline, through the time delay relay coil, to the other side of the powerline. If the delay time of the time-delay relay is exceeded (for example, if the relay is set for ten seconds of delay time before it actuates, and the switches are in opposite modes for 12 seconds), the time-delay relay will actuate. This in turn will cause the power relay to actuate, removing the AC power that is normally

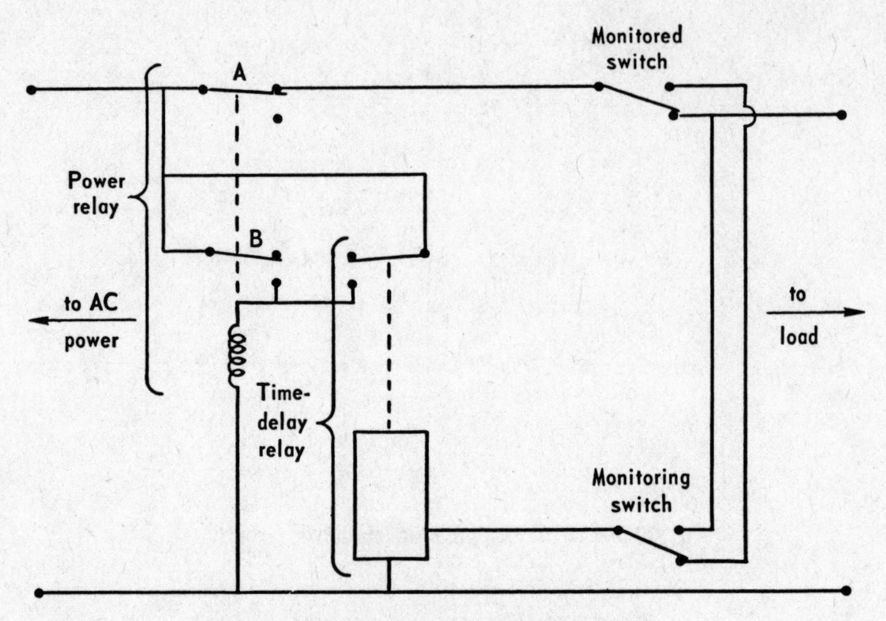

Figure 4-13. Switch failure monitoring device.

applied to the load through contact A of the power relay. Contact B of the power relay is a self-latching contact that will keep the power relay actuated, preventing application of AC power to the load, until AC power is removed from the power relay and the trouble corrected.

Transistor Ignition

In a transistor ignition system, using the circuit shown in Figure 4-14, the distributor points of an internal combustion engine are represented by S. As S opens and closes momentarily, the transistor conducts heavily for a short period of time and allows current to flow through the primary of ignition coil T. The inductive kick is boosted by the ignition coil. The resulting high voltage is fed through the distributor to the spark plugs. This type of ignition system eliminates arcing at the distributor points because the heavy flow of current from the battery is through the transistor emitter-collector path instead of through the points.

A Triac Switch

A triac is a three terminal semiconductor device capable of opening and closing high current circuits. Unlike an SCR, a triac is a bidirectional device for use in AC circuits. In Figure 4-15, a triac is shown connected in series with load RL. When switch S is open, the

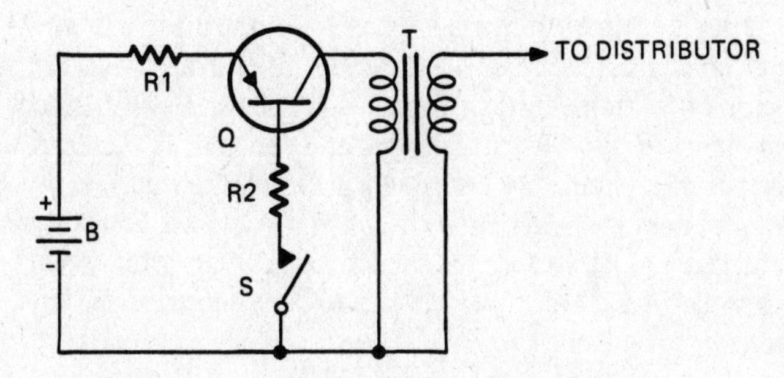

Figure 4-14. Transistor ignition system circuit.

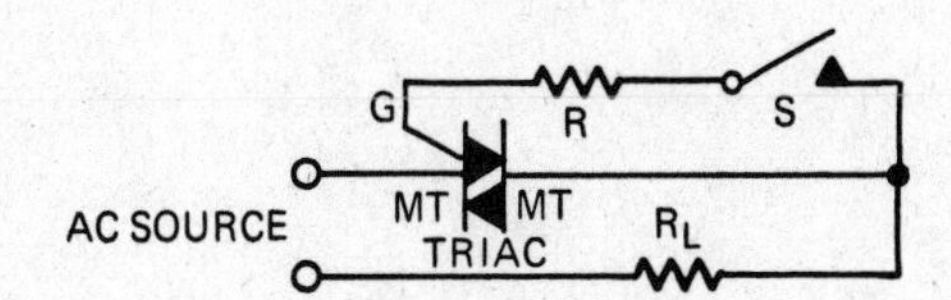

Figure 4-15. Triac switch.

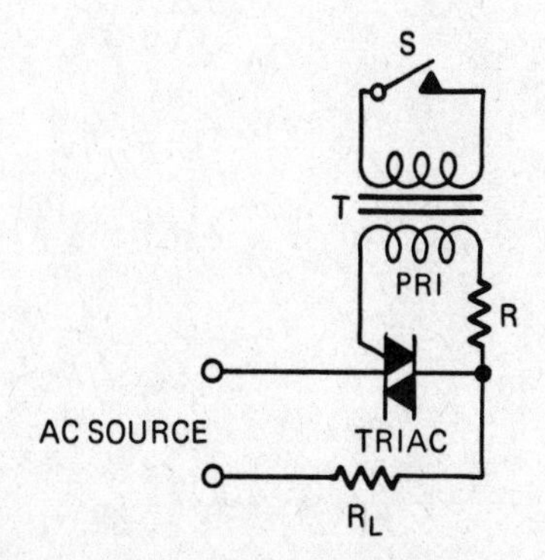

Figure 4-16. Triac switch with gate control.

triac acts as an open switch and blocks passage of current through the load. When S is closed, a small current flows to the gate of the triac through resistor R which causes the triac to switch to its conducting state and allow current to flow through the load. The triac is either open or closed. When closed, the voltage drop between its two main terminals (MT) is very small. The advantage of using this circuit is that a very light duty switch (S) can be used for controlling high current. Note in the diagram that the gate current is obtained from the load side of the circuit. This means that if there is no load, the triac will not be gated into conduction.

Triac Switch with Gate Control

In the triac switch circuit shown in Figure 4-16, the manual control switch can be at a considerable distance from the triac and its load. Gate current flows through resistor R and through the primary of transformer T which may be a 6.3 volt filament transformer or a doorbell transformer. If the resistance of R is great enough so that its resistance combined with the reactance of the transformer primary will keep gate current low enough, the triac will not be triggered into conduction until switch S is closed. Since switch S is connected across the secondary of T, closing the switch causes the primary reactance to drop low enough to allow sufficient triac gate current to flow to cause the triac to switch to the conducting state.

5

Coupling Circuits

A coupling circuit enables transfer of electrical energy from a source to a load or between stages of an electronic device. All coupling circuits introduce an "insertion loss" unless they incorporate an "active" amplifying device. For example, a step-up transformer is a "passive" coupling device which provides voltage gain, but introduces a power loss because a transformer is not 100 percent efficient. Passive filters introduce a larger insertion loss at some frequencies than others. Described and illustrated in this chapter are numerous AF and RF coupling and interfacing circuits.

Antenna Coupling, Capacitive

A receiving antenna can be coupled to the input of a receiver through a capacitor. As shown in Figure 5-1, the antenna is coupled through C1 to a tap on RF autotransformer T. The impedance between the tap and ground is lower than the impedance across C2. If the antenna were connected through C1 to the ungrounded end of T, the resonant circuit would be loaded down by the antenna. This would decrease the selectivity since the Q of T and C2 would be lowered.

Antenna Coupling, Inductive

In the circuit illustrated in Figure 5-2 RF current flowing through the primary of RF transformer T induces a voltage in the secondary through inductive coupling. The secondary is made resonant at a specific frequency by adjusting variable capacitor C. At the resonant frequency, the secondary impedance is high. And because of the reflected-impedance effect, the primary impedance is highest at the resonant frequency of the

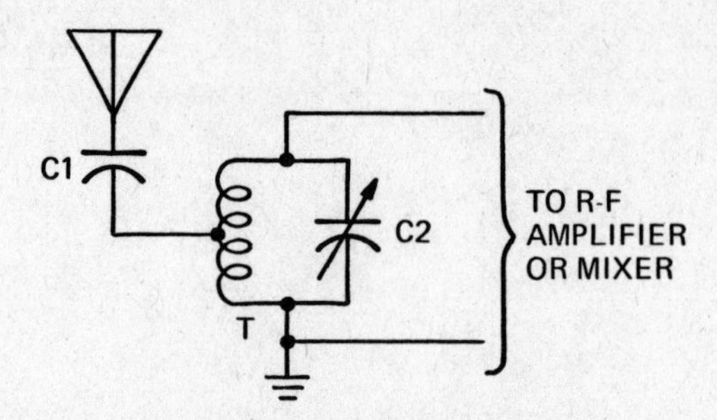

Figure 5-1. Antenna coupling, capacitive.

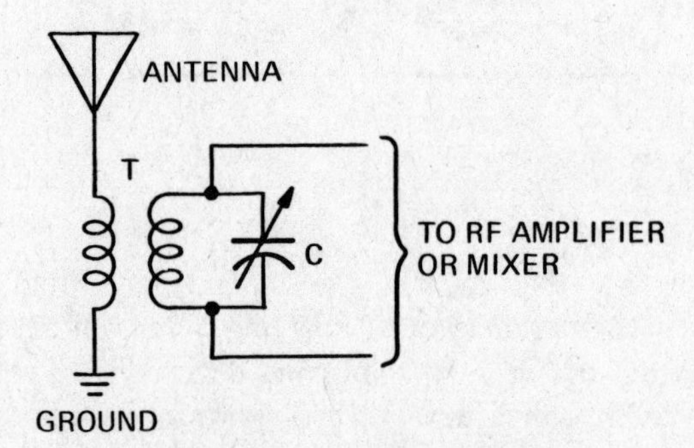

Figure 5-2. Antenna coupling, inductive.

secondary and C. Therefore, there is a greater signal voltage drop across the primary at that frequency.

Antenna Matcher

Figure 5-3 illustrates how an antenna matcher can be used to use the low-impedance output of a transmitter to drive a relatively high impedance transmission line. Capacitor C1 is tuned so that the input impedance of the matcher is 50 ohms. Capacitor C2 is tuned to the frequency of the transmitter, and the impedance of the secondary circuit can be adjusted to match that of the transmission line by selecting the taps on the transformer secondary.

Antenna Multiplexer

A single long-wire antenna can serve several receivers simultaneously with negligible interaction. The diagram shows an antenna multiplexer circuit that can be used at a receiving station to enable monitoring at several different frequencies simultaneously. X,

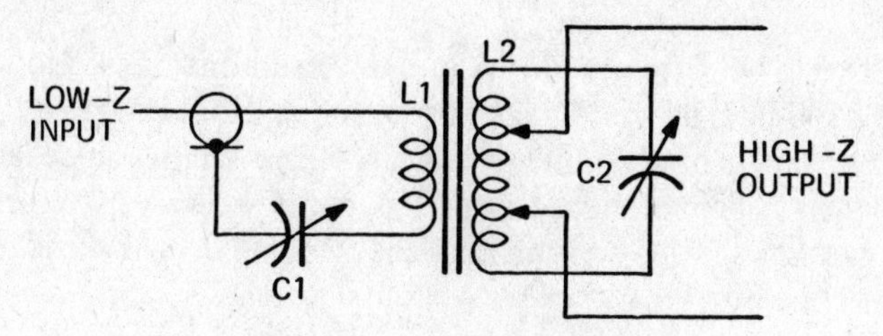

Figure 5-3. Antenna matcher.

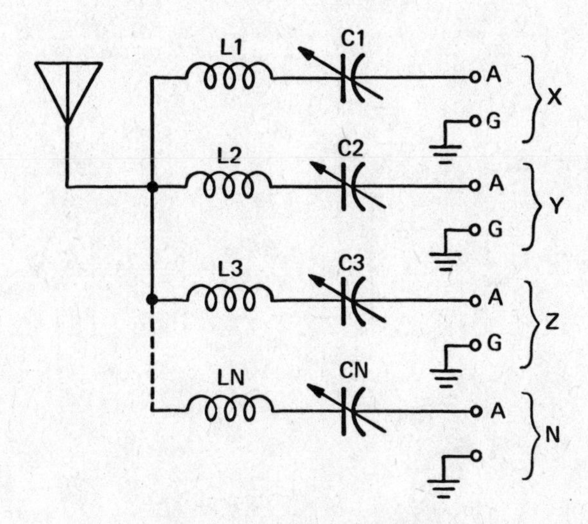

Figure 5-4. Antenna multiplexer.

Y, and Z represent the inputs of three receivers, each tuned to a different frequency. N represents the inputs of additional receivers for each of which the same type of input filter (LN-CN) must be provided. An example of the frequencies to be monitored is listed below.

Receiver input	*Frequency*
X	500 kHz
Y	2,182 kHz
Z	2,500 kHz
N1	5,000 kHz
N2	27,065 kHz

L1-C1 would be tuned for series-resonance at 500 kHz, L2-C2 at 2182 kHz and so on. These band-pass filters (L-C) readily pass signals at their resonant frequencies and attenuate signals at all other frequencies.

Antenna Splitter

To connect two television or FM radio receivers or one of each to the same antenna equipped with a 300-ohm balanced transmission line, a "splitter" (also called "coupler") is required. The function of the splitter is to maintain the required impedance relationships and to route signals to both of the receivers. Figure 5-5 shows a splitter circuit employing four 150-ohm resistors. The transmission line sees a 300-ohm load and each of the receiver inputs see a 300-ohm source.

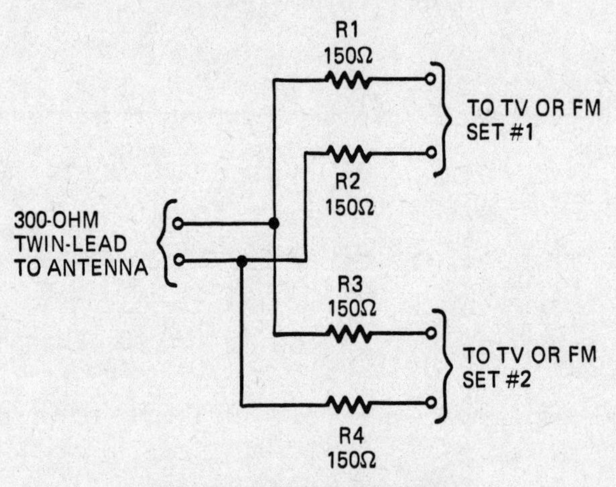

Figure 5-5. Antenna splitter.

Antenna Tuner, Capacitive

A variable capacitor, connected in series with the antenna-ground system as shown in Figure 5-6, can be used to vary the level of the signal developed across the primary of RF transformer T. Reducing the capacitance of C1 has the same effect as reducing the length of the antenna. However, when the antenna is sufficiently long and the inductance of the primary of T is adequate, C1 can be adjusted to make the antenna circuit series-resonant at the receiving frequency. In this case, the antenna and ground form a capacitor which is, in effect, in series with C1. When the antenna circuit is series-resonant at the receiving frequency, current through the primary of T is greatly increased. This causes a larger voltage to be developed across C2 and the secondary of T. Selectivity is also improved when both C1 and C2 are adjusted for maximum sensitivity at the receiving frequency.

Antenna Tuner, Inductive

A variable inductor connected in series with the antenna-ground system, as shown in Figure 5-7, can be used to vary the level of the signal developed across the primary of RF transformer T. The inductor may be a coil with an adjustable ferrite core, or a "variome-

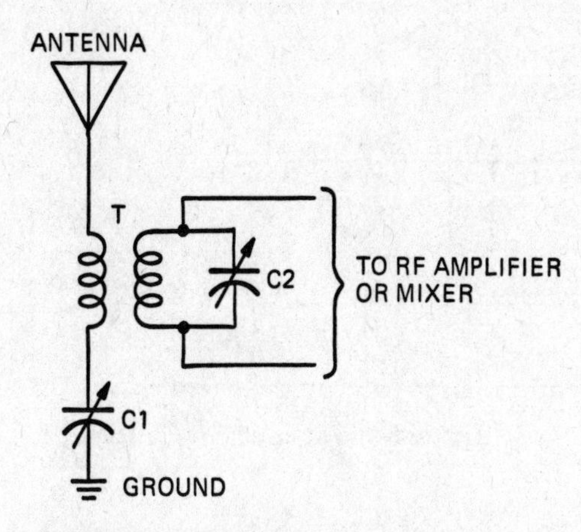

Figure 5-6. Antenna tuner, capacitive.

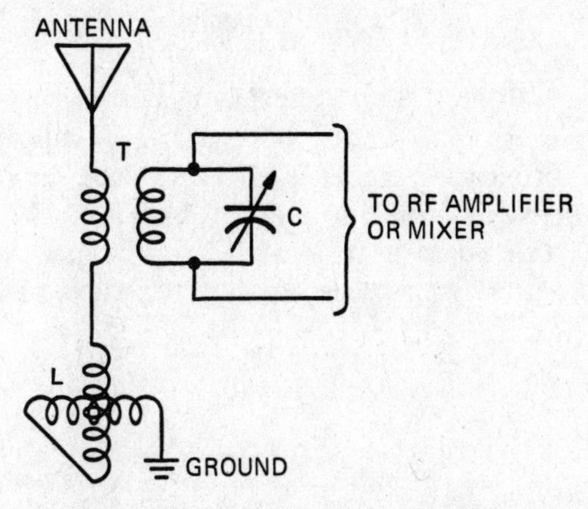

Figure 5-7. Antenna tuner, inductive.

ter," as illustrated. Increasing the inductance has the same effect as increasing the length of the antenna. Since the antenna and ground form a capacitor, adjusting L to make the antenna circuit series-resonant at the receiving frequency causes a sharp increase in current through the primary of T. Selectivity is also improved when both L and C are adjusted for maximum sensitivity at the receiving frequency.

Balanced Loop Antenna

Figure 5-8 illustrates a balanced loop antenna, useful in direction finders and other applications where an antenna connected to an unbalanced line is undesirable. The antenna is balanced through capacitors C2 and C3 and the transformer primary, and tuned by

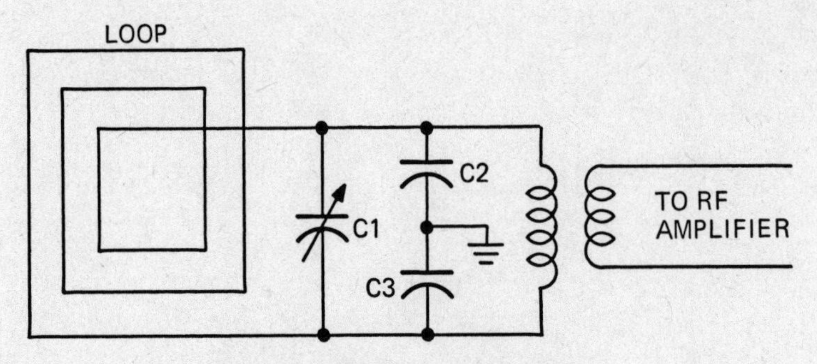

Figure 5-8. Balanced loop antenna.

means of capacitor C1 to resonance with the incoming signal. The received signal is coupled through the transformer to the first RF amplifier stage of the receiver.

CB Antenna Coupler

The circuit shown in Figure 5-9 is of a coupler that permits sharing of a CB antenna by a 27-MHz band citizens radio transceiver and an AM automobile radio. At the 11-meter citizens band frequencies, the series-resonant combination of L1-C1 provides a low-impedance path from the antenna to the CB transceiver. At AM broadcast band frequencies (540-1600 kHz), however, the small value of C1 appears as a high impedance compared to R1 and the broadcast band signals are routed through R1 to the auto radio.

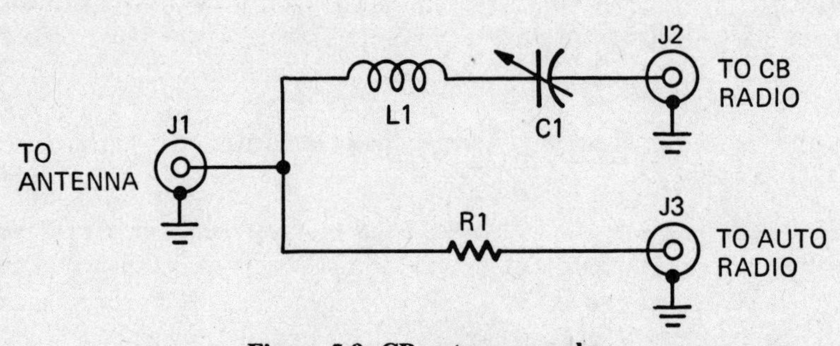

Figure 5-9. CB antenna coupler.

When the operator is transmitting with the CB transceiver, the signal from the transceiver is fed through the low-impedance network L1-C1 to the antenna. Although some of the signal gets to the auto radio through R1, it should not harm the receiver. The signal current is greatly reduced by R1 which also, because of its relatively high resistance, causes an insignificant loading effect on the transmitter.

Crystal Lattice Filter

In Figure 5-10 crystals Y1, Y2, Y3, and Y4 form a lattice filter. Two pairs of crystals are used, each pair resonant 1 kHz apart when a 2-kHz bandpass is required. The filter must be built so as to respond to a specfic band of frequencies. In an SSB transmitter application, for example, a DSB signal is fed into the filter which will pass only one of the sidebands. The filter would have a center frequency of 9.001 MHz in order to pass an upper sideband (USB) which extends from 9 to 9.002 MHz.

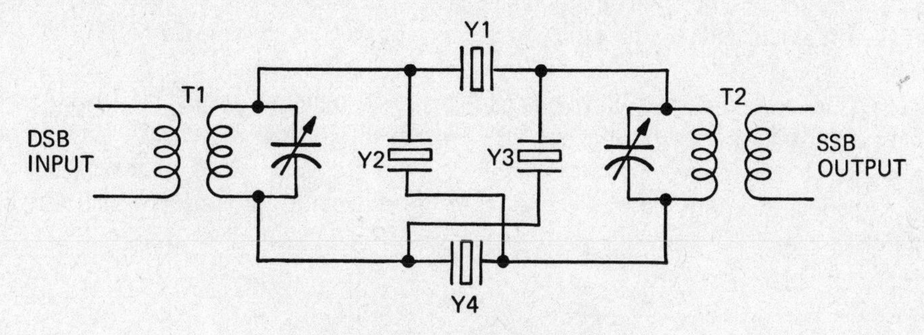

Figure 5-10. Crystal lattice filter.

Double-Tuned Transformer

Both the primary and secondary of double-tuned transformers, illustrated in Figure 5-11, are tuned either by variable capacitors, as shown in A, or by varying the inductance of the windings, as shown in B. Standard IF (intermediate frequency) transformers for use in receivers employing tubes are available in either type. Generally, when capacitance tuning is used, the transformer contains two mica compression trimmer capacitors, one shunted across each of the coils. Those employing inductive tuning usually have two adjustable ferrite cores, one within each of the windings, each of which is shunted by a fixed capacitor. In both types, both windings are usually wound next to each other on the

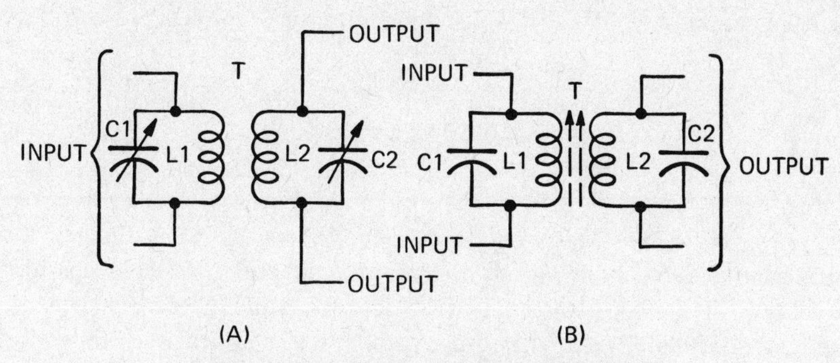

Figure 5-11. Double-tuned transformers.

same coil form. For maximum selectivity (and gain), the primary and secondary are usually tuned to the same frequency (peaked). When wider bandpass is required, the primary and secondary are usually tuned to slightly different frequencies (staggered).

Ferrite Core Antennas

For the sake of compactness, most small AM broadcast band transistor radios contain a tiny ferrite core antenna (loopstick) instead of a larger flat loop antenna. Both perform the same function and both are directional. Ferrite core antennas are also used with larger AM/FM hi-fi tuners and receivers to enable pickup of the signals of nearby AM broadcast stations.

Most ferrite core antennas have two windings, as shown in Figure 5-12. The coils are wound on a form which surrounds the ferrite core. One winding is tuned by a variable capacitor to obtain resonance at the desired receiving frequency. The other winding has fewer turns and is connected to the low input impedance of the transistor mixer or RF amplifier.

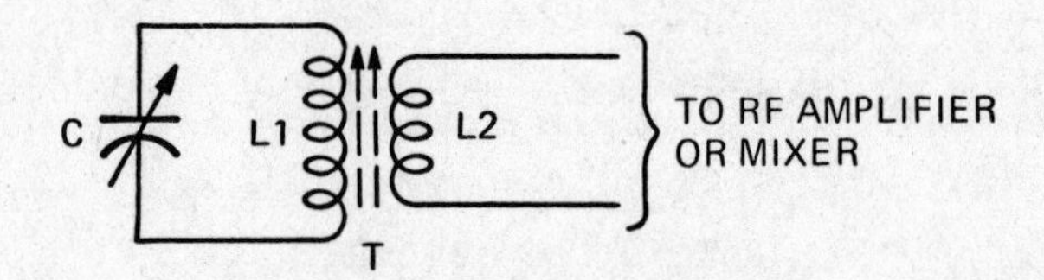

Figure 5-12. Ferrite core antenna.

A ferrite core antenna is most sensitive when either end of the core is pointed in the direction of the station being received. It is least sensitive when it is broadside to the station being received. Because of its directional characteristics, most low cost radio direction finders employ a rotatable ferrite core antenna mounted above a compass rose calibrated in degrees of azimuth. The antenna is rotated to obtain a "null" (weakest signal) since the null is more pronounced than the maximum signal point. As the ferrite rod is rotated to obtain a bearing, two nulls, 180 degrees apart, will be obtained. Therefore, the navigator must either know which null indication is correct or refer to a magnetic compass.

Loop Antenna

A loop antenna consists of a number of turns of wire wound on a flat piece of insulating material (such as Masonite) or on a wooden frame. It is a directional antenna which is most sensitive to radio signals coming from directions perpendicular to its core. The inductance of the loop antenna, in conjunction with the capacitance of tuning capacitor C, as shown in Figure 5-13, form a parallel-resonant circuit which is tuned by C. This parallel-resonant circuit is connected directly to the input stage of the receiver.

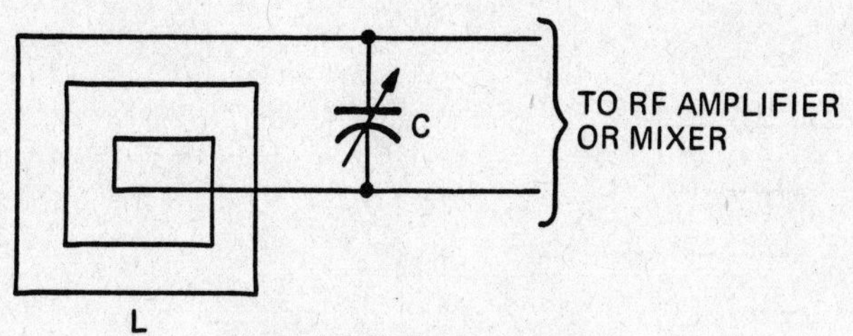

Figure 5-13. Loop antenna.

Loop Antenna Coupling

The sensitivity of a radio receiver employing a loop or ferrite rod antenna can be greatly increased by utilizing an external antenna and an earth ground. When it is inconvenient to connect the external antenna and ground to the receiver, the inductive coupling technique shown in Figure 5-14 can be used. The external antenna and ground are connected to the terminals of a flat loop antenna (L). The receiver is placed close to this external loop antenna so that energy is transferred inductively from it to the loop antenna within the receiver. For effective results, the ground connection must be made so that RF current flows through L.

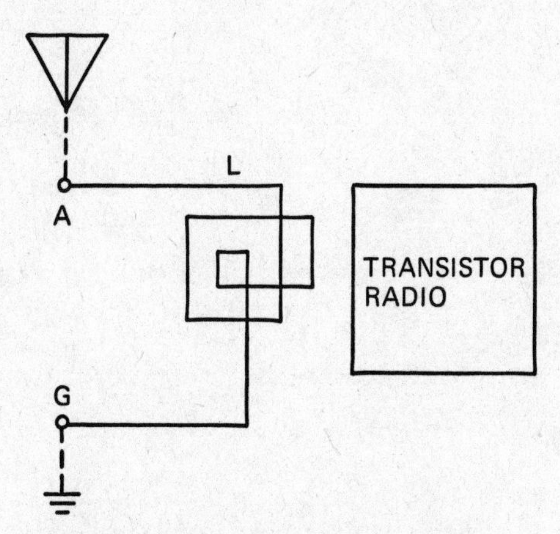

Figure 5-14. Loop antenna coupling.

Loop Antenna, Inductively Coupled

A fewer number of turns are required when the loop antenna is inductively coupled to the receiver input, as shown in Figure 5-15. The loop antenna (L) may consist of two or three turns of wire to form a rectangular coil of large diameter. It is inductively coupled by

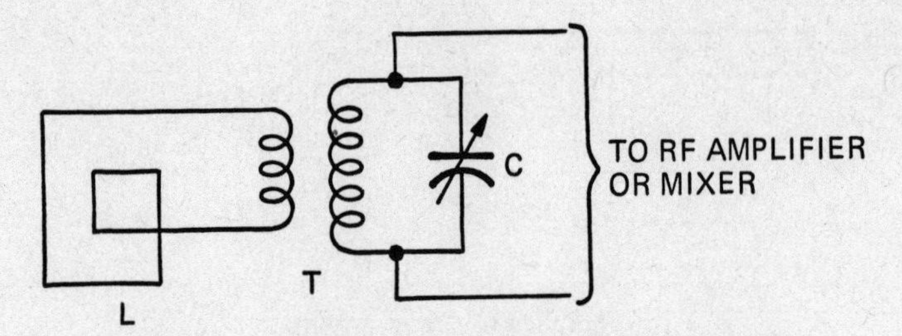

Figure 5-15. Loop antenna, inductively coupled.

RF transformer T to the receiver input. The secondary of T and tuning capacitor C form a tunable parallel-resonant circuit. The reflected resonant frequency of the loop antenna is adjusted by varying the capacitance of C.

L-Type LC Filters

Figure 5-16A is a schematic of a low-pass, L-type, LC (inductance-capacitance) filter. The components are arranged in the shape of the letter "L," and the filter will pass signals below its designated cut-off frequency.

Figure 5-16B is a schematic of a high-pass, L-type, LC filter. A high-pass filter passes signals above its cut-off frequency.

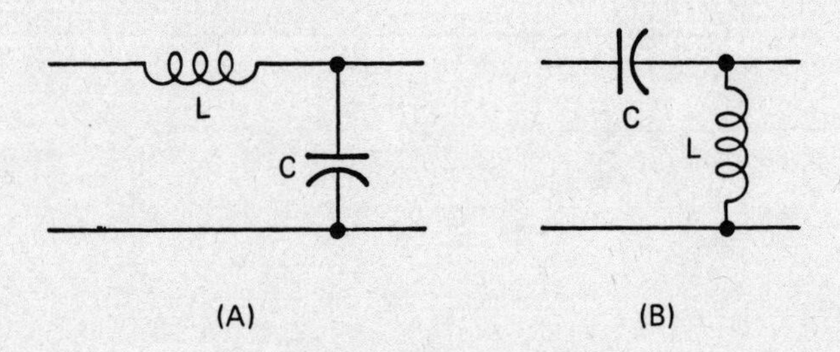

Figure 5-16. L-type LC filters.

L-Type RC Filter

Figure 5-17A is a schematic of a low-pass, L-type, RC filter. It is called an L-type because the components are arranged in the shape of the Letter "L." A low-pass filter passes signals which are below a specified frequency, while attenuating those above it.

Figure 5-17B illustrates a high-pass, L-type RC filter. A high-pass filter passes those signals which are above its cut-off frequency.

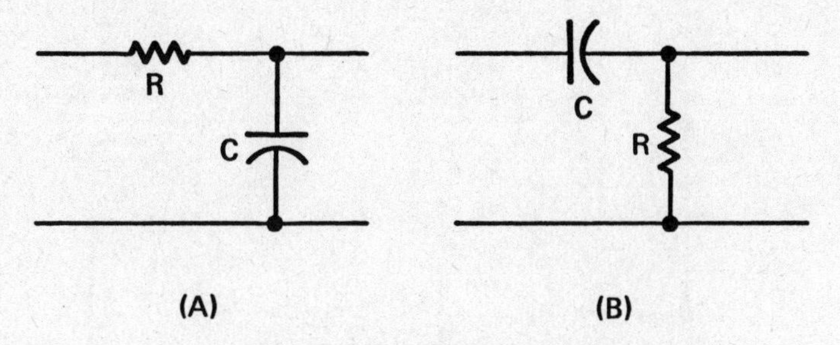

Figure 5-17. L-type RC filters.

Pi-Network Coupler

Figure 5-18 illustrates a pi-network, so named because of the resemblance of the schematic diagram of the network to the Greek letter π. In practice, in a transmitter, capacitor C1 is tuned to make the circuit resonant at the transmitting frequency and C2 is tuned to match the network to the load impedance. The pi-network acts as the tank circuit, a low-pass filter and an impedance matching network. It can be used to match a high impedance source to a low impedance load and vice versa.

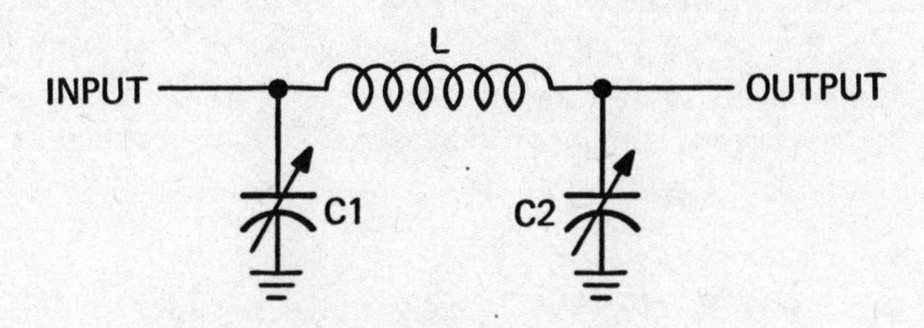

Figure 5-18. Pi-network coupler.

Preemphasis and Deemphasis Networks

Figure 5-19 shows simple preemphasis networks used in the audio circuitry of an FM radio transmitter. Preemphasis is used to provide more amplification of the higher audio frequencies than of the lower audio frequencies. This is desirable in FM transmission because, although noise is distributed randomly throughout the spectrum, voice and music frequencies are not. The higher audio frequencies do not have the amplitude of the lower frequencies. Hence, when transmitted without preemphasis, the level of unwanted noise with respect to signal level decreases at the higher frequencies.

The simple preemphasis network shown in A consists of a resistor (R) and an inductor (L). Since the reactance of the inductor rises with frequency, while the resistance

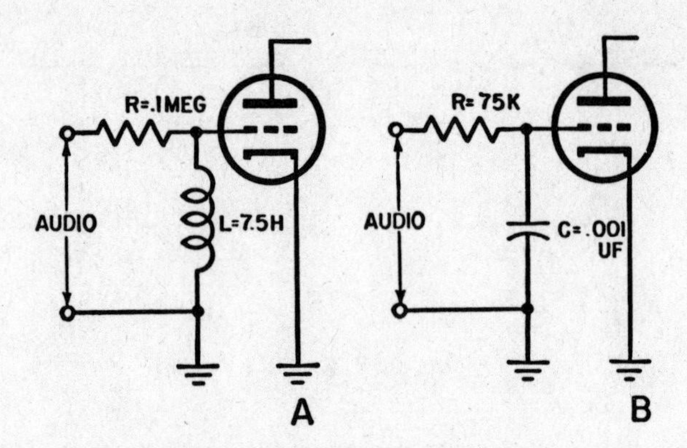

Figure 5-19. Simple emphasis networks.

stays the same, the voltage developed at the grid of the tube is greater at higher frequencies. The time constant of the network can be computed as follows:

$$\text{Time Constant in microseconds} = \frac{\text{L in henrys}}{\text{R in megohms}} = \frac{7.5\text{ H.}}{0.1\text{ M.}} = 75\text{ microseconds}$$

At the receiver, the reverse reciprocal of preemphasis is used so that the normal balance between high and low audio frequencies is restored. This is accomplished by making the time constant of the resistor and capacitor of the deemphasis network, shown in B, the same as that of the preemphasis network. Since capacitive reactance decreases with increasing frequency, the voltage applied to the grid of the tube through the deemphasis network decreases as frequency increases. The time constant of the circuit shown may be calculated using the formula:

$$\text{Time Constant in microseconds} = \text{R (ohms)} \times \text{C (microfarads)} = 75{,}000 \times .001 = 75\text{ microseconds.}$$

The standard time constant for preemphasis and deemphasis in FM broadcasting is 75 microseconds in the United States and 50 microseconds in Europe.

Repeater Coil

Figure 5-20 is the schematic symbol of an AF transformer referred to by the telephone industry as a repeater coil. It is commonly used in telephone, broadcast studio, and phone patching applications. The impedance of each winding is 150 ohms. When windings L1 and L2 are connected in series, and L3 and L4 are also connected in series, as shown in A, the input sees a 600-ohm load and the output sees a 600-ohm source. Both the impedance ratio and the turns ratio (same as voltage ratio) are 1:1.

When windings L1 and L2 are connected in series, and L3 and L4 are connected in parallel position, as in B, the input sees a 600-ohm load and the output sees a 150-ohm

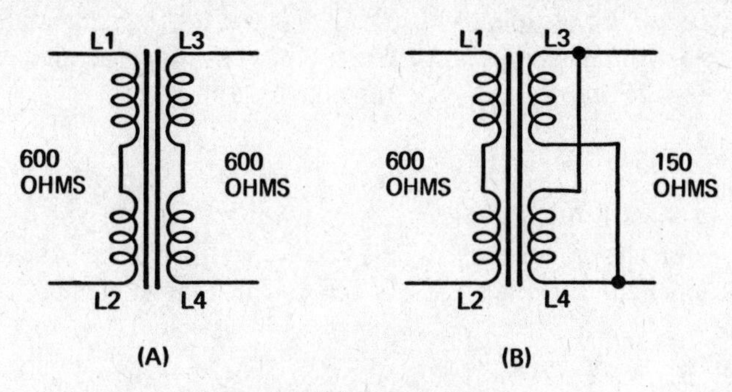

Figure 5-20. Repeater coil.

source. The impedance ratio is 4:1 but the voltage ratio is only 2:1. Windings L1 and L2 can be paralleled so that the input will see a 150-ohm load. If L3 and L4 are also paralleled, the output will see a 150-ohm source.

The transformer allows AF signals to pass through it but blocks passage of DC from one side to the other.

Selectivity Filter

The AM superheterodyne receiver circuit shown in Figure 5-21 utilizes a ceramic bandpass filter between the mixer and integrated circuit IF amplifier to provide excellent selectivity. Both the input of the RF amplifier and the input of the mixer are tuned to the frequency of the signal to be received and the local oscillator is tuned 455 kHz above or

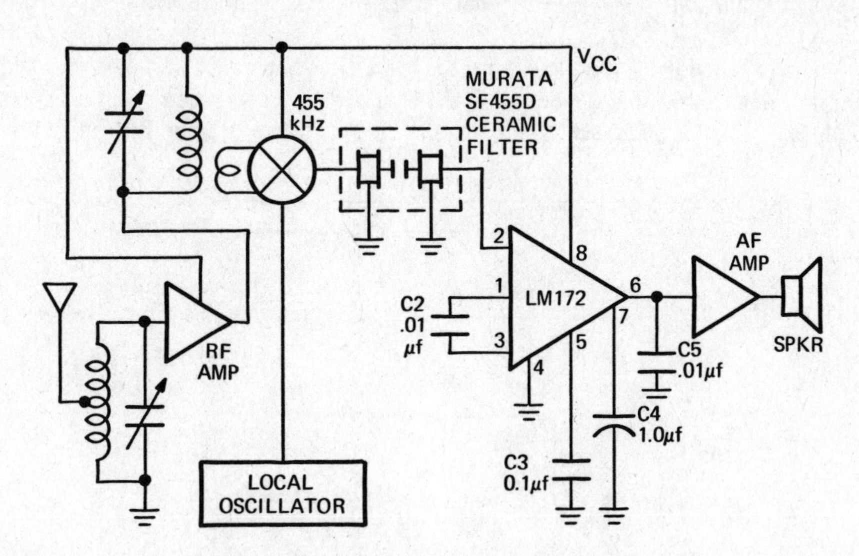

Figure 5-21. AM superheterodyne receiver circuit.

below that frequency. The resulting 455-kHz IF signal is fed from the mixer output to the IF amplifier input through the ceramic filter. This one filter eliminates the need for adjustable interstage IF transformers and requires no adjustment.

T-Network Filter

The T-Network shown in Figure 5-22 is a low-pass filter. When used as a transmitter output network, one leg of the network is tuned to match the transmitter impedance. The other leg is tuned to match the antenna resistance. Thus, the T-Network can be used to match a transmitter to a given antenna.

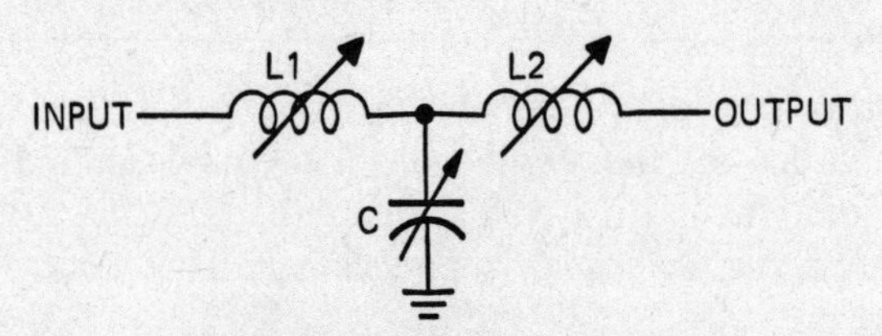

Figure 5-22. T-Network.

Transformer-Coupled Mixer-Balun

Figure 5-23 illustrates how the unbalanced outputs of two microphone or phonograph pickup preamplifiers can be mixed and fed to a balanced line. The signals from the two amplifiers are fed to the primaries of transformers T1 and T2 respectively, and are isolated from each other at this point. The secondaries of T1 and T2, however, are connected in series.

Variable T-pad R1 is used to control the level of input signal 1 at the output, and R2 the level of input signal 2 at the output. If R1 is turned all the way down and R2 is turned up, only signal 2 will be present at the output, and vice versa. R1 and R2 can be used to

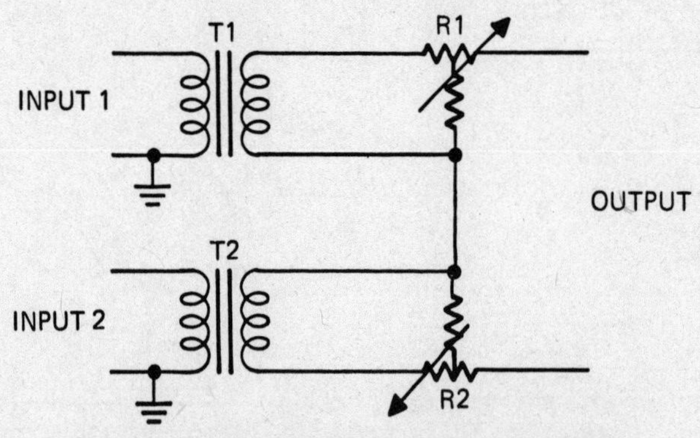

Figure 5-23. Transformer-coupled mixer-balun.

blend signals 1 and 2 in any desired manner. Since a T-pad presents a constant resistance to both the source and the load, adjustment of R1 will have no affect on the adjustment of R2 nor the impedances seen by the two input signal sources and the impedance seen by the output line.

Transmission Line Connections

Typical FM broadcast band radio receivers and television receivers have 300-ohm balanced input terminals. As shown in Figure 5-24A, the antenna can be connected to these input terminals through 300-ohm unshielded twinlead (ribbon cable) or 300-ohm shielded twinlead transmission line. The center tap of input RF transformer T may be grounded as shown. (Some receiver input transformers do not have a center tap.)

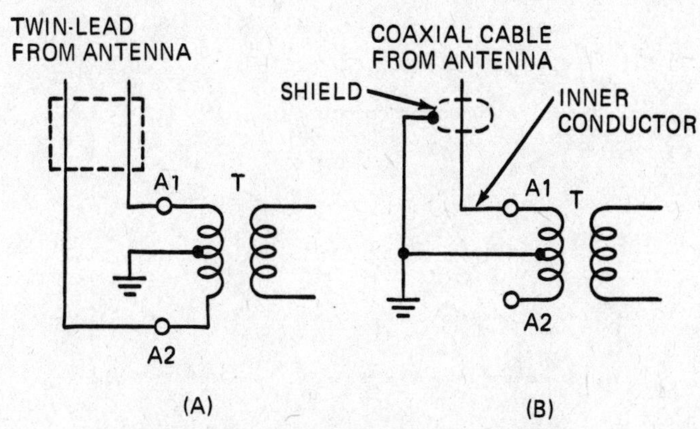

Figure 5-24. Transmission line connections.

To use 75-ohm coaxial cable as the transmission line a "balun" (matching transformer) can be used to match the 75-ohm unbalanced coaxial cable to the 300-ohm balanced receiver input. Or, as shown in B, the coaxial cable can be connected across one half of the 300-ohm primary of the input transformer. The center conductor of the coaxial cable can be connected to either antenna terminal (A1 or A2), and the shield of the coaxial cable is connected to the center tap of the input transformer primary. The impedance across A1 and A2 is 300 ohms, but, the impedance between either A1 or A2 and the center tap of the input transformer primary is 75 ohms.

Varactor-Tuned Bandpass Filter

In Figure 5-25, a voltage variable capacitor (varactor diode) is used as part of a parallel-resonant bandpass filter. The filter is tuned by the variable capacitor C1 and also by the varactor D. Varying the amount of bias voltage on the varactor varies its capacitance, and hence, the resonant frequency of the tank circuit.

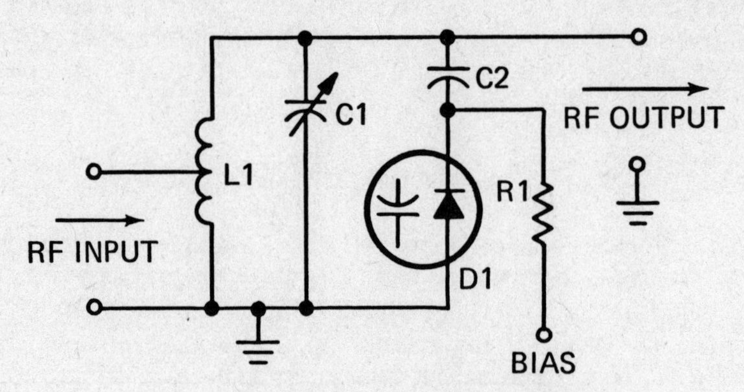

Figure 5-25. Varactor-tuned bandpass filter.

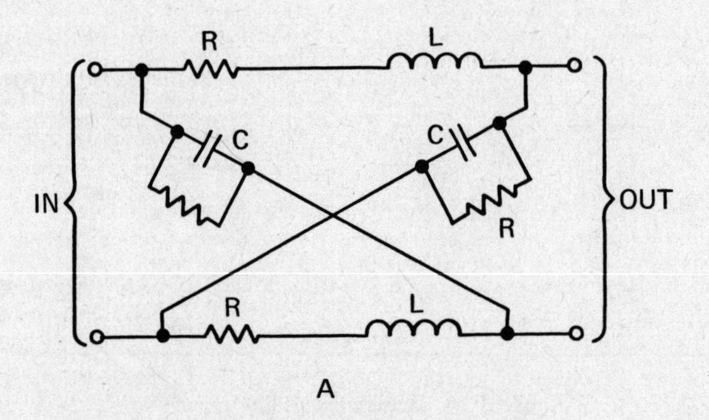

Figure 5-26A. Attenuator equalizer.

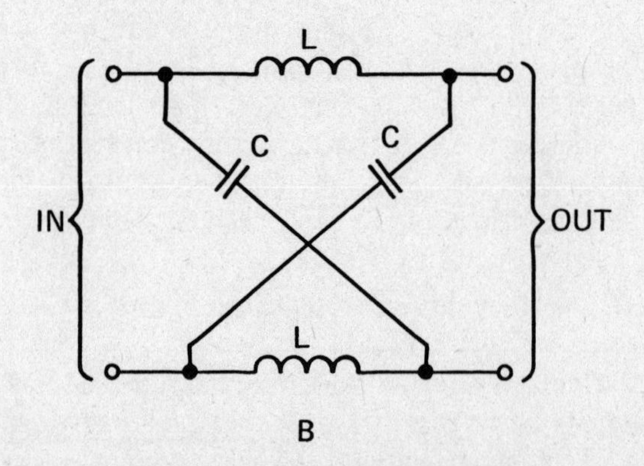

Figure 5-26B. Phase equalizer.

Video Equalizers

Equalizers are used with wideband video-pair transmission lines such as used in closed circuit television systems. There are two main types attenuation and phase equalizers. The high video frequencies are attenuated much more by the cable than the low frequencies because of cable capacitance. Therefore, a passive network or amplifier is used which has a gain frequency characteristic which peaks at the high end of the transmitted band of frequencies.

Lattice networks (bridged T) are used for equalization. These networks have a constant resistive characteristic impedance and in proper combination nearly any desired amplitude vs frequency response can be obtained. A simple example of an attenuator equalizer is shown in Figure 5-26A. A phase equalizer is shown in 5-26B.

Lattice networks are often used and designed to provide constant characteristic impedances and flat response at all frequencies. Such all-pass networks may be designed to have nearly any desired phase shift or time delay characteristic shape desired. A correctly designed all-pass network may be combined with an amplifier. This technique is often used is television systems.

(Note: The phase equalizer has a flat passband for all frequencies, but the attenuation equalizer does not provide uniform phase delay at all frequencies. Therefore, phase equalization may still be required to correct for non-linear phase shift in the amplitude equalizers as well as in the transmission line.)

Wavetrap, Absorption Type

An external loop antenna and a variable capacitor can be used as a wavetrap for a receiver employing an internal loop antenna. As shown in Figure 5-27, the external loop antenna is placed close to the receiver (the closer, the better). This loop antenna is tuned to resonate at the frequency of an interfering signal by adjusting variable capacitor C. At this frequency, the wavetrap absorbs energy.

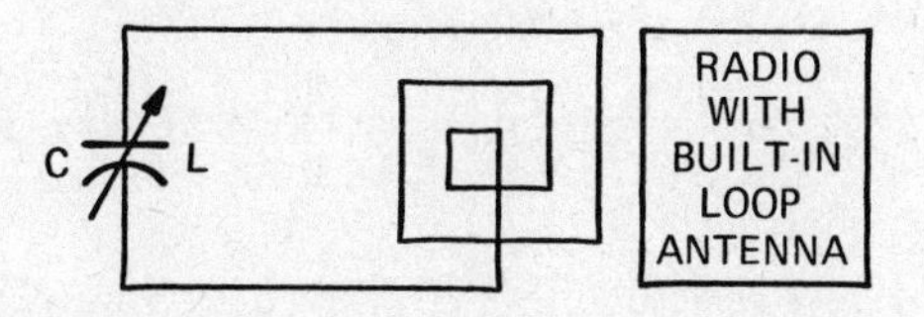

Figure 5-27. Radio with built-in antenna.

Wavetrap for Balanced Inputs

Two parallel-resonant wavetraps can be used in series with the balanced antenna transmission line and the balanced input of an FM broadcast band radio receiver or television receiver, as shown in Figure 5-28A. Both L1 and L2 are tuned to resonate at the

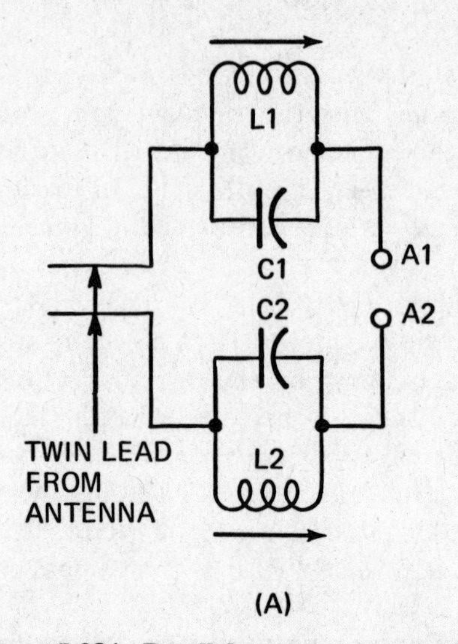

Figure 5-28A. Parallel-resonant wavetraps.

frequency of the interfering signal. At that frequency, L1-C1 and L2-C2 exhibit extremely high impedance and reject signals at that frequency. The attenuation of signals at other frequencies is very low since these signals pass easily through C1 and C2 or L1 and L2.

A single series-resonant wavetrap can be connected across the balanced receiver input as shown in 5-28B. The core of variable inductor L is adjusted so that L and C are series-resonant at the frequency of the interfering signal. At that frequency, the wavetrap acts as a short circuit. At all other frequencies, the inductive reactance of L and/or capacitive reactance of C cause very little attenuation.

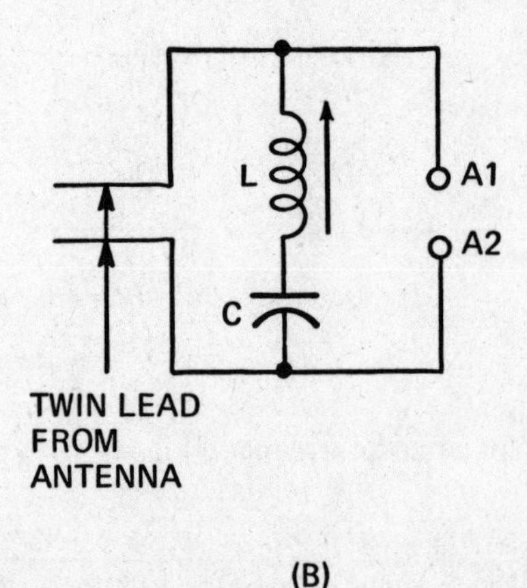

Figure 5-28B. Single series-resonant wavetrap.

Wavetrap for Unbalanced Inputs

A parallel-resonant wavetrap connected in series with the antenna terminal of a receiver, as shown in Figure 5-29A, attenuates signals at its resonant frequency. When the core of variable inductor L is adjusted to make L and C resonant at the frequency of an unwanted signal, the wavetrap exhibits an extremely high impedance at that frequency and causes relatively small insertion loss at all other frequencies.

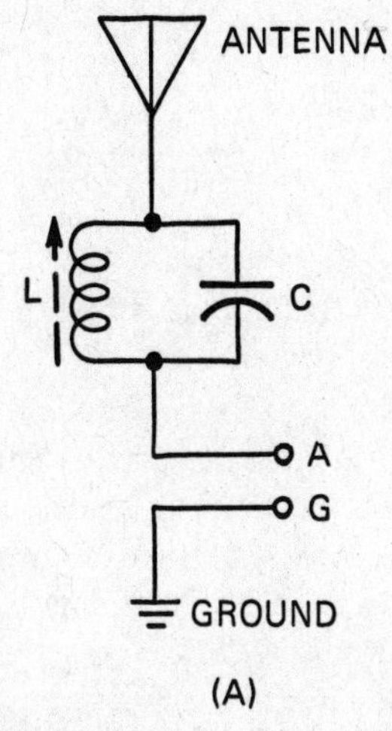

Figure 5-29A. Parallel-resonant wavetrap connected to antenna terminal of a receiver.

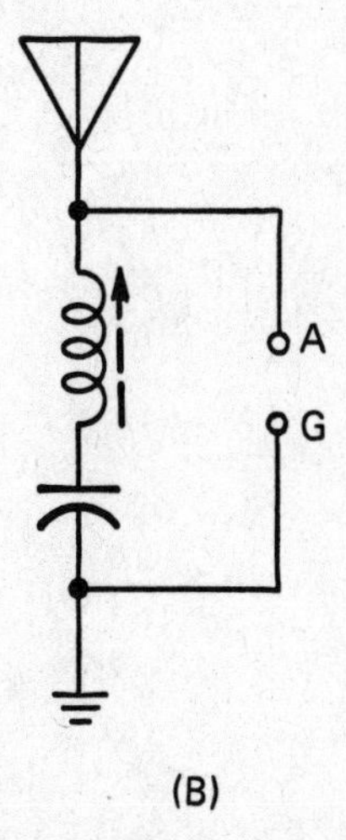

Figure 5-29B. Series-resonant wavetrap connected across the antenna and ground terminals of a receiver.

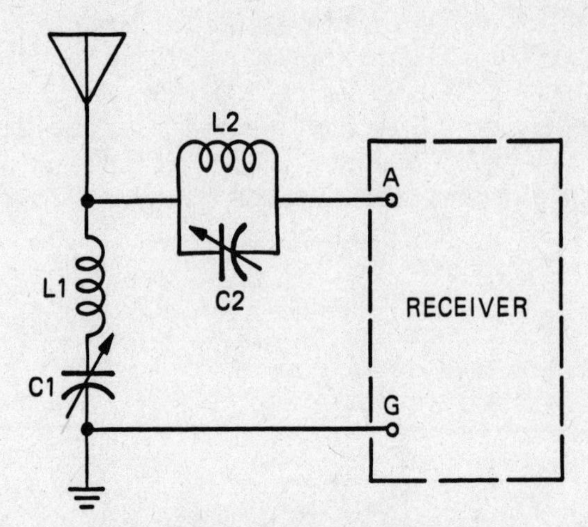

Figure 5-29C. A dual wavetrap circuit.

A series-resonant wavetrap connected across the antenna and ground terminals of a receiver, as shown in 5-29B, bypasses interfering signals to ground when tuned to the frequency of the interfering signals. At its resonant frequency, the wavetrap exhibits a very low impedance. At all other frequencies, it exhibits a much higher impedance.

A dual wavetrap circuit is shown in 5-29C. The series-resonant wavetrap (L1-C1) provides a low impedance path to ground for the interfering signal. The parallel-resonant wavetrap (L2-C2) causes a high insertion loss at the interfering signal frequency. Both wavetraps can be tuned to the same frequency to provide maximum attenuation at that frequency. Or, to attenuate two interfering signals at differing frequencies, one wavetrap can be tuned to attenuate one of the signals and the other to attenuate the second interfering signal.

In A and B, the inductor is shown as being variable and in C, the capacitors are shown as being variable. In any of these wavetrap circuits, either or both the inductor and capacitor can be variable.

6

Frequency Conversion and Multiplication Circuits

Perhaps the earliest frequency converter was the DC to AC motor generator set which consisted of a DC motor whose input voltage frequency is 0 Hz, and an alternator (AC generator) whose output voltage frequency was usually 25, 50, or 60 Hz. A rotary frequency converter usually consists of a 60-Hz AC motor and a 400-Hz (or higher frequency alternator).

In an electronic frequency converter, two AC signals of differing frequency are fed to a nonlinear "mixer" in which the sum and difference beat frequencies of the two signals are generated. In most applications, only one of the beat frequencies is utilized and the other is filtered out.

The frequency of an AC signal can be increased harmonically by an electric frequency multiplier which may consist of a diode, an electron tube, or a transistor which distorts the input signal waveform so that harmonics of the input signal frequency are generated.

Described and illustrated in this chapter are numerous circuits for changing signal frequencies.

Beam-Deflection Mixer

Figure 6-1 shows an unbalanced mixer circuit in which the input signal is applied to the control grid in the usual manner. The local oscillator signal is applied to one of the deflection plates of the beam deflection tube. The stream of electrons from the cathode is alternately switched from one plate to the other plate by the oscillator voltage applied to the deflection plates. The sum or difference beat frequency of the input and local oscillator

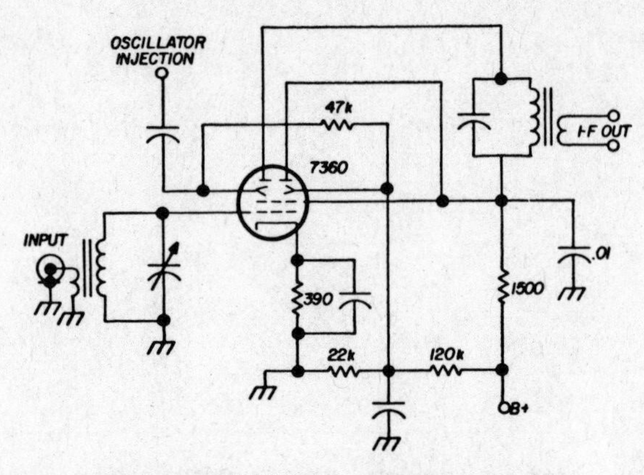

Figure 6-1. Unbalanced mixer circuit.

signals is selected by the tuned transformer connected in series with one of the plates. The other plate has a DC voltage applied to it, but is bypassed to ground at the signal frequencies by a capacitor. (Illustration courtesy of *Ham Radio*)

Bipolar Transistor Frequency Multiplier

Figure 6-2 shows a frequency multiplier circuit employing a single parallel-resonant output circuit. The input pi-network is tuned to the frequency of the input signal to present the correct load impedance for the signal source. The collector tank is tuned to the desired harmonic frequency. Attenuation of the fundamental and unwanted harmonic frequencies is provided by the collector tank circuit. In this circuit, the output signal is obtained from a tap on the secondary of the output tank circuit to permit interfacing with a low impedance load. (Illustration courtesy of *Ham Radio*.)

Harmonics of the input frequency are generated because the transistor is not biased

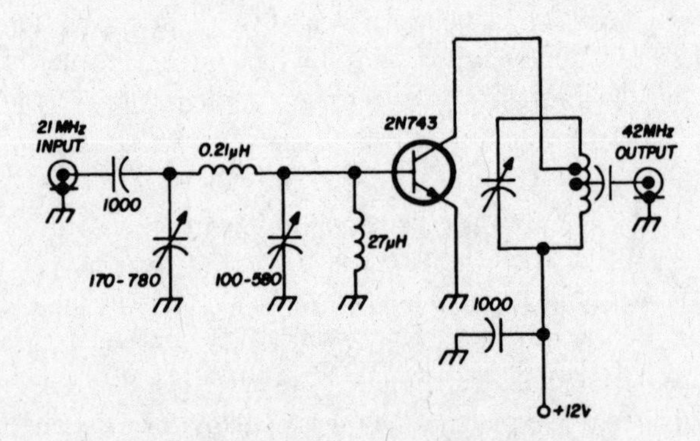

Figure 6-2. Frequency multiplier circuit.

for linear operation and the transistor operates as a Class C amplifier. In fact, when no input signal is present, there is no forward bias since the base and emitter are shorted to each other through the very low DC resistance of the 27-uH RF choke coil. When there is an input signal, the RF choke coil connected from the base of the transistor to its emitter has a relatively high reactance at the signal frequency. When the input signal makes the transistor base positive, the NPN transistor is forward biased and collector current flows through the tank circuit. Contrarily, when the input signal makes the transistor base negative, the transistor is reverse-biased and collector current is zero. Collector current flows only when the transistor is forward biased. However, RF current flows alternately in different directions within the tank circuit because of the "flywheel effect" even during the periods when no collector current flows.

In this and other frequency multipliers, the desired harmonic is selected by tuning the output tank circuit to its frequency.

Cathode-Injection Triode Mixer

The local oscillator signal in the circuit shown in Figure 6-3 is injected into the cathode circuit of the triode tube. The radio signal is fed to the grid. Cathode bias is used to make the tube operate in a non-linear manner so that heterodyning of the two signals will occur. The oscillator tank coil is inductively coupled to a coil connected in series with the cathode. The AC voltage induced into this coil alternately bucks and boosts the cathode bias voltage and thus modulates the electron stream.

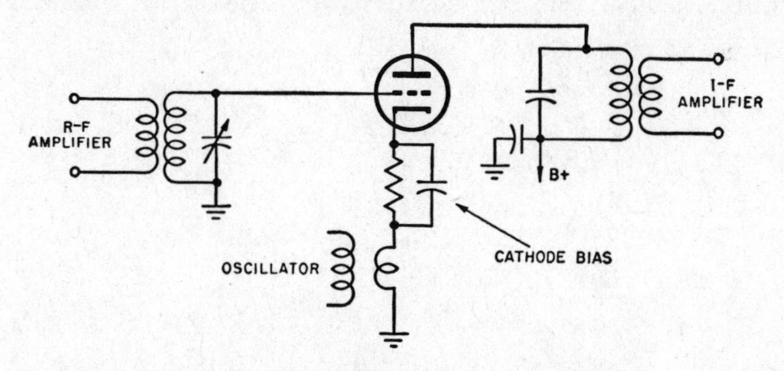

Figure 6-3. Cathode-injected triode mixer.

Dual-Diode Mixer

The mixer circuit shown in Figure 6-4 uses two diodes to develop a beat frequency. The radio signal is fed directly to the anode of one diode and the cathode of the other diode. The local oscillator LO signal is fed through RF transformer T1 to the diodes. The resulting beat frequency signal is obtained from the center tap of the secondary of T1.

The local oscillator signal alternately switches both diodes on and off during each

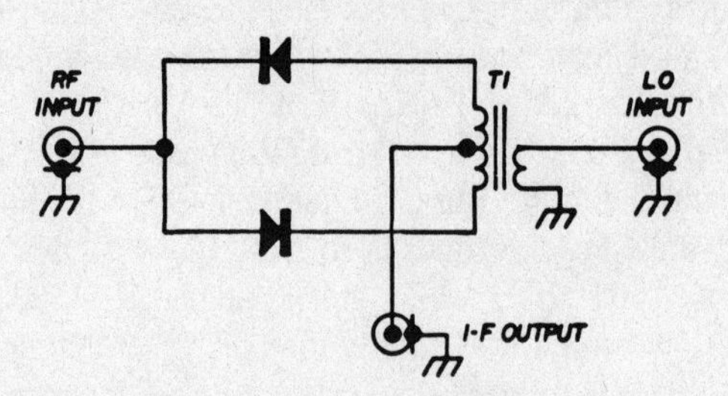

Figure 6-4. Dual-diode mixer.

cycle. The radio signal alternately reduces current flow through one diode and increases current flow through the other diode. Since the two signals interact in a non-linear manner, beat frequencies are generated, of which one is selected by the IF amplifier. (Illustration courtesy of *Ham Radio*.)

FET Frequency Multiplier

In the FET frequency multiplier circuit shown in Figure 6-5, the cutoff bias and the amount of RF drive voltage determine the conduction angle of the FET (field effect transistor) and hence the harmonic that will be generated by the multiplier. The input tank is tuned to the input frequency and the output tank coil is tuned to the frequency of the desired harmonic. Harmonic number versus conduction angle is given in the table below.

Harmonic Number	*Conduction Angle*
2	90 - 120°
3	80 - 120°
4	70 - 90°
5	60 - 72°

(Illustration courtesy of *Ham Radio*.)

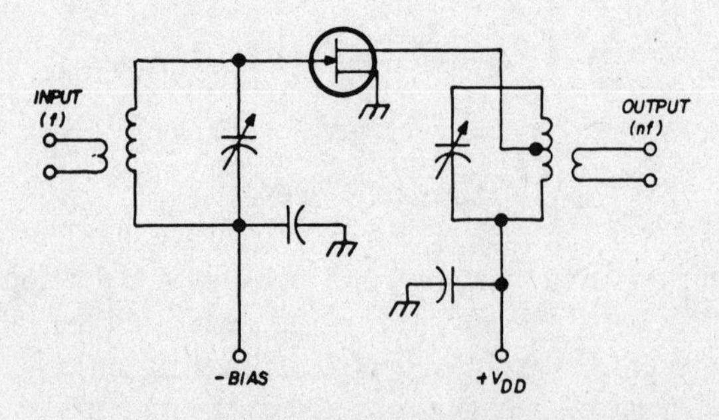

Figure 6-5. FET frequency multiplier circuit.

FET Oscillator and Frequency Doubler

Three FETs (field effect transistors) are used in the frequency-generating circuit shown in Figure 6-6. The oscillator Q1 is crystal-controlled. The oscillator output signal, at the fundamental frequency, is coupled through T1 to the push-push frequency multiplier stage. The gates of the Q2 and Q3 are connected in push-pull and their drains are connected in parallel for push-push operation. A single potentiometer is used as the source bias resistor for both Q2 and Q3. By adjusting this potentiometer the drain currents of these two transistors can be equalized.

The push-push amplifer automatically generates the second harmonic of the input signal frequency in essentially the same manner that the ripple frequency of a full-wave rectifier is equal to twice the source voltage frequency. (Illustration courtesy of *Ham Radio*.)

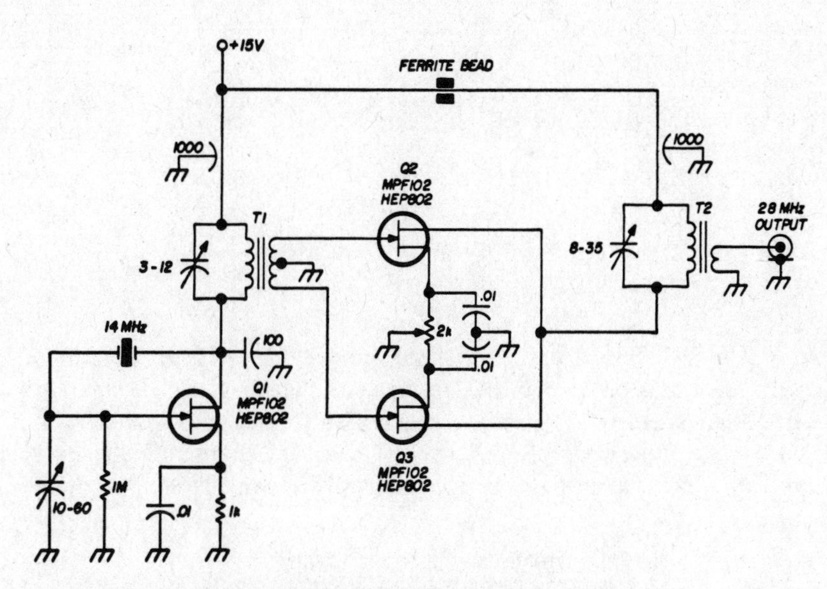

Figure 6-6. Frequency generating circuit.

Frequency Converter

The circuit in Figure 6-7 can be used to expand the tuning range of a receiver. For example, 150-174 MHz VHF band signals can be translated so they can be received with a communications receiver tuned to 10.7 MHz. The VHF antenna is connected to J1 and the receiver is connected to J2. Stations are tuned in by adjusting variable capacitor C2 which determines the frequency of local oscillator V1B. Sensitivity is improved by adjusting variable capacitor C1 to make L1 resonant at the frequency of the intercepted signal so that maximum signal voltage is fed to the grid of mixer V1A. (V1A and V1B comprise a

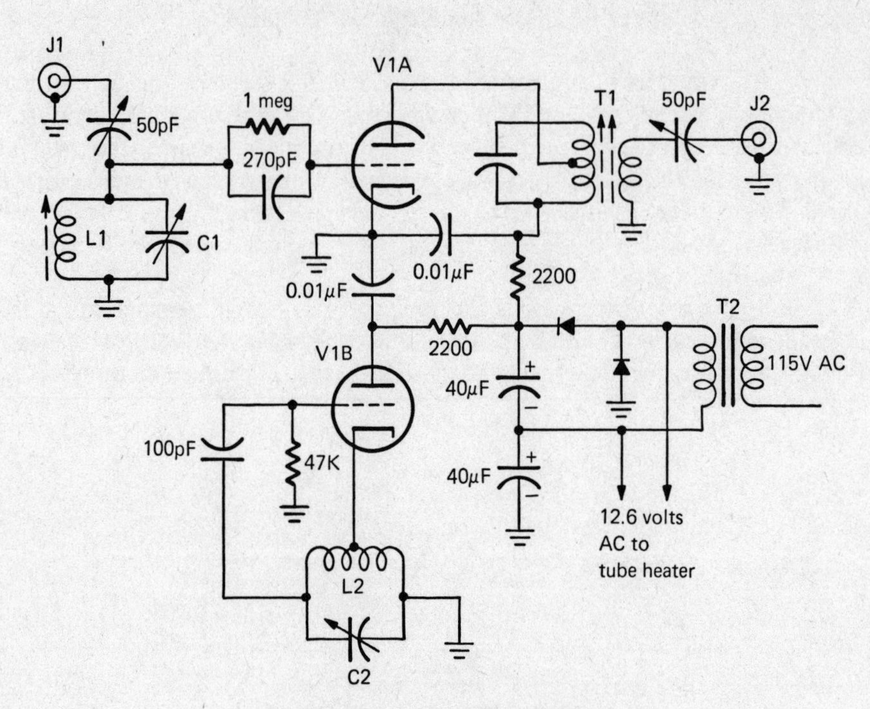

Figure 6-7. Frequency converter.

12.6-volt dual triode.) T1 is an IF transformer tuned to 10.7 MHz (or to whatever frequency the receiver is tuned). The local oscillator frequency must be equal to the sum or difference of the intercepted signal frequency and the frequency to which T1 is tuned.

The power supply consists of a 12.6-volt filament transformer (T2) and a voltage-doubler rectifier which supplies approximately 30 volts to the plates of the triodes.

Frequency Multiplier Trap

The use of a series-resonant wave trap at the input side of a transistor frequency multiplier, as shown in Figure 6-8, can improve the multiplier's efficiency. The input tank is tuned to the frequency of the input signal, but the wave trap connected across the base and emitter of the transistor is tuned to output frequency. This makes the multiplier more efficient by reducing the effect of base-to-collector capacitance. (Illustration courtesy of *Ham Radio*.)

Frequency Multiplier with Dual Pi-Networks

Figure 6-9 shows a frequency multiplier circuit employing a bipolar transistor and input and output pi-networks. The pi-network consisting of L1 and its associated

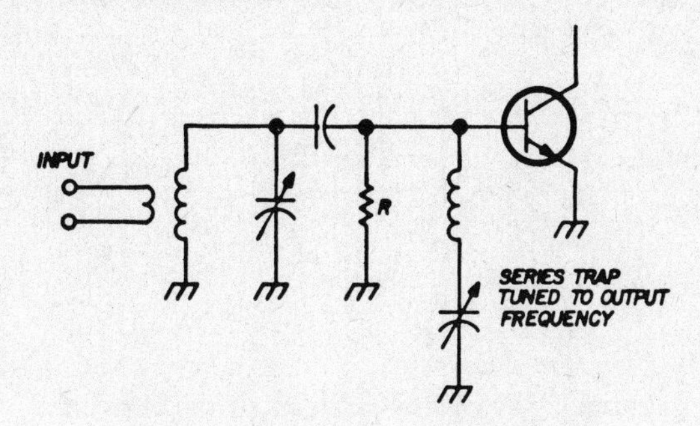

Figure 6-8. Frequency multiplier trap.

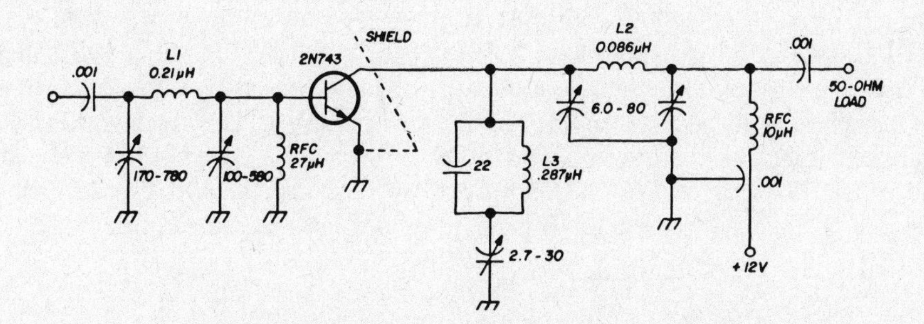

Figure 6-9. Frequency multiplier with dual pi-networks.

capacitors is tuned to the input signal frequency. The output pi-network consisting of L2 and its associated capacitors is tuned to the desired harmonic frequency. When used as a frequency doubler, it is necessary to include a wave trap tuned to the input signal frequency. This wave trap consists of L3, a shunt fixed capacitor, and a series variable capacitor. If the circuit is used as a frequency tripler, two wave traps would be required, one for the input frequency and one for the second harmonic. (Illustration courtesy of *Ham Radio*.)

Grid-Injection Triode Mixer

The mixer circuit shown in Figure 6-10 uses grid-leak bias. Both the local oscillator signal and the radio signal are fed to the grid of the triode tube. The RF transformer at the left is tuned to the frequency of the radio signal and the IF transformer at the right is tuned to the sum or difference beat frequency of the input signals. Since the grid-leak bias causes the triode to operate in a non-linear manner, it is a "detector." When the superheterodyne receiver was first developed, the mixer was called the "first detector" and the demodulator was called the "second detector."

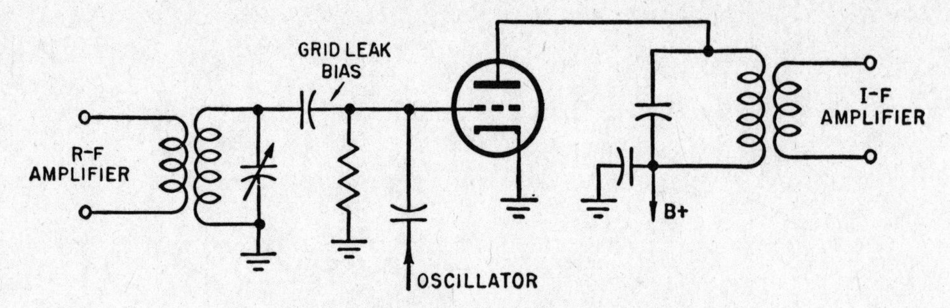

Figure 6-10. Grid-injector triode mixer circuit.

Hot Carrier Diode Mixer

Four "hot carrier" diodes are used in bridge-type mixer circuit shown in Figure 6-11. In a radio receiver application, the intercepted signal is fed in at R and the local oscillator signal at L. If only the local oscillator signal is present, it is balanced out and does not appear at output I. When an intercepted radio signal is also present, the bridge is unbalanced as the diodes are switched on and off by the signals. As a result of this action, an IF signal will be present at I. The frequency of the IF signal is equal to either the sum or difference of the input frequencies, depending upon the primary impedance of T2.

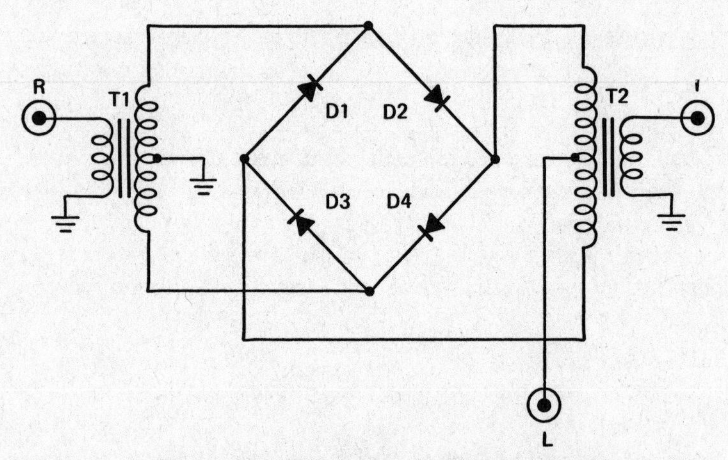

Figure 6-11. Hot carrier diode mixer circuit.

Low Impedance Frequency Doubler

Shown in Figure 6-12 is a frequency doubler circuit using a bipolar transistor, a pi-network input circuit, and a double-tuned output coupling transformer circuit. The

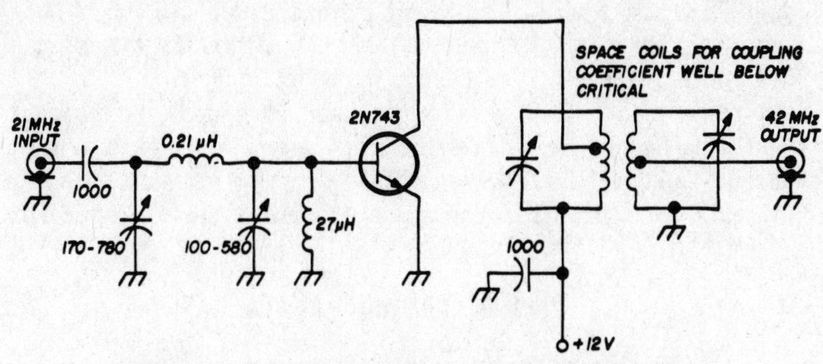

Figure 6-12. Low impedance frequency doubler circuit.

input pi-section, consisting of the two variable capacitors and an inductor, is tuned to the frequency of the input signal, providing the correct load impedance for the input signal source. Both windings of the output transformer are tuned to twice the input frequency. The coupling of the windings is critical since too close coupling will result in the passing of spurious harmonics, and too loose coupling will result in inadequate output level. The output signal is obtained from a tap on the secondary winding of the output transformer to enable feeding the output signal to a low impedance load. (Illustration courtesy of *Ham Radio*.)

Pentagrid Converter Circuit

The circuit shown in Figure 6-13 is of a pentagrid converter using a five-grid tube as a combination local oscillator and mixer in a radio receiver. The radio signal from the previous RF amplifier or from the antenna is applied through a tuned RF transformer to the

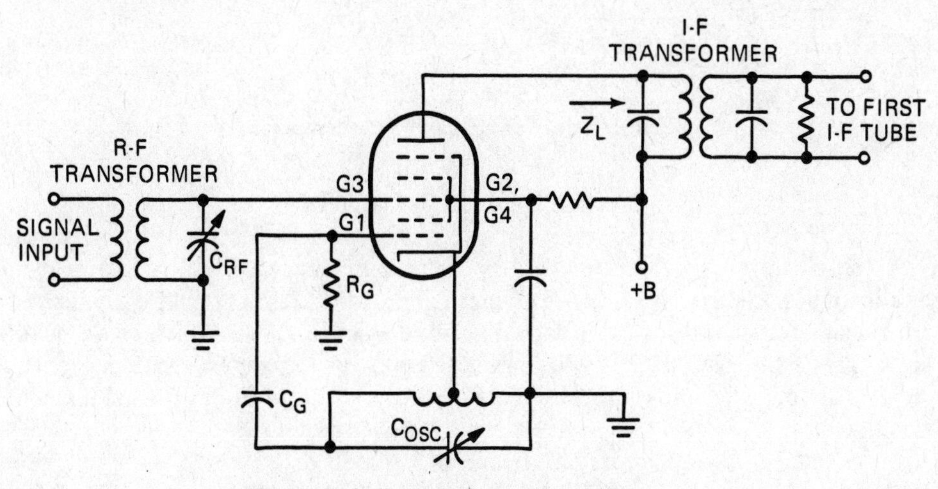

Figure 6-13. Pentagrid converter circuit.

injection grid G3 which is surrounded by the two screen grids (G2 and G4) which are connected together. The Hartley oscillator uses the control grid G1, the cathode, and one screen grid G2 as a triode tube. Grid G2 acts as the triode plate.

Plate current is modulated by both the oscillator signal at G1 and the radio signal at G3, but at differing frequencies. The result is plate current flow that contains both of the input frequencies and their beat frequencies. The IF transformer in the plate circuit filters out the unwanted frequencies and the only signal remaining is the selected beat frequency.

Pentode Autodyne Converter

A single pentode tube is used in this simple autodyne converter circuit. The presence of L3 in the cathode circuit causes oscillations at the frequency to which L2 and C1 are tuned. Potentiometer R2 is used to set the screen grid voltage at the point where stable oscillations are produced. Since L2 and C1 are not tuned precisely to the receiving frequency, the resulting intermediate frequency must be fairly low. For example, if the IF is 50 kHz, and the receiving frequency is 710 kHz, L2-C1 would be tuned for resonance at either 760 kHz or 660 kHz. Although some loss of sensitivity results, this circuit is a simple and low-cost way of obtaining frequency conversion.

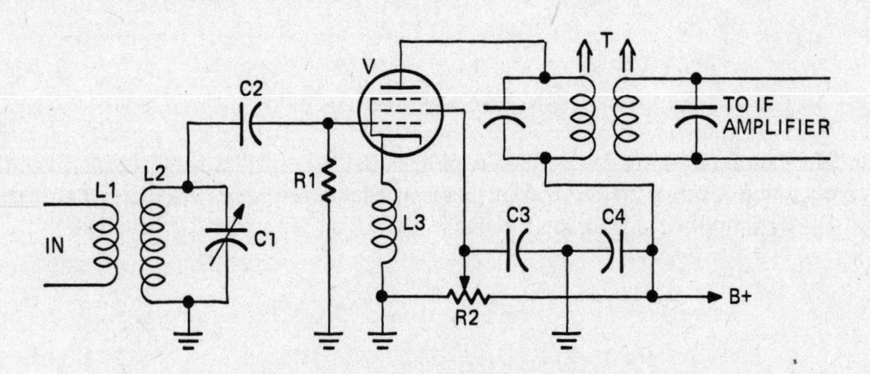

Figure 6-14. Pentode autodyne converter.

Push-Pull Frequency Multiplier

The circuit shown in Figure 6-15 is a push-pull frequency multiplier employing electron tubes. The circuit closely resembles a push-pull, Class C amplifier, except that cross-neutralization capacitors are not used since the output-input frequencies are not the same. The input tank is tuned by the balanced capacitors to the frequencies of the input signal. The output tank is tuned by its associated capacitors to the desired harmonic frequency.

The highest obtainable harmonic frequency is largely determined by the bias applied to the tubes. They are biased for Class C operation with the conduction angle of each tube

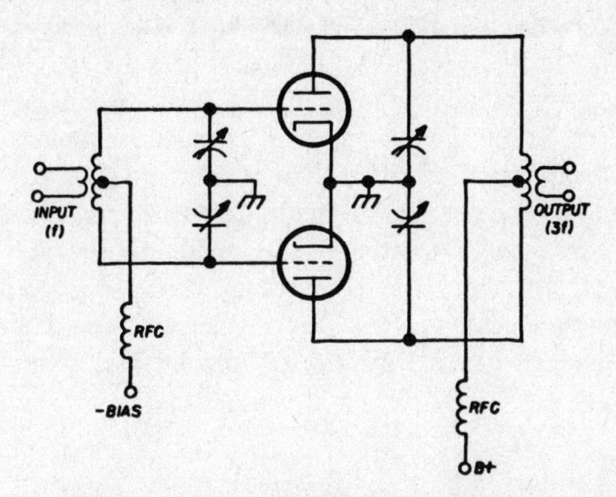

Figure 6-15. Push-pull frequency multiplier circuit.

considerably less than 180 degrees. The higher the frequency multiplication factor, the narrower the conduction angle must be for high efficiency. This type of circuit is useful only for generating odd harmonics because the push-pull action cancels out the even harmonics. (Illustration courtesy of *Ham Radio*.)

Push-Pull Transistor Frequency Multiplier

Figure 6-16 illustrates the use of two NPN transistors in a push-pull frequency multiplier circuit. The input tank is tuned to the frequency of the input signal, and the output tank is tuned to the desired harmonic frequency. The two transistors are connected in a common-emitter configuration and are biased for Class C operation during each cycle of input signal. Unless the input drive level is high enough, a small forward bias voltage must be provided. (Illustration courtesy of *Ham Radio*.)

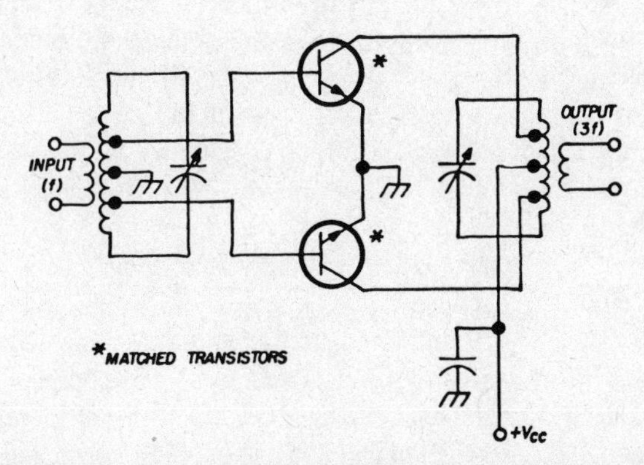

Figure 6-16. Push-pull transistor frequency multiplier circuit.

Push-Push Transistor Frequency Multiplier

A push-push transistor frequency multiplier circuit is shown in Figure 6-17. This type of frequency multiplier cancels the fundamental and odd harmonics, so is useful as a frequency doubler, quadrupler, and so on. The input tank is tuned to the input signal frequency and the output tank is tuned to the desired harmonic. The circuit shown uses two NPN transistors connected in the common-emitter configuration, but PNP transistors could be used as well. The emitter-base junctions of the transistors are connected in push-pull, but the base-collector junctions are connected in parallel. Unless the input signal level is high enough a small amount of forward bias is required. (Illustration courtesy of *Ham Radio*.)

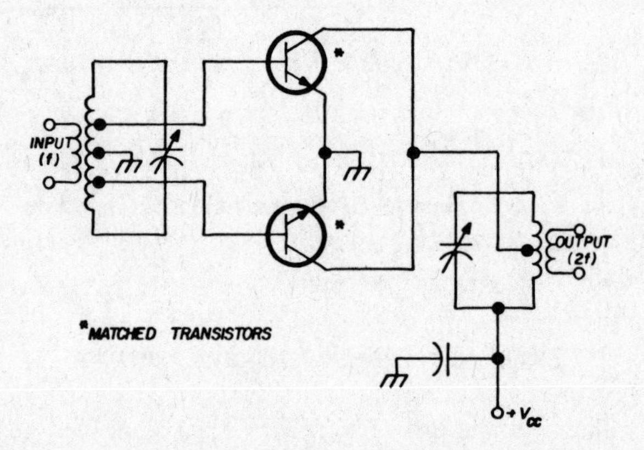

Figure 6-17. Push-push transistor frequency multiplier circuit.

Second Harmonic Frequency Converter

The first portable superheterodyne receiver, now a museum piece, was constructed by Harry Houck for Major Edwin H. Armstrong, the inventor of the superheterodyne receiver. The superheterodyne development depended upon the invention of the triode tube by Dr. Lee De Forest. This particular receiver (still operable) employs a "second harmonic" frequency converter whose circuit is shown in Figure 6-18, shown for historical reasons. Today, the use of this kind of circuit is seldom necessary (but by no means is the circuit obsolete) in most receivers because the IF is far removed from the local oscillator frequency.

When this circuit was conceived, a relatively low IF was used in order to obtain adequate gain and selectivity because higher-frequency IF transformers were not available. Assume that 50 kHz is the IF (intermediate frequency) and that reception of WOR on 710 kHz is required. This means that the local oscillator should operate at 660 kHz or 760 kHz to obtain a 50-kHz IF signal. Since the local oscillator signal frequency is so close to the receiving frequency that serious "image" frequency interference and local oscillator locking to the receiving frequency could result, Major Armstrong decided to tune the local

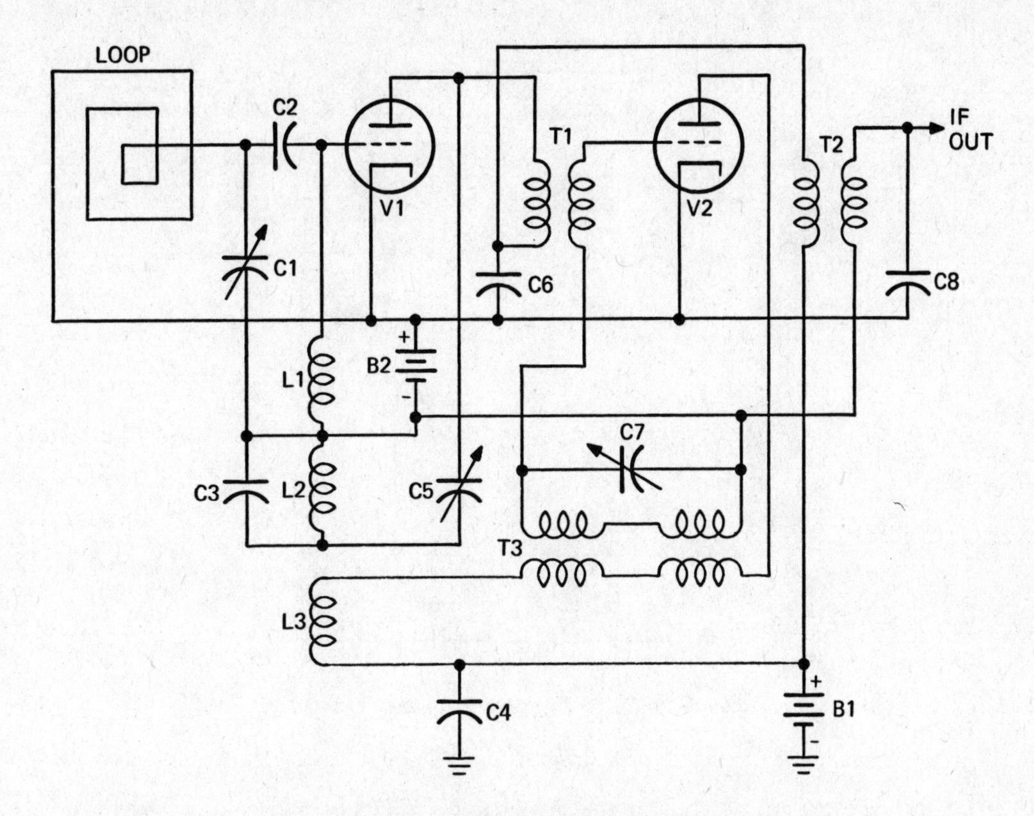

Figure 6-18. Second harmonic frequency converter.

oscillator to a frequency one octave lower than the required injection frequency.

By tuning the local oscillator V2 to 330 kHz or 380 kHz to receive WOR on 710 kHz (as an example), its second harmonic can be used as the 660-kHz or 760 kHz as the injection signal. In the circuit shown in the diagram, C1 is adjusted to make the loop antenna resonant at the receiving frequency, and C7 is adjusted to make T3 resonant at half the required injection frequency. The signal generated by V2 is inductively coupled from L3 to L2 and is tuned to its second harmonic with C5. IF transformer T2 is tuned to the IF which is equal to the receiving frequency plus or minus the second harmonic of the local oscillator frequency.

In modern VHF and UHF communications receivers, the same principle is used. Harmonics of the local oscillator crystal frequency are utilized as the required injection frequency signal.

Series-Injection Frequency Converter

In most frequency converters the local oscillator signal is usually injected into the base or emitter of a mixer transistor or the grid of a mixer tube, or electron-coupled as in a

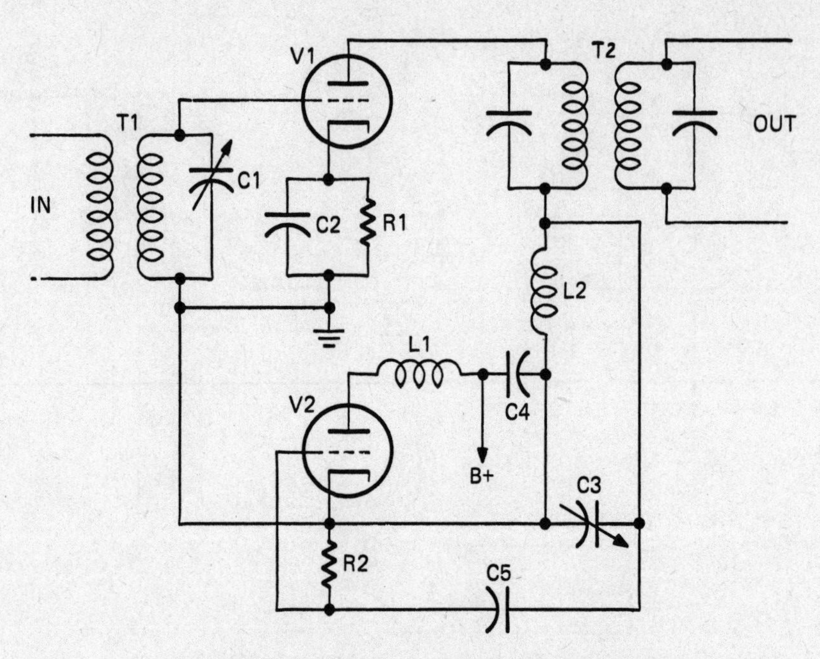

Figure 6-19. Series-injection frequency converter.

pentagrid tube converter circuit. In the circuit shown in Figure 6-19, the local oscillator plate-modulates the mixer. In this circuit, triode V1 is the mixer which is cathode-biased for non-linear operation, and its input is tuned to the receiving frequency with C1. Triode V2 is the local oscillator whose frequency is adjusted with C3 which, with L2, forms the oscillator tank circuit. Plate voltage to V1 is routed through L1 (oscillator feedback coil), the plate-cathode path of V2, L2, and the primary of IF transformer T2. As V2 plate current varies at the oscillator frequency rate, so does the instantaneous voltage at the plate of V1. Since the incoming radio signal frequency is modulated by the oscillator frequency, an IF signal is generated by the heterodyning of the two signals.

Transistor Autodyne Converter

A single transistor (Q1) serves as both the mixer and local oscillator in the superheterodyne receiver circuit shown in Figure 6-20. L1 is the ferrite rod antenna which intercepts radio signals and L2 represents the two-winding local oscillator coil, both of which are tuned by a two-gang variable capacitor. The intercepted radio signal is fed from L1 through C1 to the base of Q1. The local oscillator signal, which is generated by Q1, is injected into its emitter through C2. Oscillations are produced because the tickler coil of L2 is in series with the collector circuit of Q1 and is inductively coupled to the tank portion of L2, causing positive feedback.

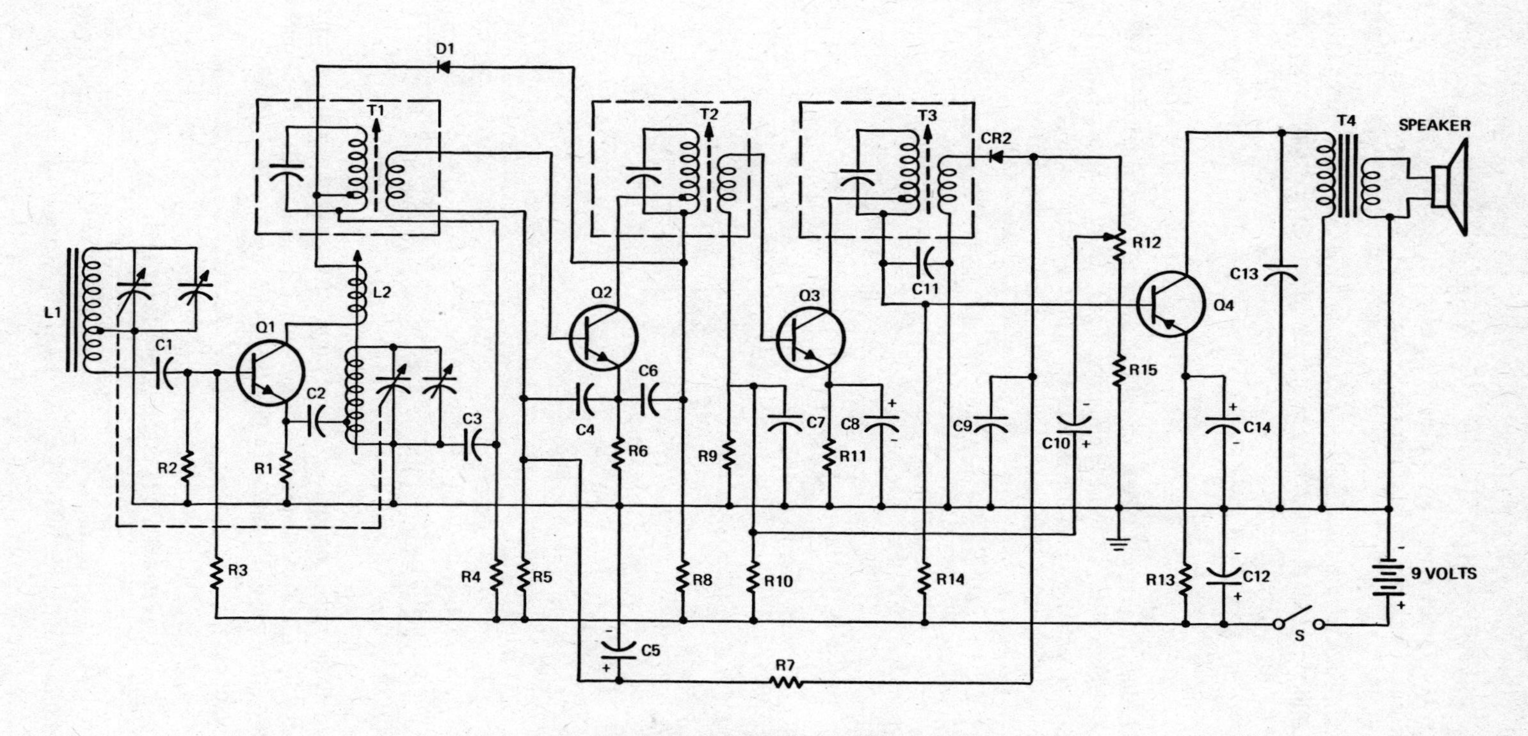

Figure 6-20. Transistor autodyne converter.

If the receiver's IF is 455 kHz and the receiver is tuned to a station operating on 1000 kHz, Q1 will generate oscillations at 1455 kHz. As the tuning dial is rotated, the two-gang tuning capacitor tracks so that the oscillator signal will be 455 kHz higher in frequency than the receiving frequency.

In this circuit, IF transformer T1 acts as a bandpass filter that passes the 455-kHz IF signal and rejects the local oscillator and intercepted radio signals. IF transformers T2 and T3 further increase the selectivity.

Transistor Parametric Multiplier

Figure 6-21 illustrates a circuit for a typical transistor parametric amplifer. This type of multiplication makes use of the base-collector depletion capacitance, and must be designed carefully to achieve optimum efficiency. The parametric effect only becomes important when the RF level is high enough to swing a large portion of the collector-to-base voltage.

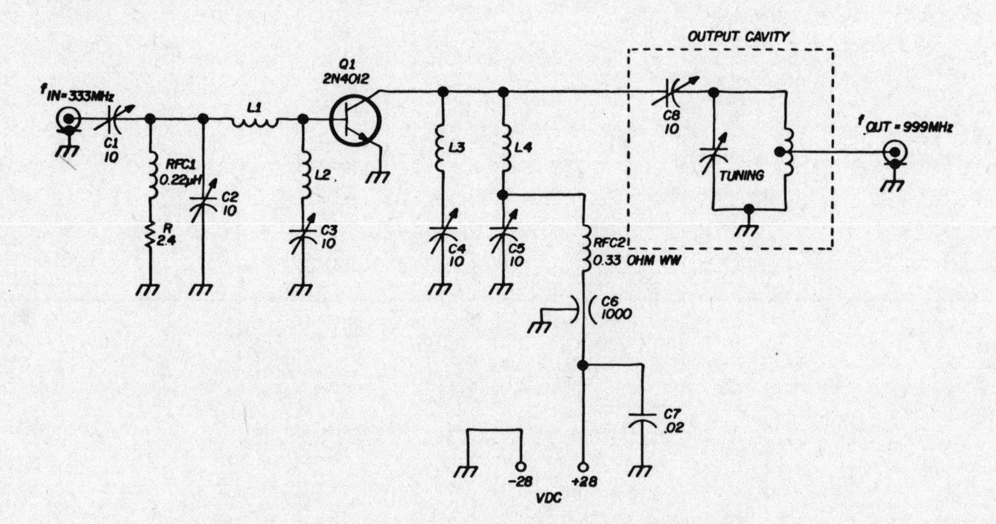

Figure 6-21. Transistor parametric multiplier circuit.

In a transistor parametric multiplier, idler circuits are used in the collector circuit to increase efficiency by reflecting undesired harmonics back to the collector-base capacitance. These idler circuits are illustrated by the networks L3-C4 and L4-C5. (Illustration courtesy of *Ham Radio*.)

Tunnel Diode Frequency Converter

A tunnel diode is used as both the mixer and local oscillator in the frequency converter circuit shown in Figure 6-22. The local oscillator frequency is determined by the

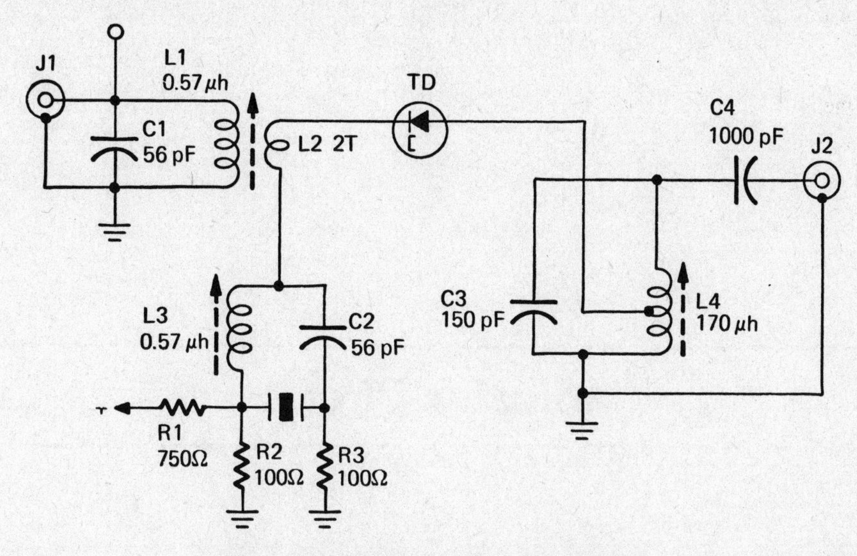

Figure 6-22. Frequency converter circuit.

crystal and its associated components. The IF signal is developed across L4 and C3 and is fed out through C4 to jack J2. Using the indicated values of components, this converter can be used to translate 27-MHz band signals to the AM broadcast band. In such an application, the output from J2 would be fed to the antenna jack of an AM broadcast band auto radio or to the antenna terminal of a communications receiver. The power source can be a 1.5 volt flashlight cell. (Illustration courtesy of General Electric Company.)

7

Indicator Circuits

The most commonly used status indicators are incandescent and neon lamps which are being rapidly replaced in new equipment designs by the LED (light emitting diode). Quantitative indicators include analog and digital meters and the electron ray indicator tube. Described and illustrated in this chapter are various kinds of electrical and electronic indicator circuits.

AC Ammeter

A DC milliammeter or ammeter can be used as an AC ammeter by rectifying the AC and measuring the resultant DC. In Figure 7-1 T is a step-down transformer whose low-voltage secondary is connected in series with the load (between the AC input plug and the AC output receptacle). When load current is zero, no voltage is induced into the primary of T, and meter M indicates zero. When load current flows through the secondary of T a voltage is indicated into the primary. This AC voltage is rectified by diode CR and capacitor C becomes charged. The DC voltage across C is measured by the DC voltmeter consisting of multiplier resistor R and meter M (a DC microammeter, milliammeter, or ammeter).

If T has a 115-volt primary and a 2.5-volt secondary and its secondary current rating is 10 amperes, the AC voltage across its primary, when used in this circuit, will be 115 volts when load current is 10 amperes. When load current is lower, the primary voltage will be lower. Since the reactance of the low-voltage secondary is about 0.25 ohm, the AC load voltage will not be significantly lower than the AC line voltage. When load current is 10 amperes, the voltage loss will be about 2.5 volts and will be less at lower load currents.

This is neither a direct-reading nor an accurate AC ammeter. It gives relative indications, but can be calibrated to give quantitative indications. It is intended to be an inexpensive troubleshooting tool.

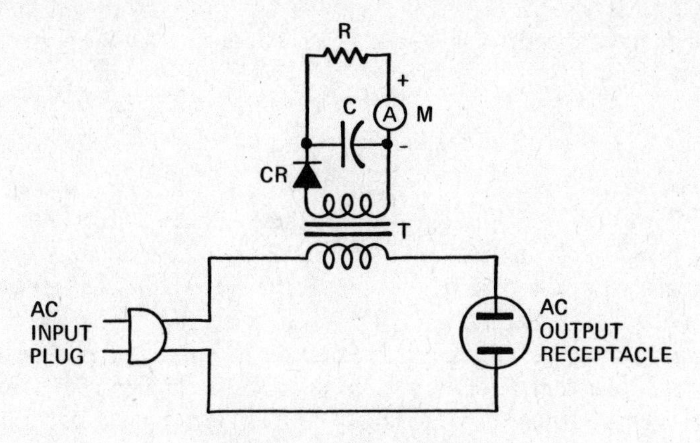

Figure 7-1. AC ammeter.

Ammeter

The magnitude of an electric current is often measured with an ammeter. It may be a meter so designed that all of the current measured actually flows through the meter winding. Too, it can be a meter which is connected into the circuit through a "shunt," as shown in Figure 7-2. For example, if the meter is a 0-1 DC milliammeter and its internal resistance is 100 ohms, its range can extend to 1 ampere full scale by using a 0.1-ohm resistance as a shunt. If the shunt resistance is 100 ohms, the full-scale range is doubled since half of the current flows through the meter and half flows through the shunt. The full-scale range can be computed by:

$$I = E / \frac{Ri + Rs}{Ri \times Rs}$$

where:

Ri is the internal resistance of the meter,
Rs is the resistance of the shunt,
E is the voltage drop across the meter when indicating full-scale, and
I is the current flowing through Ri and Rs in parallel at full-scale meter reading.

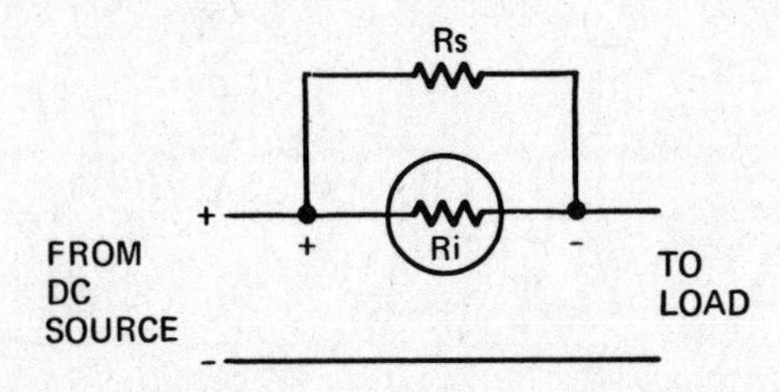

Figure 7-2. Meter connected into circuit through a "shunt."

E must be computed first. It is equal to RI, with R being the known meter resistance and I the full-scale rating of the meter movement. In the case of a 0-1 DC milliammeter whose internal resistance is 100 ohms, E is equal to 0.1 volt (100 x 0.001).

Amplified S-Meters

Figures 7-3A and 7-3B illustrate how inexpensive meters can be made to serve as signal strength indicators in radio receivers. In Figure A, a single stage of transistor amplification is used to boost the weak AVC (automatic volume control) signal to a level that can be used with an inexpensive 0-1 ma meter. In addition, the AVC operation of the receiver is affected very little by the addition of this circuit because of the relatively high input resistance. Potentiometer R5 is used to set the meter to zero with no signal applied, and potentiometer R3 reduces the sensitivity so that the strongest signal applied causes maximum meter deflection. More isolation and sensitivity is available by using two stages of amplification, as shown in Figure B.

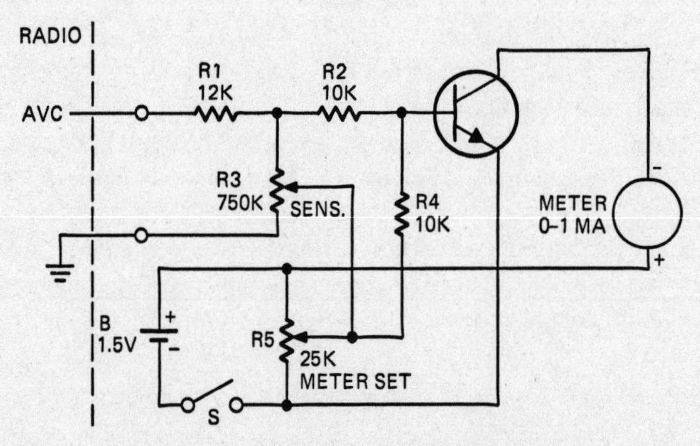

Figure 7-3A. Single stage of transistor amplification used to boost weak AVC.

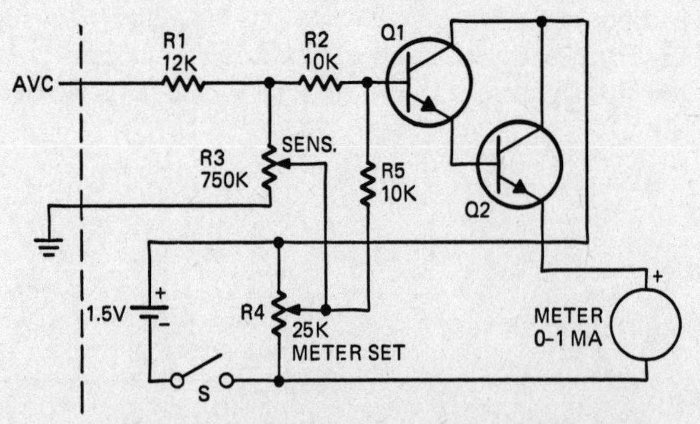

Figure 7-3B. Use of two stages of amplification.

Bridged S-Meter

Figure 7-4 illustrates a meter circuit that can be used to read the strength of the signal being processed by a receiver. The input to C1 is from one of the IF stages, and C1 is tuned to the IF of the receiver. The signal is detected by Q1 and its associated circuitry, and fed to bridge amplifier Q2. With no signal applied, resistor R2 is adjusted so as to balance the bridge (causing no reading on meter M). The received signal then causes the bridge to unbalance and the meter to register commensurate with the strength of the signal.

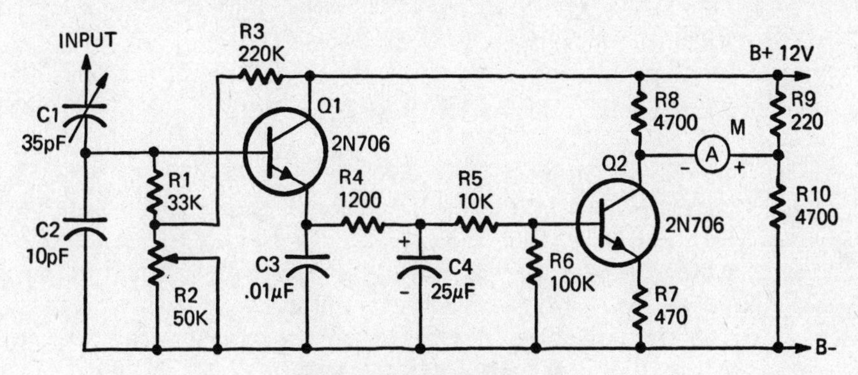

Figure 7-4. Bridged S-meter.

Capacitor Tester

An ordinary low-wattage incandescent lamp, such as a pilot lamp, can be used in an electrolytic capacitor testing circuit, as shown in Figure 7-5. B can be a 6-volt battery, I a No. 47 lamp and S a three-position SPDT switch. When a large value capacitor (C) is connected across X and Y, setting S to the CH (charge) position will cause the lamp to flash momentarily as C charging current flows through both C and I. Then, when S is set to the DIS (discharge) position, the lamp will again flash as C discharges through I.

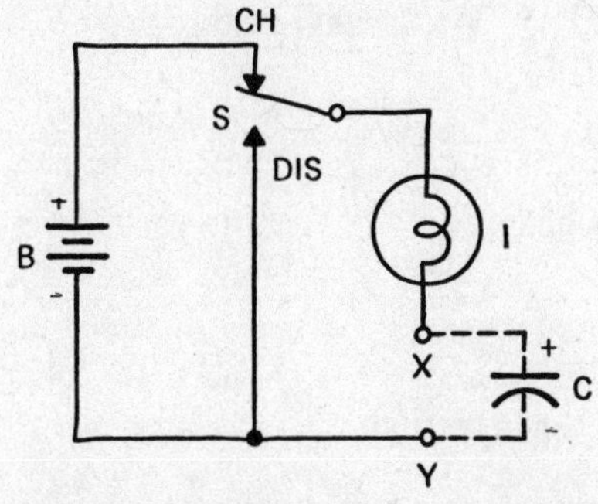

Figure 7-5. Low wattage incandescent lamp used in an electrolytic capacitor circuit.

The brilliance and duration of the flashes depends upon the time constant (TC) of C and I. The "hot" resistance of a No. 47 lamp is approximately 42 ohms. But, its "cold" resistance is probably less than 5 ohms. If the capacitance of C is 100 mf, for example, TC = 100 mf times 0.000005 megohms (5 ohms) or 500 microseconds (0.0005 second) when the lamp filament is cold or around 4200 microseconds when the filament is hot. When testing a capacitor in this circuit, the time constant is between these two extremes since the lamp filament resistance changes during the course of the test.

If the lamp glows continuously when S is in the CH position, the capacitor is shorted. If the lamp does not flash at all, the capacitor is open or dehydrated, or its capacitance is too low for this test circuit. The advantages of this tester is that it permits GO-NO/GO tests to be made quickly with only a low DC voltage applied.

Crystal Markers

A simple method for generating marker frequencies is shown in Figure 7-6. In this circuit, four crystals (any number can be used) are shown connected across the output of a sweep generator which is connected to the input jack. As the sweep generator frequency varies, energy at the series-resonant frequency of each crystal is shunted to ground as its frequency is passed. A suck-out or glitch indication will appear for each crystal frequency on the swept oscilloscope display. At other than its series-resonant frequency, each crystal looks like an open circuit. But, at its series-resonant frequency, each crystal short circuits the sweep generator output. Commercial sweep generation systems often use a variation of this technique to place markers on their swept displays to indicate frequency. (Illustration courtesy of *Ham Radio*.)

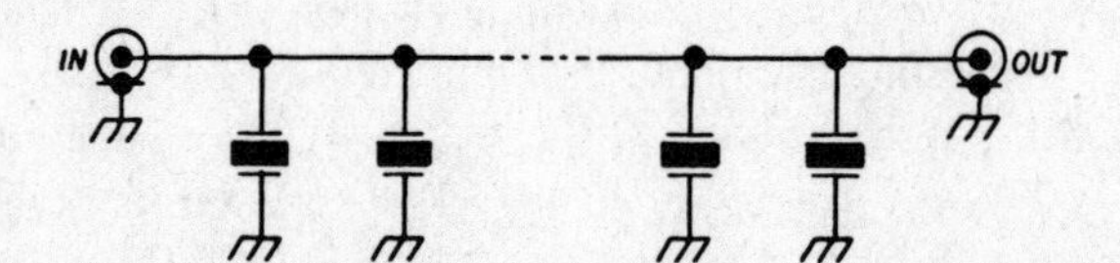

Figure 7-6. Simple method for generating marker frequencies.

Dry Battery Tester

Testing dry cells and dry batteries with a DC voltmeter can lead to erroneous conclusions. The tests should be made with a "load" applied. Although the tester circuit shown in Figure 7-7 does not yield quantitative indications (actual volts), it does yield yes-no indications. It consists of a normally-open push button switch (S), a transformer (T), and a neon lamp (I). The transformer can be a 6.3-volt filament transformer with a 115-volt primary. The battery (or cell) to be tested is connected to points X and Y. When S is closed momentarily, current flows through the low-resistance secondary of T which acts as a load. When S is opened, current flow ceases and an inductive kick is developed across the primary of T and the neon lamp glows momentarily. If the neon lamp does not glow, the battery (or cell) is no longer useful.

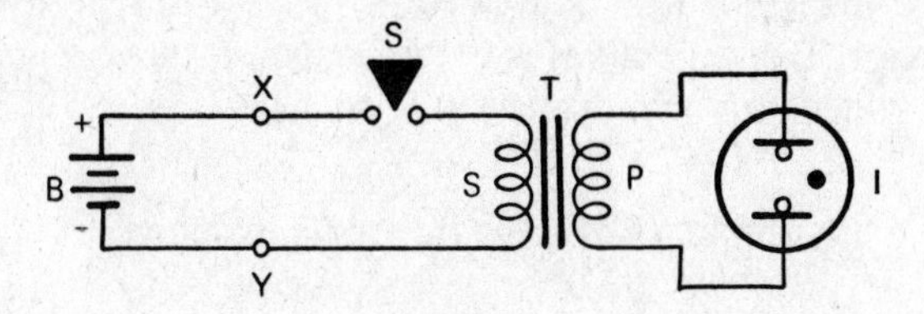

Figure 7-7. Dry battery tester circuit.

Dual Electron Ray Indicator Tube

A circuit employing a dual electron ray indicator tube (V) is shown in Figure 7-8. The voltages at its ray control electrodes (G1 and G2) are controlled independently by triode-connected remote cutoff pentodes V1 and V2. When the control grid of V1 is made more negative, its plate voltage and G1 voltage rise and cause more of the target area of V to glow. The same is true with respect to V2.

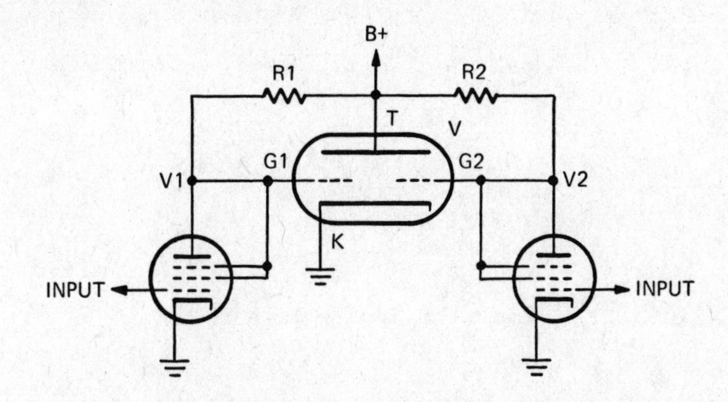

Figure 7-8A. Circuit employing a dual electron ray indicator tube.

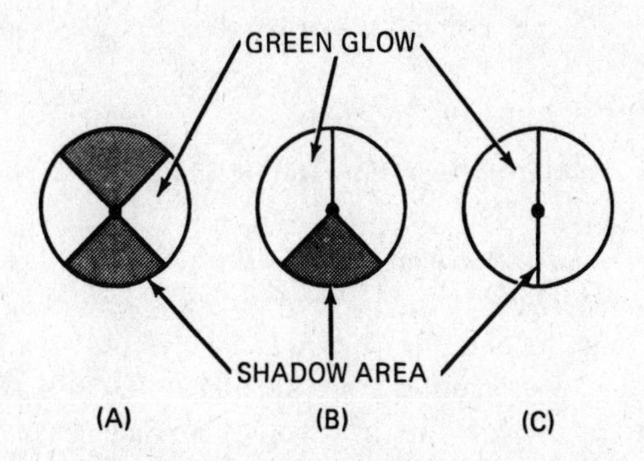

Figure 7-8B. Typical target flow conditions.

Typical target flow conditions, illustrated in Figure 7-8 are: (A) both sections indicate no input voltage; (B) both sections indicate medium input voltage; (C) both sections indicate full input voltage for zero-degree shadow areas.

Electron Ray Tube Power-On Indicator

Because of its relatively large diameter screen (target) and bright green glow, an electron ray indicator tube (6E5, etc.) is sometimes used as a pilot lamp or status indicator. In the circuit shown in Figure 7-9, its target and plate (through R1) are connected to a positive DC voltage source (usually around 250 volts). Its heater (leads X and Y) are connected to a source of 6.3 or 12.6 volts AC or DC, depending upon rated heater voltage. With both voltages applied, and after the cathode has reached operating temperature, the entire target area will glow green because of cathode bias developed across R2 which makes the control grid negative with respect to the cathode.

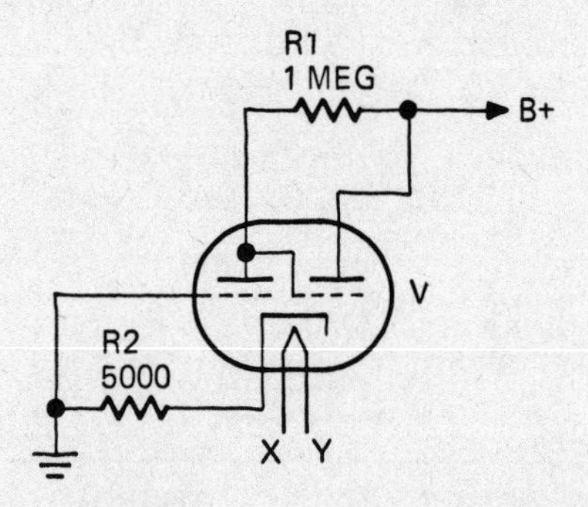

Figure 7-9. Electron ray tube power-on indicator.

This kind of indicator is used in high power transmitters, RF induction heaters, and other apparatus which should not be made operational until after a brief warm-up period. The green "eye" will not glow until the indicator tube has warmed up. Nor will it glow if the required DC plate and target voltage is not present which could be an indication of equipment malfunction.

Electron Ray Tube Tuning Indicator

The electron ray indicator tube (6E5, etc.) is used as an audio level indicator in tape recorders, and was very widely used in radio receivers as a tuning indicator prior to World War II in a circuit similar to the one shown in Figure 7-10. In this circuit, V1 is a remote cutoff pentode IF amplifier whose gain is controlled by AVC voltage, V2 is a duo-diode-triode combination detector and AF amplifier and V3 is an electron ray indicator tube (magic eye or electron eye).

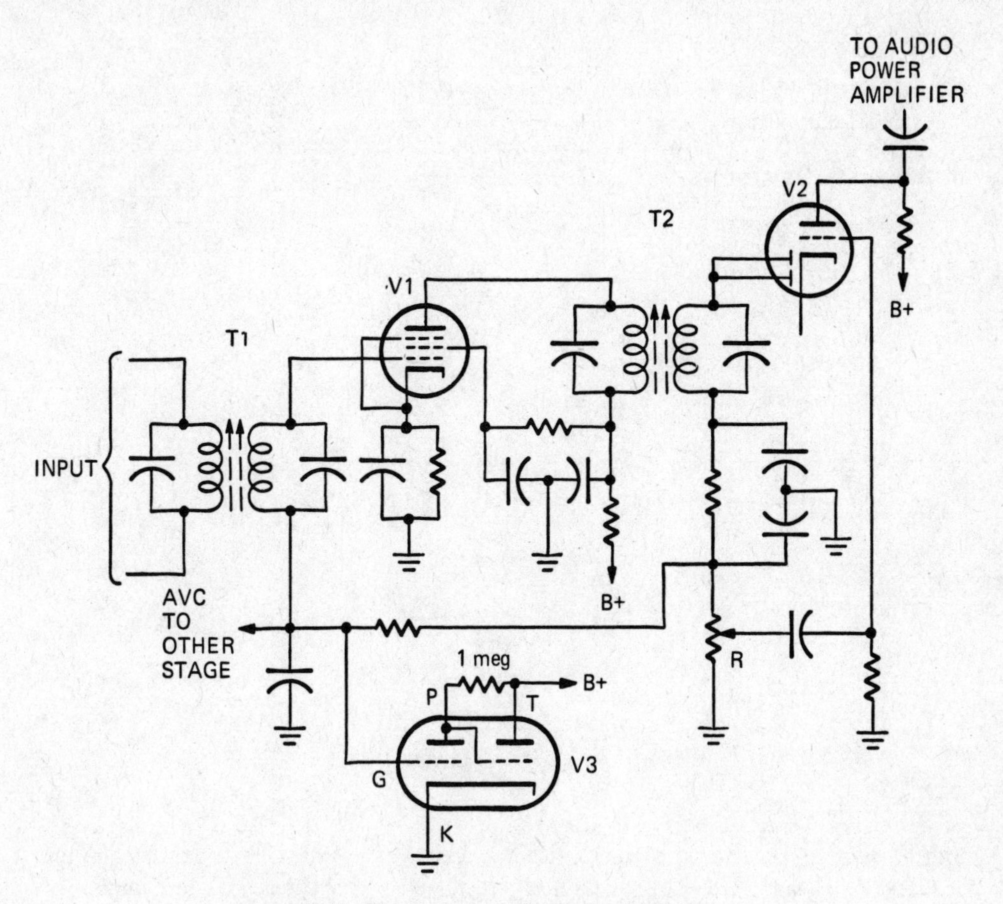

Figure 7-10. Electron ray tube tuning indicator.

The control grid of V3 is connected to the AVC buss. When no radio signal is being received, AVC voltage developed across volume control R is very low and the shadow angle of V3 is approximately 100 degrees. When a signal is being received, AVC voltage rises (becomes more negative) and V3's shadow angle narrows or is reduced to zero degrees, depending upon the strength of the signal. To tune the receiver so that the intermediate frequency will be in the center of the IF amplifier passband, the tuning dial is rotated to obtain minimum V3 shadow angle.

Electronic Voltmeter

Figure 7-11 shows how a meter and two commercially available IC's are connected for use as a high-impedance electronic voltmeter. The high input impedance is assured by connecting the input to the non-inverting input of the first integrated circuit amplifier.

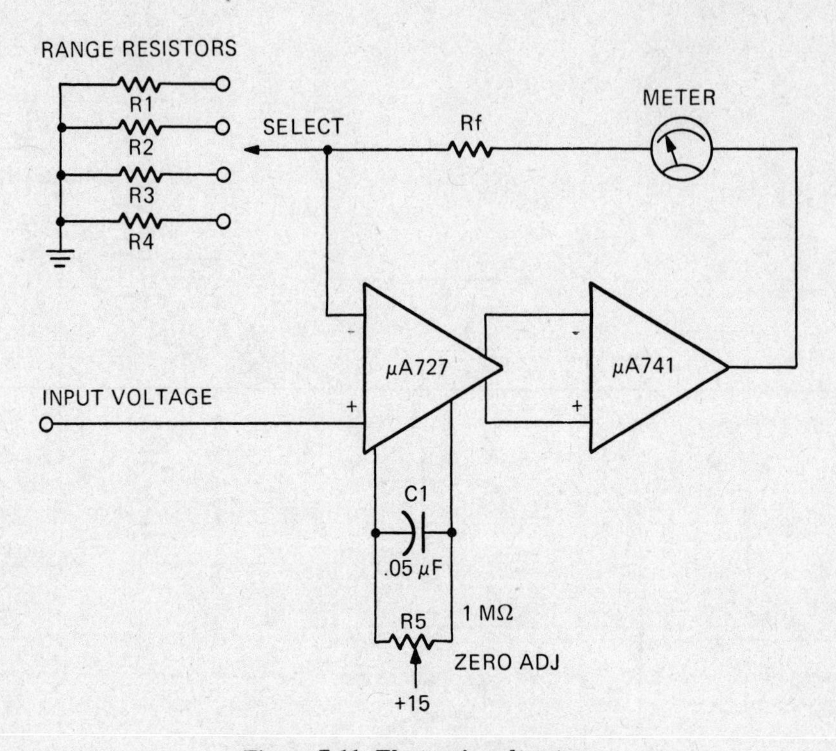

Figure 7-11. Electronic voltmeter.

With no signal input, the zero adjust potentiometer, R5, is set so that the meter in the feedback loop reads zero. The position of the select switch determines the amount of feedback signal that is shunted to ground.

In-Line S-Meter

Figure 7-12 shows the S-Meter of an AM radio receiver. It indicates the relative strength of intercepted radio signals. In the diagram, V1 represents a remote cutoff IF or RF amplifier pentode tube whose gain is controlled by a negative AVC voltage fed to its control grid. Tube V2 represents an AF power amplifier pentode or beam power tube. Meter M is usually a 0-1 DC milliammeter with a scale calibrated in arbitrary S units from 1 to 9 and dB above S9. Resistor R2 is a current limiting resistor, and potentiometer R3 is used for setting the meter at zero when no signal is being received.

The DC voltage at the cathode of V1 rises as AVC falls, and decreases as AVC voltage becomes greater. The DC voltage at the cathode of V2 remains steady. Potentiometer R3 is adjusted when no signal is being received so that the DC voltage at the + terminal of the meter is the same as at the cathode of V1. No current flows through the meter. When a radio signal is received, AVC voltage rises and V1 cathode voltage falls

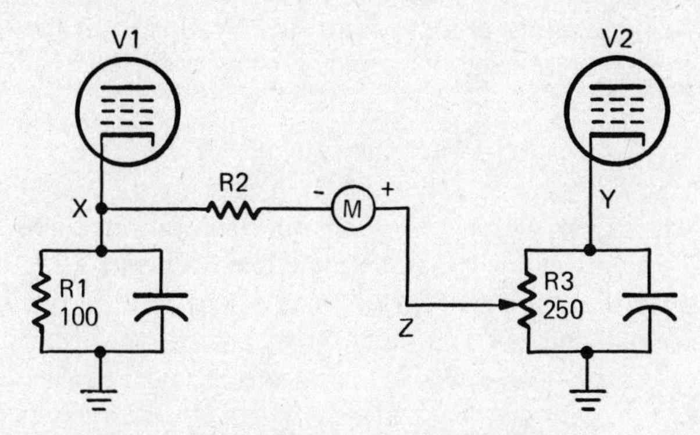

Figure 7-12. S-meter of an AM radio receiver.

(because cathode current is less). Current now flows through the meter. The lower the V1 cathode voltage, the higher the meter indication.

Multimeter

The circuit shown in Figure 7-13 is that of a multirange voltmeter/milliammeter employing a 0-1 DC milliammeter as the indicator. An 8-position rotary switch is used to select voltage and current ranges (10, 100, and 1000 volts AC, or DC, and 10 and 100 milliamperes DC). The top three resistors are the AC voltage multipliers, the next three

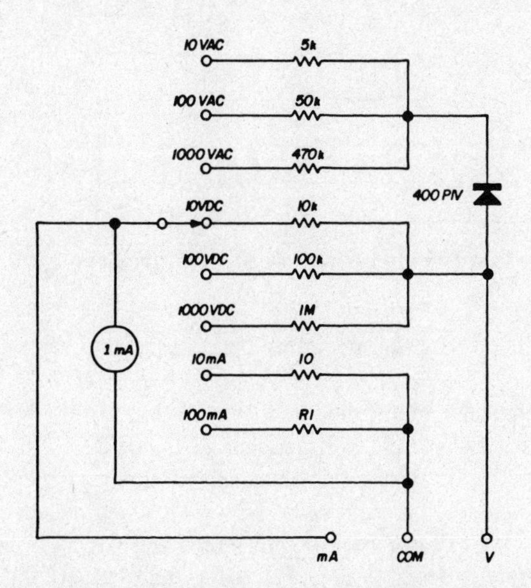

Figure 7-13. Multi-range voltmeter/milliammeter circuit.

are the DC voltage multiplier resistors, and the lower two resistors are the current shunts. When measuring AC voltage, the diode rectifies the AC voltage and allows pulsating DC to flow through the meter movement. (Illustration courtesy of *Ham Radio*.)

Neon Indicators

Figure 7-14 shows two configurations in which neon lamps are often found in appliances and electronic equipment. Neon glow lamps operate cool, have long life, require low power and operate rapidly. In 7-14A, a neon lamp is shown connected in series with a resistor and functions as a pilot lamp to indicate when an appliance is on. In 7-14B, a neon lamp is used to indicate when a preset temperature has been reached. When the thermostat opens, current that was formerly flowing through the thermostat is now shunted through the neon lamp and its ballast resistor, causing it to glow.

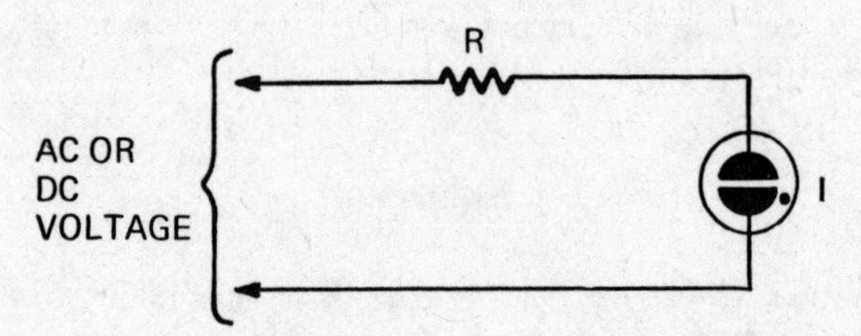

Figure 7-14A. Neon lamp to indicate when appliance is on.

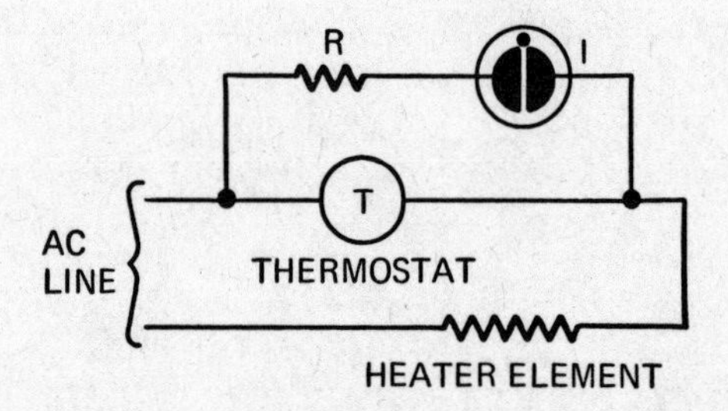

Figure 7-14B. Neon lamp to indicate when preset temperature is reached.

Overmodulation Indicator

An electron ray indicator tube (6E5, 6U5, etc.) is used in the circuit illustrated in Figure 7-15 to indicate negative AM overmodulation. Diode CR1 blocks passage of the positive plate voltage and CR2 bypasses any leakage current through CR1. Normally, the eye of the electron ray tube will remain open. However, when negative overmodulation exceeds 100 percent, CR1 will become forward-biased and CR2 will be reverse-biased. Then, a negative voltage will be fed through R2 to the control grid of the electron ray tube causing its eye to close every time negative overmodulation peaks occur.

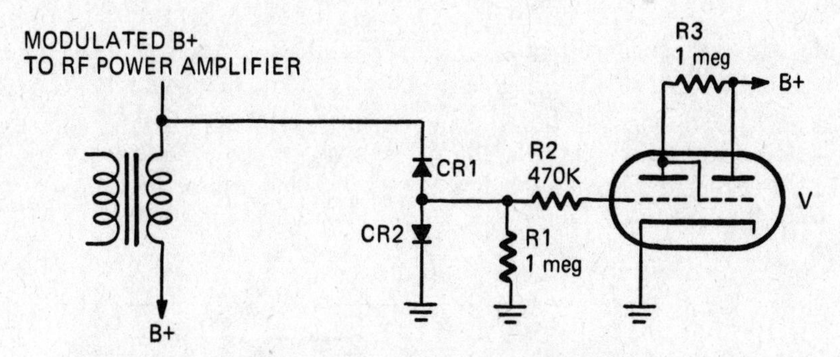

Figure 7-15. Electron ray indicator tube to indicate negative Am overmodulation.

Pilot Lamps

Figure 7-16A illustrates how a neon lamp may be used as a pilot lamp by connecting it across the AC input line through a series resistor. In B, an incandescent lamp is used. Incandescent lamps are usually placed across one of the secondaries of a power transformer, generally a low-voltage filament winding.

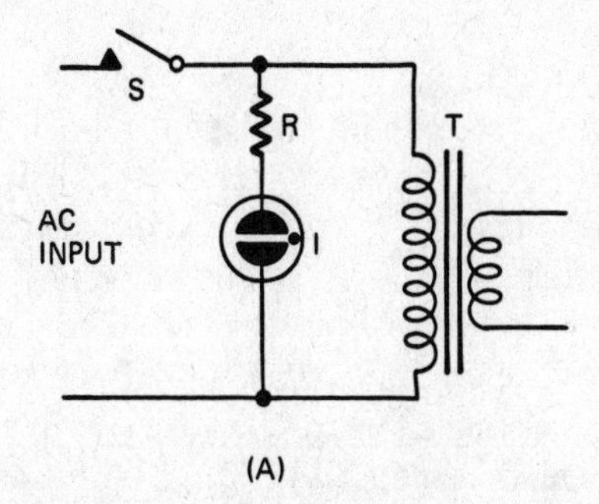

Figure 7-16A. Neon lamp used as a pilot lamp.

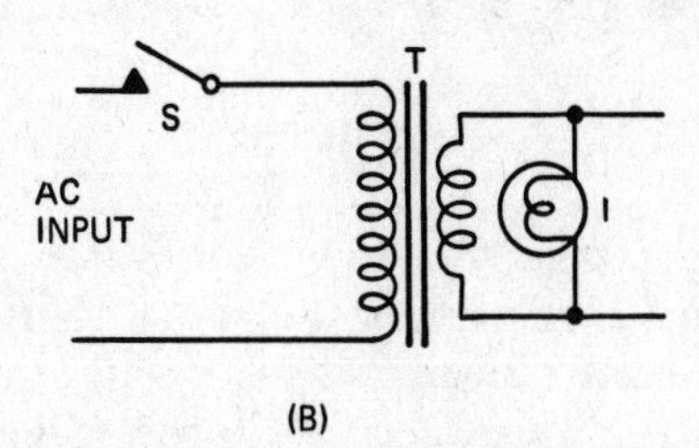

Figure 7-16B. Incandescent lamp used as a pilot lamp.

Plate Current S-Meter

Monitoring the plate current of an RF or IF amplifier stage with a DC milliammeter (A), as shown in Figure 7-17, will give an indication of the strength of the received signal. A stronger signal will cause a higher AVC voltage to be applied to the control grid of the tube and plate current will be reduced. When no signal is being received, AVC voltage will be low and plate current, as indicated by the meter, will be maximum.

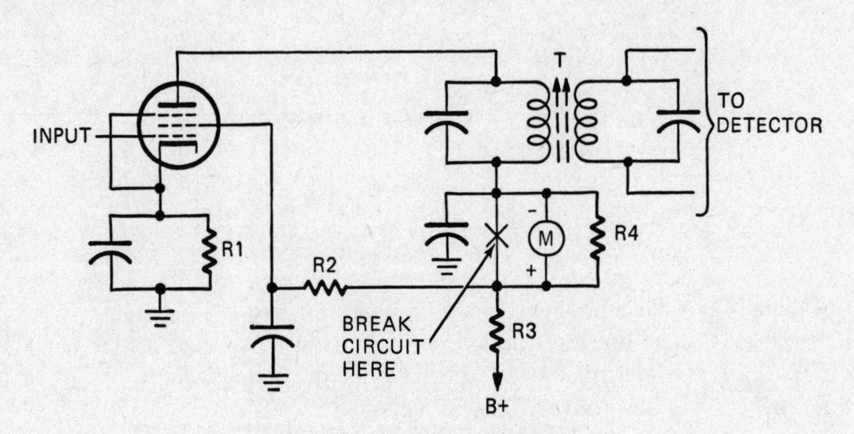

Figure 7-17. Plate current S-meter.

RF Output Meter

The relative RF power output of low power radio transmitters is often indicated by a meter that measures DC voltage obtained by rectification of the RF signal. In this circuit, the RF output signal is tapped at the antenna jack and is fed through variable capacitor C1 to shunt diode rectifier CR. The resulting pulsating DC voltage across CR is measured by a DC voltmeter circuit consisting of multiplier resistor R and DC microammeter (or milliammeter) M. The meter is shunted by C2 which bypasses RF and stabilizes the movement of the meter. This measuring circuit can be calibrated by measuring actual RF power with an RF wattmeter connected to the antenna jack and adjusting C1 to obtain relatively corresponding readings at meter M.

S-Meter Addition

An S-meter, indicating strength of received signals, can easily be added to a radio receiver as shown in Figure 7-19. Transistor Q is the last IF stage, and diode CR is the audio detector. The components surrounded by the dotted lines are added as shown, connected between ground and the point at which the AGC (automatic gain control) voltage is derived to be fed back to previous stages. A stronger signal will cause a

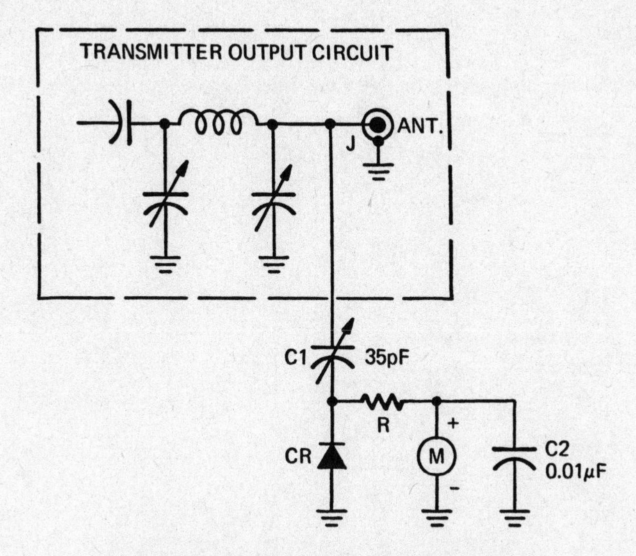

Figure 7-18. RF output meter circuit.

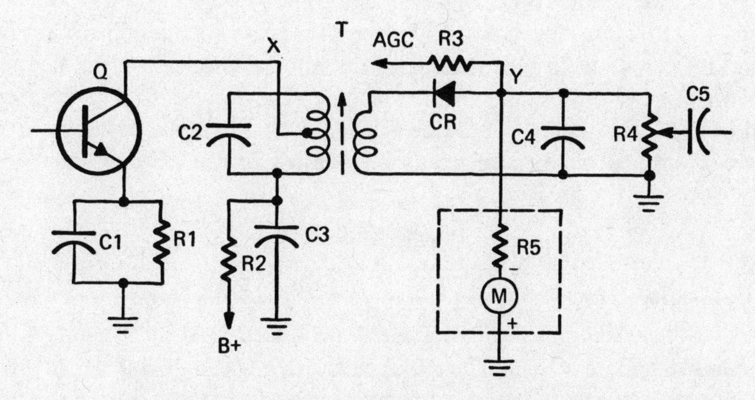

Figure 7-19. S-meter added to a radio receiver.

commensurately larger AGC to be developed, and this voltage will be indicated as a higher reading on the meter M.

Transistor Dip Meter

The circuit shown in Figure 7-20 is of a dip meter, designed by RCA, which is used for measuring the resonant frequency of tuned circuits. Frequency of oscillation of the transistor oscillator is determined by the L1-C1 tank circuit. The generated RF signal is

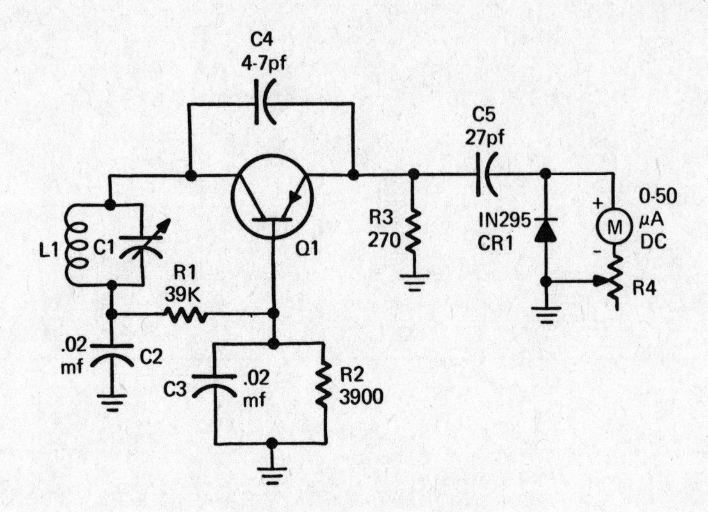

Figure 7-20. Dip meter circuit.

rectified by diode CR1 and its DC output is applied to meter M, a 0-50 DC microammeter which indicates RF signal level. Potentiometer R4 is set to obtain full-scale reading on the meter, and L1 is held next to the resonant circuit to be checked. C1 is then tuned until there is a noticeable "dip" in the meter indication. The dip results from RF energy being absorbed from L1. Generally, L1 is one of several plug-in coils, each of which permits coverage of a specific band of frequencies. (Courtesy of RCA Corporation.)

Volume Unit Meter

It is common practice in broadcast studios and recording studios to measure audio signal level with a VU meter whose scale is calibrated in volume units. A VU meter circuit is shown in Figure 7-21. When bridged across a 600-ohm line the circuit is loaded by 7500-ohms which causes insignificant drop in signal voltage. The 3900-ohm T-pad is used for setting zero VU reference level. A VU meter differs from conventional AC voltmeters in its logistical characteristics. It is designed specifically for measuring the levels of complex wave forms such as produced by speech and music.

VSWR Meter

Figure 7-22 shows the circuit of VSWR (voltage standing wave ratio) meter. It indicates both forward and reflected power when inserted in the coaxial cable transmission line between the transmitter and the antenna. Resistors R1 and RS assure that the transmitter "sees" a load of approximately 50 ohms. Diodes CR1 and CR2 rectify the RF energy for application to the meter amplifier transistor. R5 is set for minimum indication on the

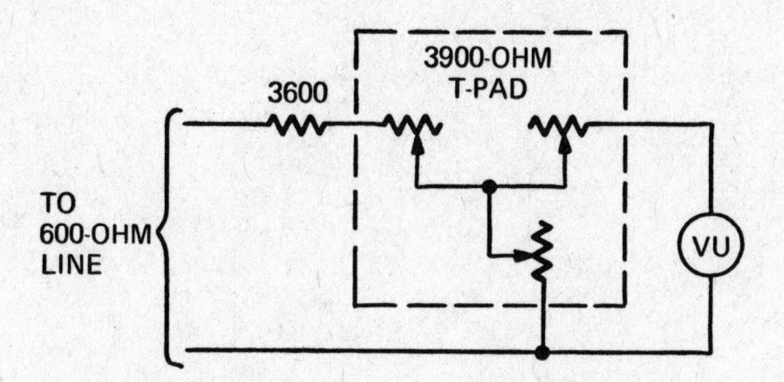

Figure 7-21. Volume Unit meter circuit.

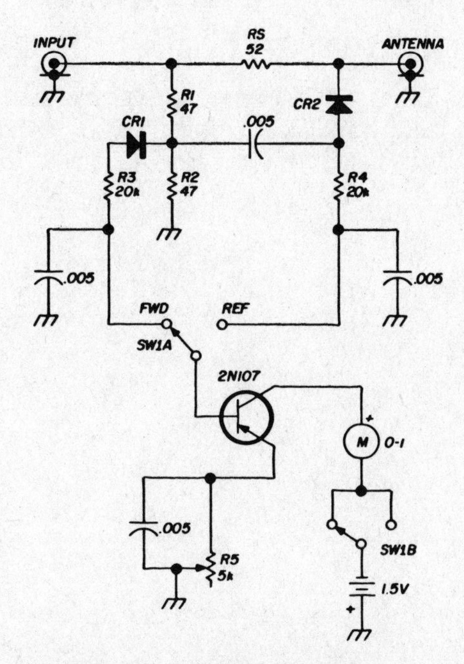

Figure 7-22. VSWR meter circuit.

meter when the transmitter is turned off. Because RS is in series with the transmitter output signal, this VSWR meter will cause an insertion loss and should be used only when checking the antenna system. (Illustration courtesy of *Ham Radio*.)

Wheatstone Bridge

Figure 7-23 shows the circuit of a Wheatstone bridge that may be used to determine the value of an unknown resistance. R1 and R2 are fixed resistors of known values, and

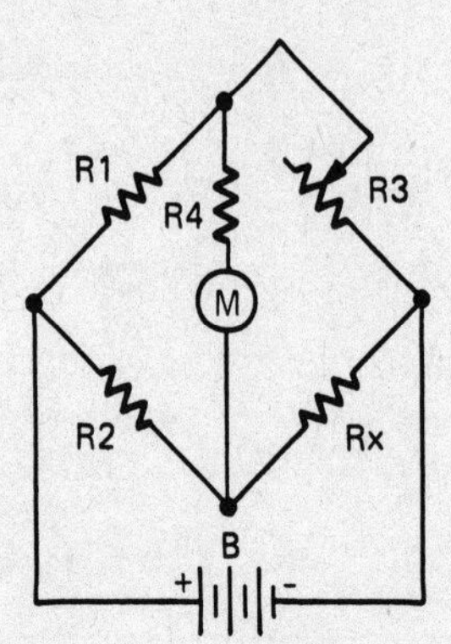

Figure 7-23. Wheatstone bridge circuit.

R3 is a calibrated rheostat. The meter is a center-zero microammeter and R4 is a multiplier resistor that converts the meter into a galvanometer. With the unknown resistor, Rx, in the circuit, rheostat R3 is adjusted until the meter indicates a null (zero).

8

Logic Circuits

Digital computers depend upon logic circuits, as do many electronic control systems. A logic circuit is one that makes a logical decision. It may be required to respond only if one or the other of two signals is present at its input, or to say "no" if a third signal is present. A logic circuit decides automatically to deliver or cut off an output signal when it senses that "all systems are A-OK" or not OK.

There are two basic types of logic, positive and negative. *Positive logic* means that a logical "1" is more *positive* than a logical "0." Thus, in a positive logic system, a logical "0" might be a potential of + 1.2 volts, while a logical "1" might be a potential of + 5.0 volts. In a *negative* logic system, a logical "0" could still be, for example, + 1.2 volts, but a logical "1" would be a more *negative* potential—say, -4.0 volts. In negative logic, the "1" is more negative than the "0."

Since there are so many types of logic circuits and applications, only the basic types are illustrated and described in this book. Those readers needing more information about logic circuits should refer to works covering this specific subject.

AND Circuit and OR Circuit Principles

An AND circuit is widely used in computers and in logic systems. An analogy of an AND circuit is shown in Figure 8-1A. To make the lamp light, both S1 and S2 must be closed. Closing only one of the switches will not make the lamp glow.

A solid state AND circuit is shown in Figure 8-1B. Closing S1 will not cause current to flow through load resistor RL because transistor Q1 cannot conduct unless Q2 is also conducting. However, when both S1 and S2 are closed, both transistors are forward-biased and conduct, allowing current to flow through RL. The circuit can be reversed, as

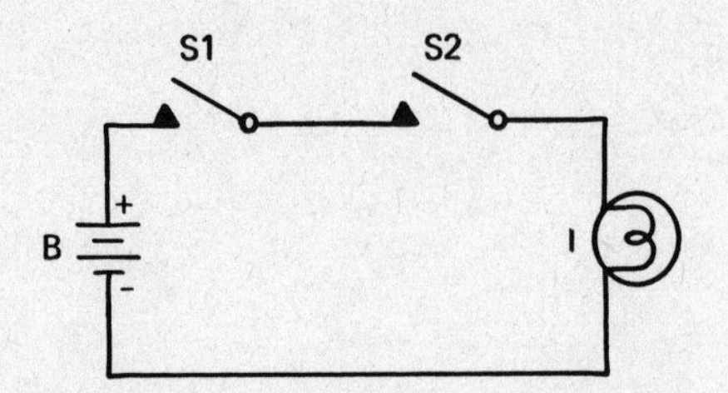

Figure 8-1A. Analogy of an AND circuit.

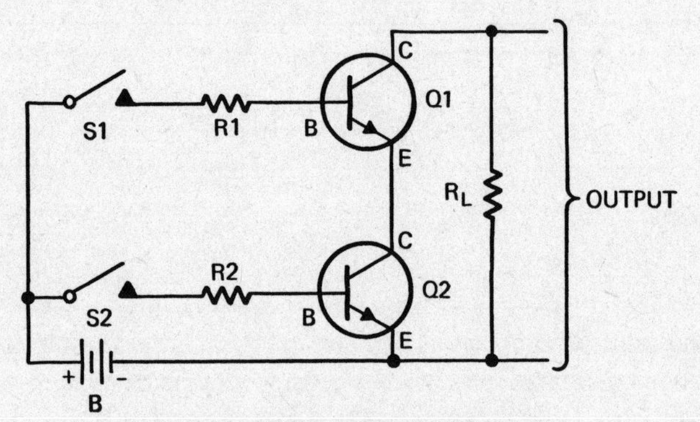

Figure 8-1B. Solid state AND circuit.

shown in Figure 8-1C. Here, the collectors and emitters of the two transistors are in parallel position. Both transistors are normally forward-biased so they are saturated (maximum collector current flows), and output voltage is almost zero. Closing S1 cuts off Q1. But since Q2 is still conducting, output voltage does not rise. However, when S2 is closed, with S1 also closed, both transistors are cut off and output voltage rises to the level of the supply voltage.

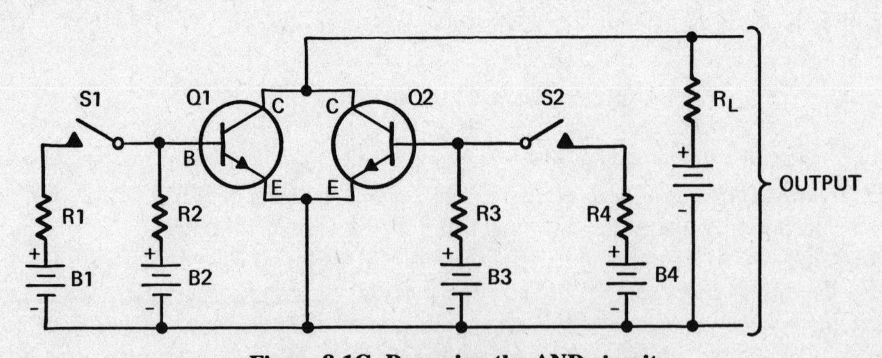

Figure 8-1C. Reversing the AND circuit.

Figure 8-1D is an illustration of the simplest type of OR circuit. Closing switch S1 *or* switch S2 will cause the lamp to light. Closing both switches will also cause the lamp to light. This may be written as the equation:

$$S1 + S2 = I1.$$

That is, S1 *or* S2 will light the lamp, I1. In logic equations the mathematical symbol "+" means *or*, and the symbol "." means *and*.

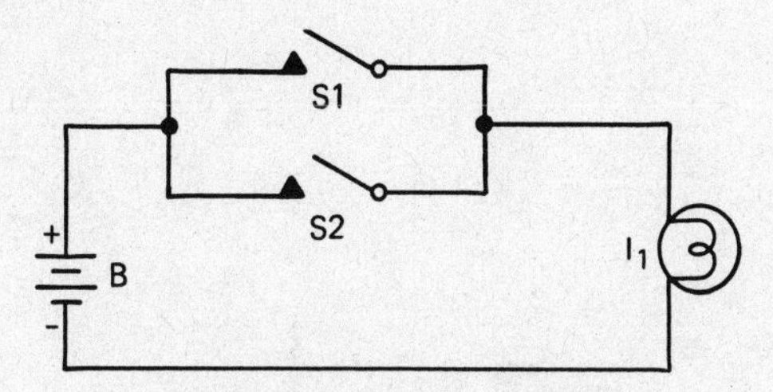

Figure 8-1D. OR circuit.

CML OR and NOR Gate

Figure 8-2 illustrates a gate circuit which provides non-inverted and inverted outputs using current mode logic. The transistors are biased by constant-current sources which keep them far out of saturation. The collectors of Q1, Q2, and Q3 are normally running at a positive potential (logical 1). A positive input on any of their bases causes the associated transistor to conduct more heavily, lowering the potential at the collectors of Q1, Q2, and Q3 to a logical 0. Thus the output of Q1, Q2, and Q3 is $a + b + c = d$. At the same time, the heavier conduction of Q1, Q2, or Q3 caused by a positive input on its base also causes the emitter of Q4 to become less negative, reducing its conduction, and causing the collector voltage of Q4 to rise to a logical 1. Thus the output of Q4 is $a + b + c = \overline{d}$.

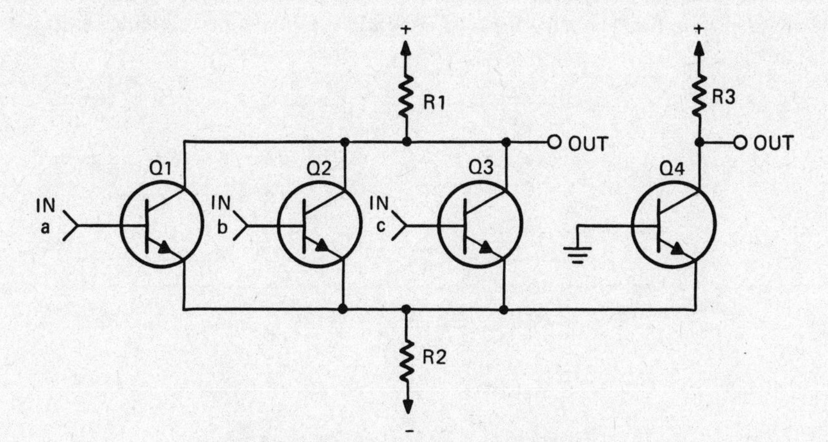

Figure 8-2. CML OR and NOR gate.

DCTL NAND Gate

In the circuit shown in Figure 8-3, direct coupled transistor logic (DCTL) is used to create a NAND condition. Normally, both transistors are cut off, and the output at the collector of Q1 is the + potential felt through R, a logical 1. A + input at a and b turns on both transistors Q1 and Q2, causing them to conduct. Their conduction causes the output at the collector of Q1 to fall to a logical 0. Thus $a \cdot b = \overline{c}$.

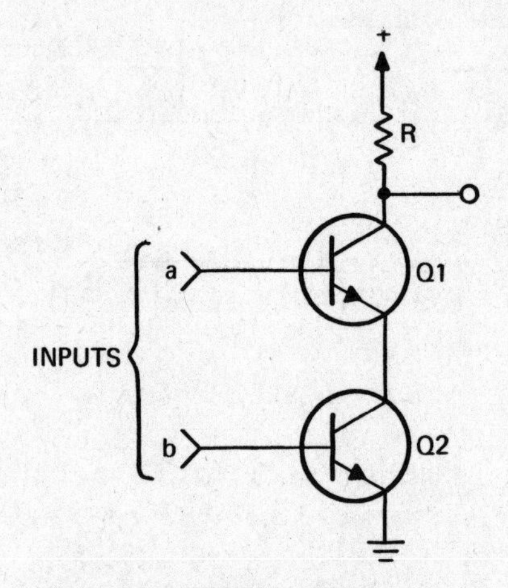

Figure 8-3. DCTL NAND gate.

DCTL NOR Gate

Figure 8-4 illustrates a direct coupled transistor logic NOR gate. Both transistors Q1 and Q2 are normally cut off, yielding a logical 1 (a positive potential) at their collectors. A + input (logical 1) at the base of Q1 *or* Q2 causes that transistor to conduct, cropping the potential at the collectors to a logical 0. Thus $a + b = \overline{c}$.

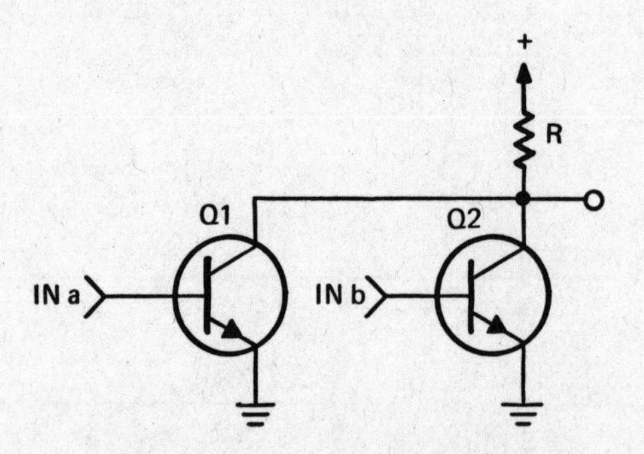

Figure 8-4. DCTL NOR gate.

DL AND Gate

Diode logic is illustrated by the AND gate shown in Figure 8-5; a negative input (logical 1) is required at a, b, *and* c. When this occurs, all three diodes (D1, D2, and D3) are cut off and the potential at the output goes negative, i.e., to a logical 1. The formula for this AND gate is:

$$a . b . c = d.$$

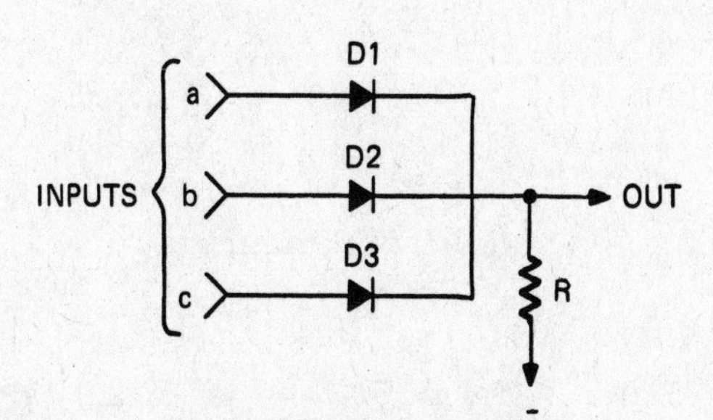

Figure 8-5. DL AND gate.

DL OR Gate

In the circuit shown in Figure 8-6, logic is performed by diodes—hence the name diode logic. With no input (logical 0) at a *or* b *or* c, the output is clamped at the level of the positive supply (logical 0). A - input at either a *or* b *or* c causes current to flow through R and the associated diode, making the output go negative (logical 1). This may be expressed as:

$$a + b + c = d.$$

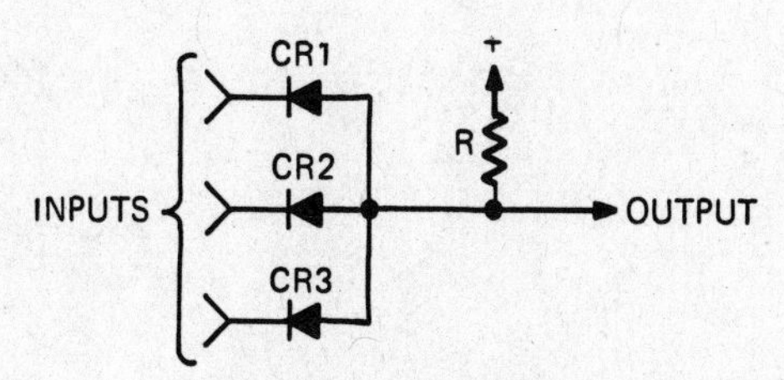

Figure 8-6. DL OR gate.

DTL NAND Gate

Figure 8-7 shows a diode transistor logic NAND gate circuit. With no input, transistor Q is cut off, and the output at its collector is a logical 1. When a *and* b *and* c receive

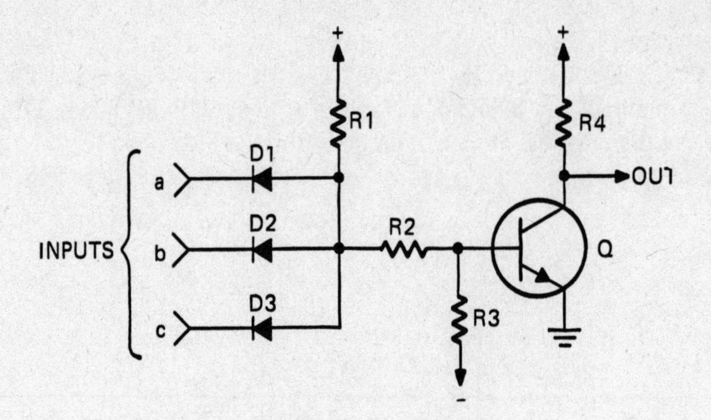

Figure 8-7. DTL NAND gate.

logical 1's and go positive, the base of transistor Q also goes positive, turning on the transistor. The collector voltage falls to a logical 0. Thus a . b . c = $\overline{\text{d}}$.

DTL NOR Gate

A NOR gate circuit using diode transistor logic is illustrated in Figure 8-8. The transistor is normally cut off, yielding a logical 1 (positive potential) at the output at the collector. If a *or* b *or* c go positive, the associated diode conducts, and the base of transistor Q swings more positive, turning it on. Collector current through the transistor causes the output voltage to fall to a logical 0. Thus a + b + c = $\overline{\text{d}}$.

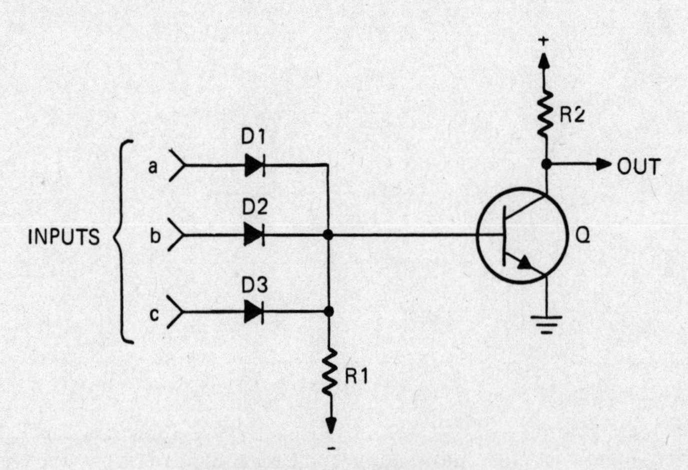

Figure 8-8. DTL NOR gate.

ECL OR-NOR Gate

The emitter-emitter coupled logic gate shown in Figure 8-9 is a type of non-saturated logic gate that provides both an OR and a NOR output. This type of gate is extremely fast, with a typical stage delay of 2 nanoseconds.

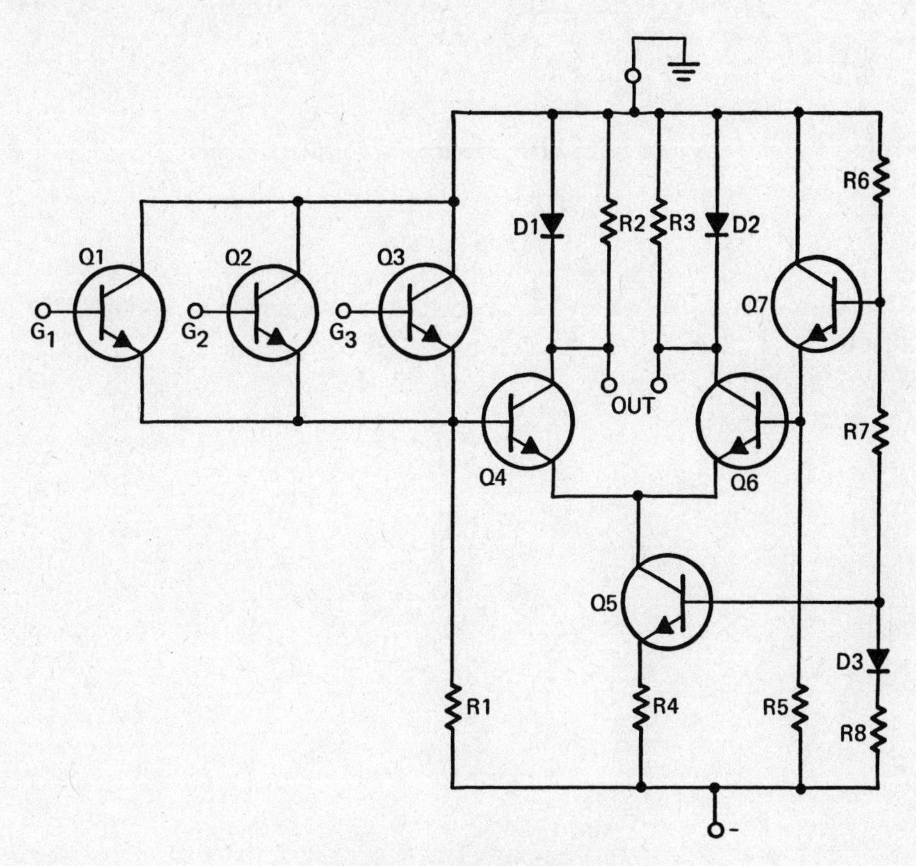

Figure 8-9. ECL OR-NOR gate.

Inverter

Figure 8-10 is a diagram of an inverter circuit using a PNP transistor. The transistor is normally cut off by the voltage provided by the divider R2 and R3. The inverter shown is operated by *negative* logic, so a logical "1"is a *negative* signal. A logical "1" applied to the base of transistor Q1 through R1 causes Q1 to conduct. The flow of collector current causes the output voltage at the collector of Q1 to become less negative—that is, it goes to a logical "0" in terms of negative logic. Thus $a = \overline{b}$. The inverter circuit may also be called a NOT circuit.

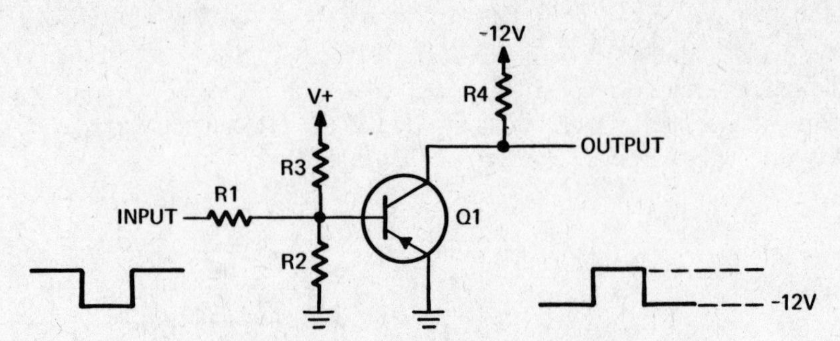

Figure 8-10. Inverter circuit using a PNP transistor.

Lamp Driver

Figure 8-11 illustrates a simple lamp driver circuit. A positive pulse (logical "1") forward-biases the transistor, causing collector current to flow through the lamp. (Illustration courtesy of *Ham Radio*.)

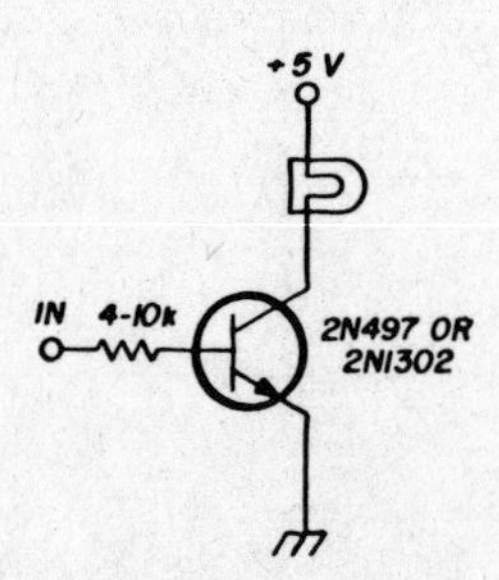

Figure 8-11. Lamp driver circuit.

Low Level Logic NOR Gate

Figure 8-12 shows low level logic circuit. Transistor Q is normally cut off, giving a logical 1 at the output. A signal (logical 1) to a or b or c will cause D4 to cut off. The full + potential is felt through R2 to the base of Q, causing Q to conduct and the output at the collector of Q to fall to a logical 0. Thus $a + b + c = \overline{d}$. This type of logic circuitry is also called current switching diode logic.

Low-Noise DTL NAND Gate

The diode transistor logic NAND gate illustrated in Figure 8-13 differs from the conventional DTL NAND gate because of the addition of diodes D4 and D5. These diodes are level shifting devices that allow a higher noise margin for this type of gate.

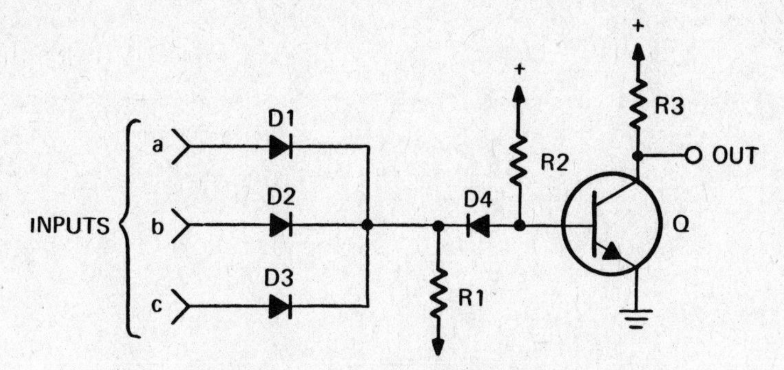

Figure 8-12. Low level logic NOR gate.

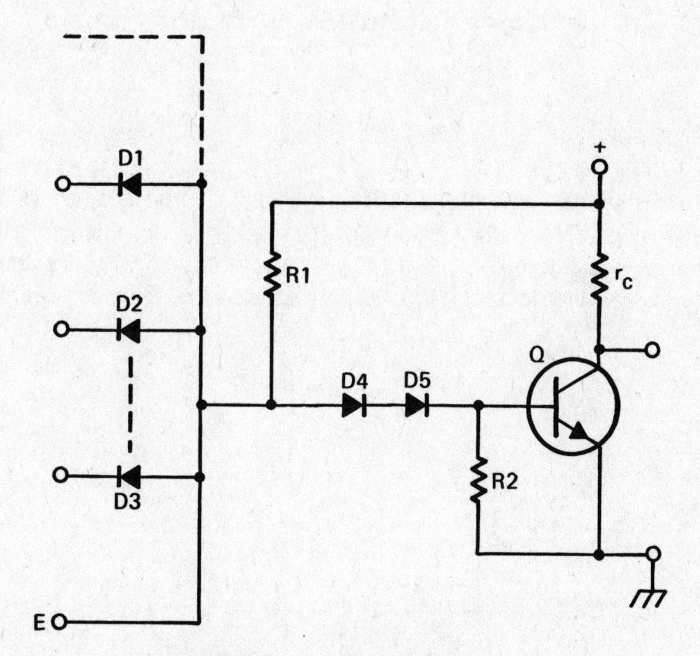

Figure 8-13. DTL NAND gate.

MOSFET NAND Gate

The NAND gate circuit illustrated in Figure 8-14 uses three metal-oxide semiconductor field-effect transistors (MOSFET's) in a negative logic system. The negative voltage establishes a logical "1" at the output under no- or one-input conditions. Thus, with either A or B having no input, there is an open in the source circuit of the output MOSFET, and the output is a negative potential (logical "1"). (Illustration courtesy of *Ham Radio*.)

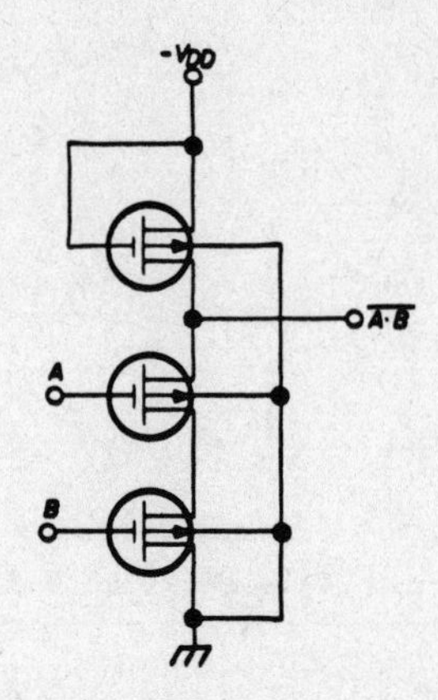

Figure 8-14. MOSFET NAND gate.

RCTL NOR Gate

Figure 8-15 shows an example of resistor capacitor transistor logic (RCTL) circuit. Normally, the transistor is cut off, producing a logical 1 (positive potential) at the output of the collector. A positive input at either a or b causes transistor Q to turn on, lowering the potential at its collector to a logical 0. The capacitors increase the base current for fast switching. Here $a + b = \overline{c}$.

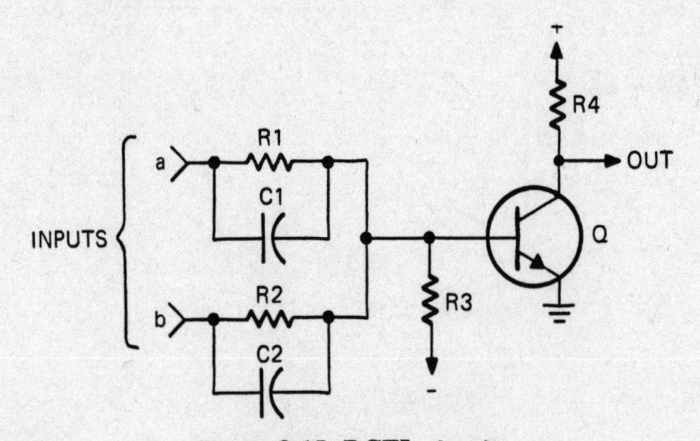

Figure 8-15. RCTL circuit.

RTL NOR Gate

Figure 8-16 illustrates a NOR gate circuit employing resistor transistor logic (RTL). Transistor Q is normally cut off with no input to the gate, and its collector is running at the

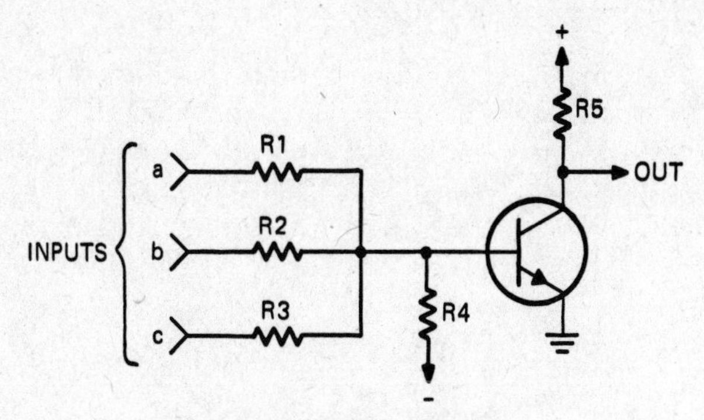

Figure 8-16. NOR gate circuit.

+ potential felt through R5 (a logical 1). With any + input at a or b or c, the transistor is forward-biased, and the flow of collector current causes the collector voltage to drop to a logical 0. Thus $a + b + c = \overline{d}$.

SCR NAND Gate

Figure 8-17 illustrates the use of silicon-controlled rectifiers in a NAND gate. With no inputs at a, b, and c, the SCR's are not conducting, and the output is the positive potential felt through resistor R4 (logical 1). When all three SCR's receive gate voltages (that is, when a *and* b *and* c are all logical 1's), they conduct, and the output falls to a logical 0. Thus, $a . b . c = \overline{d}$. The SCR's have a built-in memory and must be reset.

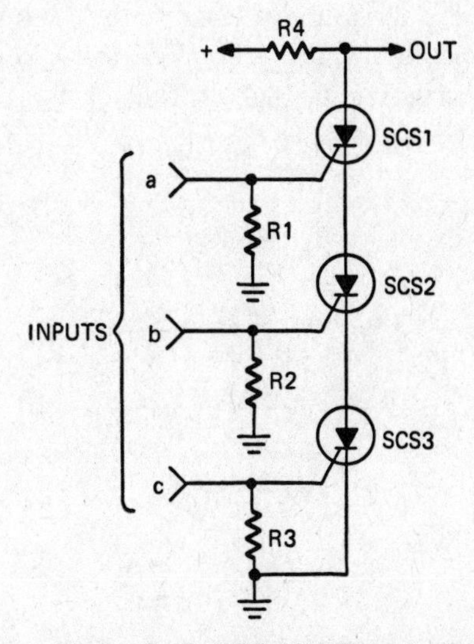

Figure 8-17. SCR NAND gate circuit.

SCR NOR Gate

Figure 8-18 illustrates the use of silicon-controlled rectifiers to create a NOR gate. With no input to the gate of any of the SCR's, they are not conducting, and the output at their anodes is a logical 1. If a *or* b *or* c receives a logical 1 (positive potential), the SCR associated with that input conducts, and the output at the anodes of the SCR's falls to a logical 0. Thus $a + b + c = \overline{d}$. The SCR's have a built-in memory and must be reset.

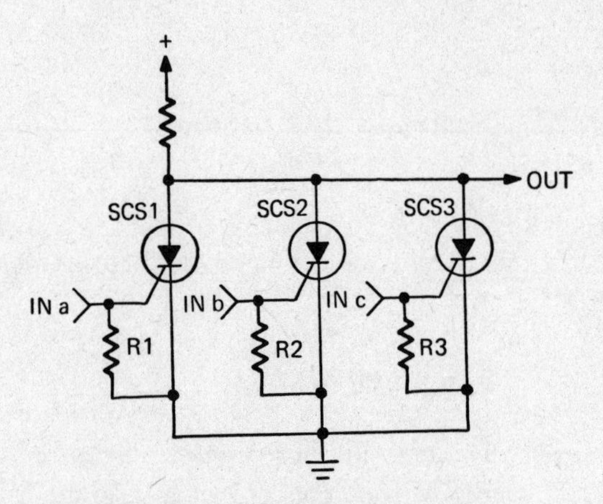

Figure 8-18. SCR NOR gate circuit.

TDL Gate

Figure 8-19 illustrates tunnel diode logic (TDL) performed by a tunnel diode switching from the low voltage to the high voltage state. To use the circuit as an OR gate the tunnel diode is biased near peak current through resistor R4. For use as an AND gate the tunnel diode is biased near ground potential.

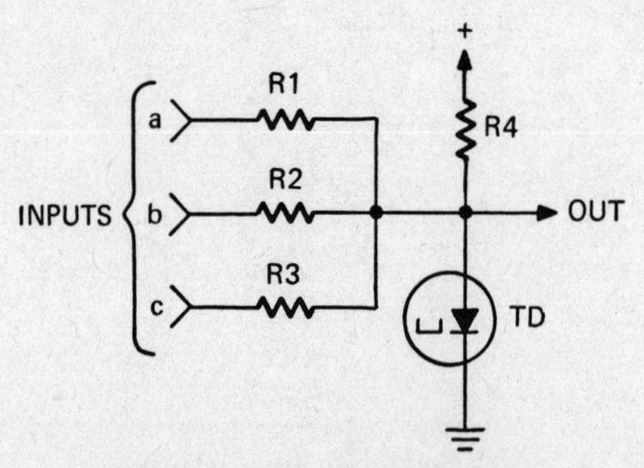

Figure 8-19. TDL gate circuit.

Transistor Flip-Flop

Figure 8-20 is a schematic representation of the operation of a transistor flip-flop. Switches S1 and S2 represent inputs. Since the two transistors will not be perfectly balanced, one of them will conduct and one will be cut off with no input applied. Assume that initially transistor Q2 is conducting and Q1 is cut off. Because of the heavy conduction of Q2, the collector voltage (and hence, the K output) will be almost zero. Likewise, since Q1 is cut off, the output at J will be a large negative voltage. This is *negative logic*, so a negative voltage represents a logical 1.

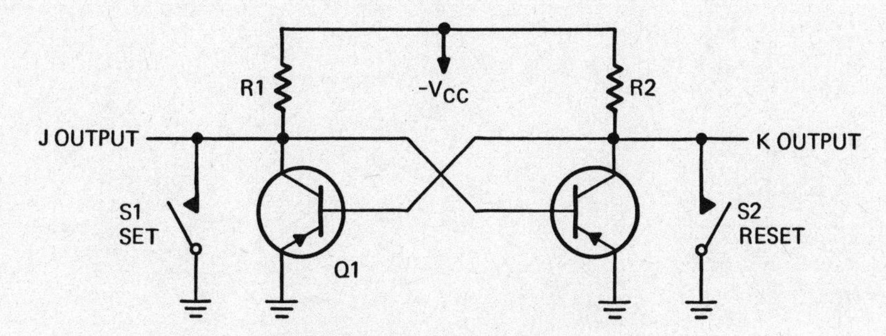

Figure 8-20. Transistor flip-flop.

If switch S1, the "Set" switch, is momentarily closed, ground potential is applied to the collector of Q1 and the base of Q2 simultaneously, removing the forward bias from the base of Q2 and cutting it off. As Q2 cuts off, the K output becomes more negative (a logical 1), and since it is also applied to the base of Q1, brings Q1 heavily into conduction. This causes the potential at the collector of Q1 to become less negative (a logical 0).

When switch S2 (simulating a "reset," or clear, input) is closed, Q1 is again cut off by the ground potential applied to its base. The collector of Q1 again becomes more negative (rises to a logical 1), in turn forward biasing Q2. As Q2 conducts, its collector, which is also the J output, becomes less negative, falling to a logical 0.

TTL NAND Gate

Figure 8-21A illustrates a NAND circuit using transistor-transistor logic (TTL or T^2L). The transistors are direct-coupled, and the input transistor has a multiple emitter, one for each input. When the input to any of the emitters of Q1 is at a logical 0 (0 volts), Q1 is turned on. The collector current of Q1 is insufficient to forward-bias Q2. When all three inputs at Q1 rise to a logical 1 (+ voltage), Q1 cuts off and Q2 turns on. The rise in emitter voltage when Q2 turns on causes Q4 to turn on, causing the output at the collector of Q4 to fall to a logical 0. This circuit has a propagation delay of 13 nanoseconds.

Figure 8-21B illustrates a variation of the circuit. This circuit has a propagation delay time of only 6 nanoseconds.

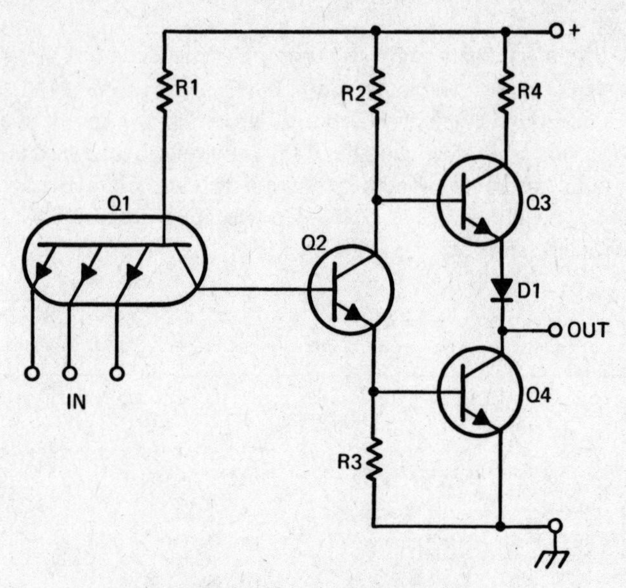

Figure 8-21A. TTL NAND gate.

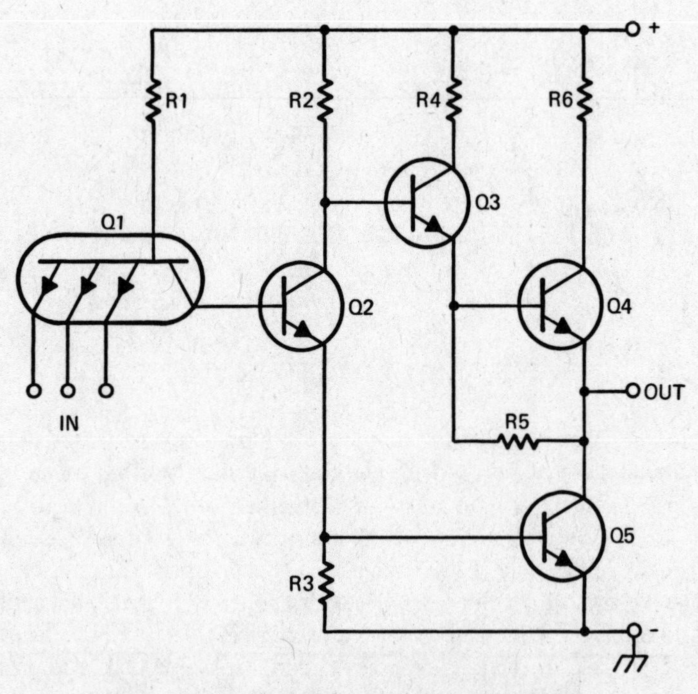

Figure 8-21B. Variation of the TTL NAND gate.

9

Modulation and Demodulation Circuits

Information is transmitted by modulating a transmission medium and is recovered by demodulating. When one is speaking into a telephone, the DC flowing through the telephone wires is modulated by the telephone transmitter (microphone). At the telephone at the other end of the telephone circuit, the undulating DC is transmogrified by the telephone receiver (earphone) from electrical into acoustic energy.

The RF output signal of a radiotelephone transmitter (including broadcast) is either amplitude modulated (AM) or frequency modulated (FM). The radiated RF signal contains information which is demodulated within a radio receiver. The receiver's demodulator is often called a detector (AM) or a discriminator (FM), but ''demodulator'' is a more descriptive term. In a radio receiver intercepting an AF-modulated RF signal the output of the demodulator is known as the ''recovered audio'' signal.

The RF output signal of the picture transmitter of a television station is amplitude modulated by the composite video signal and, when broadcasting in color, by a phase modulated subcarrier. The RF output signal of the sound transmitter of a television station is frequency modulated by the audio signal.

A multiplexed FM stereo broadcast transmitter is frequency modulated by the main audio program signal, representing the left plus right (L + R) audio channels and simultaneously by a 38-kHz amplitude modulated stereo subcarrier which is modulated by the L-R signals of the stereo audio program and a 19-kHz pilot signal. In addition, some FM broadcast transmitters are modulated by SCA (Subsidiary Carrier Authorization) FM subcarrier signals at 67 kHz and/or other frequencies above 53 kHz and below 75 kHz.

The main audio program information is recovered by the FM demodulator of an FM receiver which also recovers the stereo and SCA subcarriers, but not their modulating signals. Additional subcarrier demodulators recover the subcarrier information.

A frequency division multiplexed (FDM) FM microwave communications transmittor is simultaneously modulated by several FM, AM, or SSB (single sideband) subcarriers, each modulated by its own audio signals. An FDM FM microwave television transmitter is simultaneously modulated by a video (picture) signal and an FM subcarrier which is modulated by an audio (sound) signal. The video signal is recovered by the FM demodulator of a microwave receiver which also recovers the subcarrier signal(s) which, in turn, are demodulated by subcarrier demodulators.

Information is also transmitted via radio by on-off modulation of the carrier signal as when transmitting messages in International Morse Code or when keying the transmitter on and off with a teletypewriter. This kind of signal is recovered by sensing its presence or absence. When FSK (frequency shift keyed) modulation is used, the transmitter carrier frequency or the frequency of a subcarrier transmitter is shifted when it is keyed. The coded intelligence is recovered by a discriminator.

Digitally coded information, representing data or voice, is transmitted through on-off or FSK modulation by a train of pulses that vary in amplitude, width, and time with respect to each other, or in coding format. Again, at the receiving end of the circuit, appropriate subcarrier receivers are required.

Described and illustrated in this part of the book are numerous modulator and demodulator circuits.

Absorption Modulator

The RF power output of a radio transmitter is varied when it is amplitude modulated. A very simple way of producing an AM signal is to connect a carbon microphone to a one- or two-turn coil (L2) placed near the transmitter output tank coil (L3) as shown in Figure 9-1. When this link coil is terminated in a resistance, some of the RF energy is absorbed by the resistance. In this circuit, the resistance is variable and the amount of power absorbed by it varies with the resistance of the carbon microphone element. As sound waves are picked up by the microphone, the resistance of its carbon-granule element

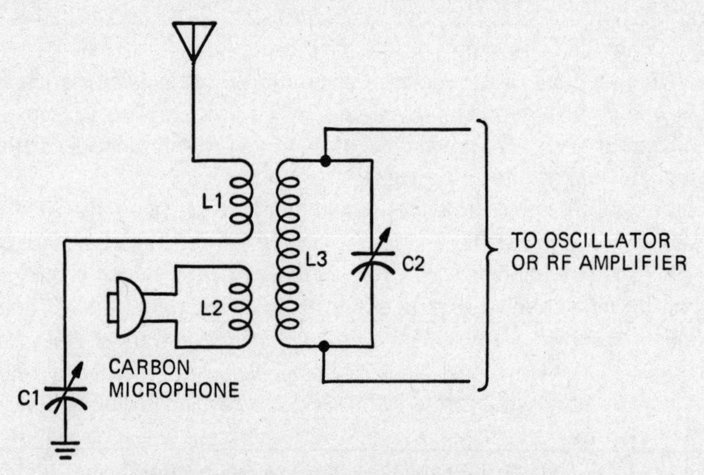

Figure 9-1. Absorption modulator circuit.

varies. This action reduces the RF power output of the transmitter and causes the output signal to be amplitude modulated. This circuit is suitable only for use with low-power transmitters (less than 1 watt) since the microphone could be damaged by excessively strong RF energy. Furthermore, this method for producing an AM signal is inefficient as it reduces the amount of power that is transmitted instead of increasing it as in a plate modulation system.
(Note: No DC is required for the carbon mike since it is excited by RF.)

Active Detector Circuit

The active detector built into the LM172 IC is actually a differential amplifer (Q1, Q2) with an emitter follower (Q4) providing a feedback loop, as shown in Figure 9-2. (Courtesy National Semiconductor Corp.) This type of detector will respond to very small signals. (A conventional diode detector will not.) Emitter follower Q4 is automatically forward-biased by the differential amplifier so it will be responsive to weak signals. Unlike a diode, this detector provides a voltage gain of 3 instead of a loss.

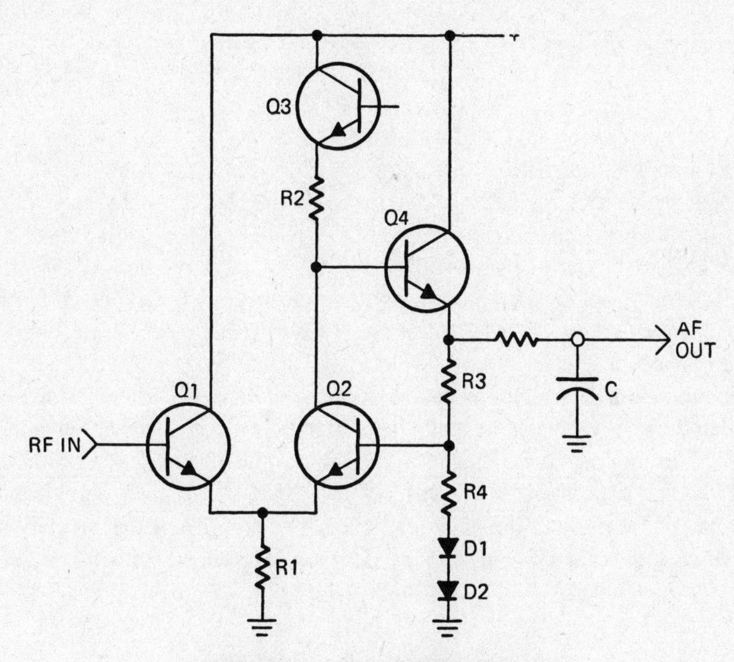

Figure 9-2. Active detector circuit.

Adjustable Phase Modulator

In the sonar phase modulator circuit shown in Figure 9-3 phase shift is introduced by the difference in phase between a constant-level signal arriving at the plate via grid-to-

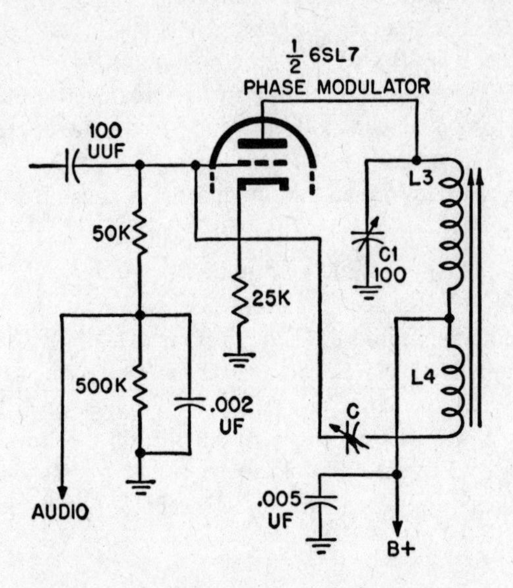

Figure 9-3. Sonar phase modulator circuit.

plate interelectrode capacitance, and a signal amplified by the tube whose plate current is modulated by an audio signal.

The plate tank coil is divided into two parts and the plate-supply voltage is introduced at a tap. The voltages across the two parts of the coil are 180 degrees out-of-phase with each other. Part of the out-of-phase voltage from the plate is fed back to the grid through capacitor C. This varies the phase and amplitude of the component passed by the Cgp of the tube, C1 varies the magnitude and phase angle of the impedance in the plate circuit through resonance.

When the modulating signal is applied to the grid there is variation in the instantaneous AC plate current of the tube, and, because the coil has a powdered-iron core, the variations in current change the magnetic saturation, thus changing the actual inductance of the coil. With the variation of this inductance caused by the audio signal, the resonant frequency and phase angle of the plate load circuit are also changed at an audio rate. This phase variation adds to that already produced by the fundamental circuit and the feedback circuit.

AM Detector and Audio Amplifier

Figure 9-4 illustrates the use of a single dual-diode/triode tube in a detector and audio amplifier circuit that is commonly used in tube-type superheterodyne receivers. The amplitude-modulated input signal from the IF amplifier is coupled through T1 and applied to the two diode plates of the tube. These diodes conduct during the positive half-cycles of the IF input signal. Capacitor C3 will charge to almost the peak value of the IF carrier

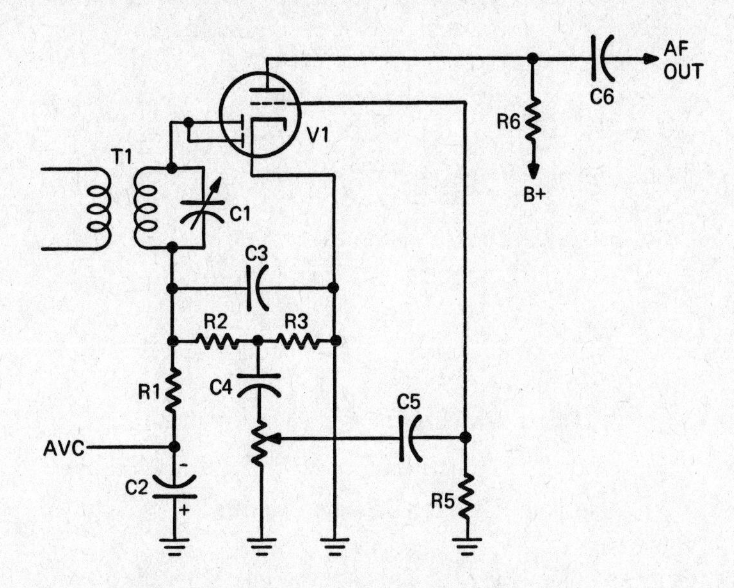

Figure 9-4. AM detector and audio amplifier.

signal. Because of the time constant of C3, R2, and R3, the charge across C3 will follow the amplitude modulation of the carrier signal.

The recovered audio is taken from the junction of R2 and R3 and applied to the grid of the tube. R4 is the volume control whose setting determines the level of audio signal to the grid. Grid-leak bias is provided by C4 and R5. The amplified audio output signal at the plate of the tube is fed through C6 to the audio power amplifier stage.

The rectified negative DC voltage at the ungrounded end of C3 varies at an audio rate, but its average value is proportional to the strength of the IF carrier signal from the IF amplifier. This negative AVC (automatic volume control) voltage is fed back to the grids of RF and/or IF stages to automatically regulate their gain. When the intercepted radio signal level rises, the average negative voltage across C3 also rises, and voltage reduces the amplification of previous stages thus keeping the signal level through the IF amplifier stages relatively constant.

Balanced Modulator

In the balanced modulator circuit shown in Figure 9-5, modulation is applied through a transformer (T2) to the center tap of the transformer T1 which feeds the RF carrier to the grids of two triodes that are connected in push-pull. The tubes are biased for non-linear operation so that an AM signal will be developed across T3. If the carrier and modulating inputs are reversed, i.e. the carrier applied through transformer T2 and modulation through T1, the output will be a double-sideband-suppressed carrier signal. Plate current flow in the push-pull arrangement is as shown by the letters i_{b1} and i_{b2}.

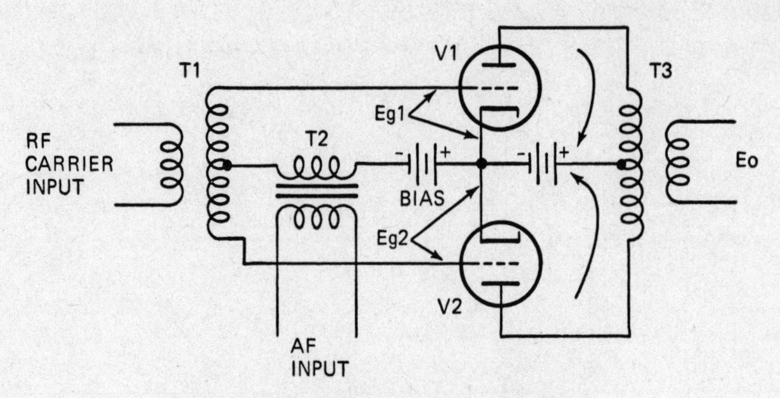

Figure 9-5. Balanced modulator circuit.

Beam Deflection Balanced Modulator

A beam deflection tube is used in the balanced modulator circuit shown in Figure 9-6. This type of circuit is used in some SSB (single-sideband) transmitters to suppress the carrier signal. The unmodulated RF carrier signal is fed through C1 to the control grid of beam deflection tube V. The plates of the tube are connected in push-pull to the output tank circuit (L1, C8A, and C8B). Since the RF signal is fed in single-ended and fed out to

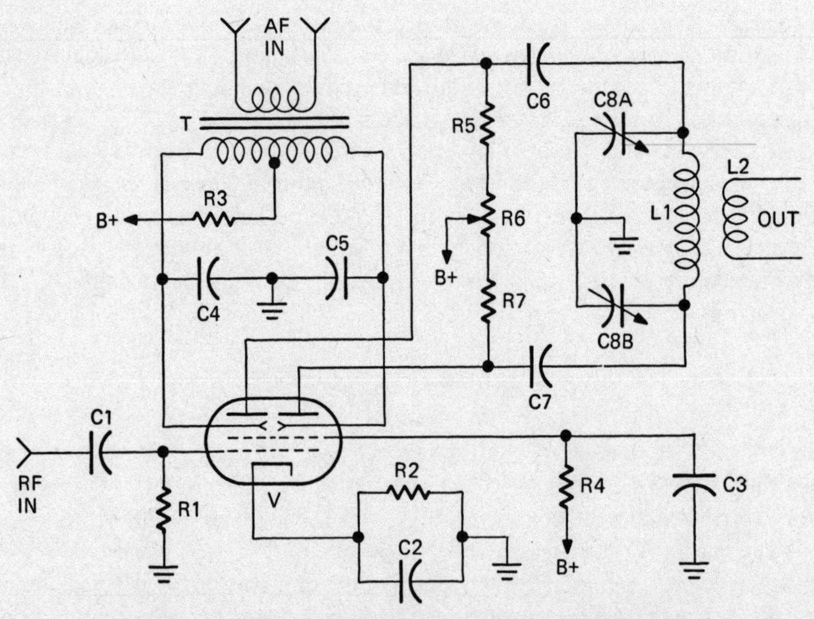

Figure 9-6. Beam deflection balanced modulator circuit.

a push-pull tank circuit, it is balanced out, and with no AF modulation applied, there is no signal across the output tank circuit. The AF modulating signal is fed through transformer T to the deflection electrodes of the tube in push-pull. The AF signal alternately pulls the electron beam to one plate or the other. As a result, the RF signal is modulated by the AF signal and two sidebands are generated. These sideband signals are developed across the output tank circuit, but the RF carrier frequency is not. R6 is a balancing control with which the circuit can be set for maximum carrier suppression. To obtain an SSB signal, a filter is connected between L2 and the next stage which eliminates one of the sidebands.

Capacitor Microphone Modulator

Figure 9-7 shows shunt-fed Hartley oscillator circuit in which the capacitor of the frequency-determining tank circuit is a capacitor-type (electro-static) microphone. When sound waves strike the microphone diaphram, the diaphragm to fixed plate capacitance changes, causing the frequency of oscillation to also change. Thus, the use of a capacitor (condenser) microphone provides a simple means of direct frequency modulation.

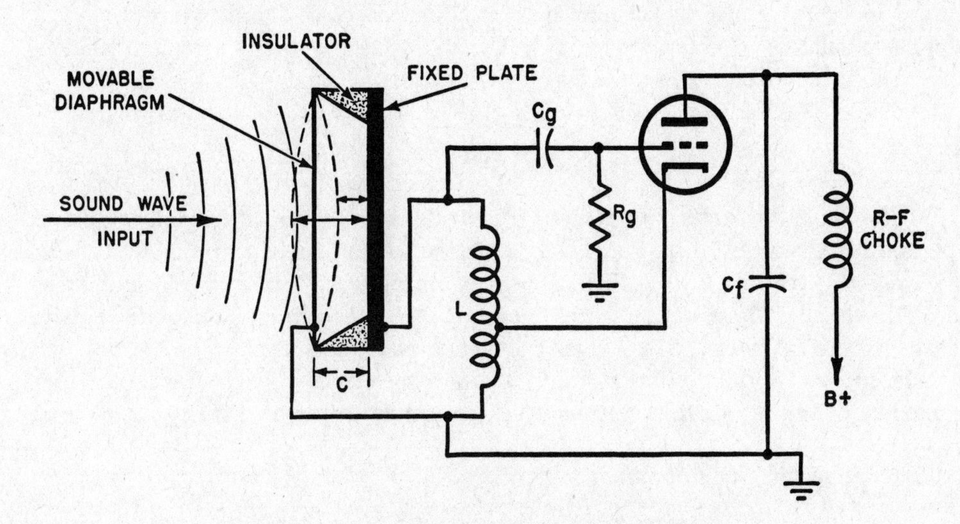

Figure 9-7. Shunt-fed Hartley oscillator circuit.

Constant-Impedance Phase-Shift Modulator

In the phase-modulator shown in Figure 9-8, a pentode tube is used as a buffer amplifier and a triode tube is used as the phase modulator. The audio signal fed to the grid of the triode tube causes its plate-cathode to change at the audio rate. A constant-impedance network is used to insure that amplitude variations are not introduced. For this reason, the triode is connected as a cathode follower whose output impedance is equal to 1/Gm. The voltage across cathode resistor Rk is used since it changes more uniformly

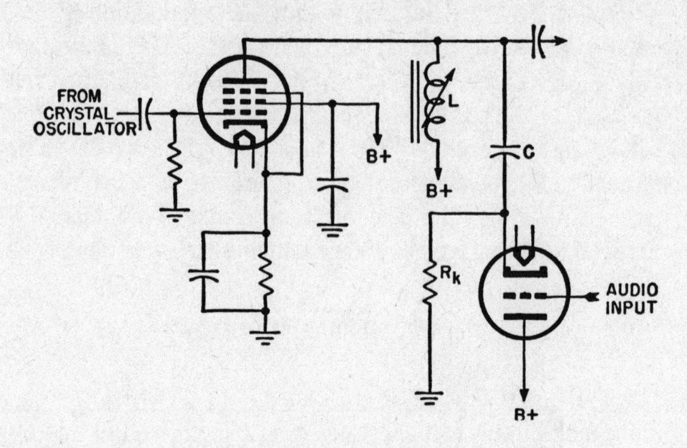

Figure 9-8. Phase modulator circuit.

with grid voltage than would the voltage across a plate resistance. The resistance between cathode and ground changes with the audio signal, and the phase of the output signal is varied. Inductor L resists any change in the total impedance and keeps the output signal amplitude constant. For any change in frequency, a change in capacitive reactance is cancelled by an opposite change in inductive reactance.

Crystal Detector

The simplest radio receiver consists of a diode, headphones, antenna, and ground, connected as shown in Figure 9-9. Before germanium and silicon diodes were developed, the diode was referred to as a "crystal." This crystal was a piece of galena or other metallic substance processing semiconductor properties. The abbreviation for diode, CR, means "crystal rectifier." Alternatively, the abbreviation D is used.

In this circuit, the positive half-cycles of the intercepted RF signal are shunted to ground by the diode. A DC voltage is developed across the diode, causing DC to flow

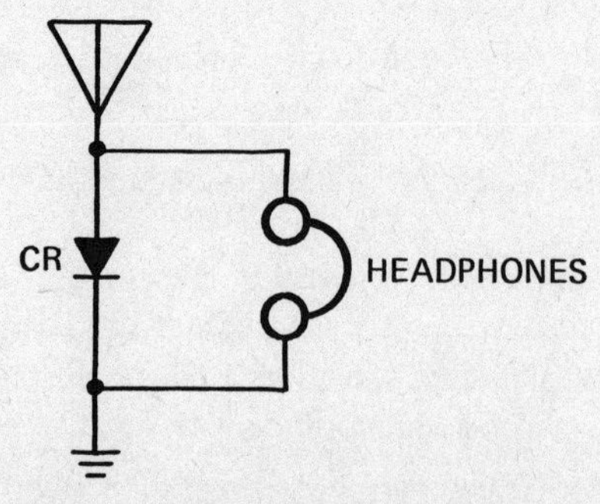

Figure 9-9. Simplest radio receiver.

through the headphones. If the RF signal is amplitude-modulated, the current through the headphones varies in amplitude at the modulating-frequency rate.

This receiver has extremely poor selectivity since it has no tuning circuits. It is most sensitive at the resonant frequency of the antenna-ground system.

Delayed AVC Detector

Figure 9-10 is the circuit of a detector/audio amplifier stage which provides delayed AVC (automatic volume control) voltage to the preceding RF and/or IF stages. The upper diode of the tube serves as the AM detector. The lower diode (AVC rectifier) is returned to ground through R6 and will be affected by the cathode bias of the triode tube. Because of the normal conduction of the triode audio amplifier portion of the tube, a positive voltage with respect to ground is developed at the cathode. The lower diode section cannot conduct until the positive peak signal applied to its plate exceeds the cathode voltage.

The bottom diode supplies the negative AVC voltage to the preceding RF and/or IF stages. Because this diode cannot conduct until there is a relatively strong signal coupled through T1, action is "delayed." Thus, no AVC voltage is present when receiving weak signals. But when receiving strong signals, AVC voltage is developed which regulates the gain of the RF and/or IF stages to keep signal level constant.

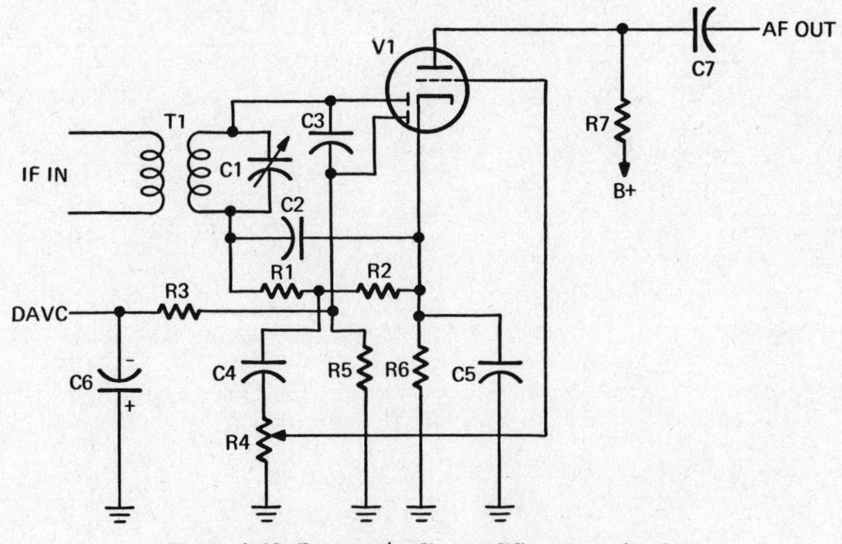

Figure 9-10. Detector/audio amplifier stage circuit.

Detector Selectivity

Figure 9-11 illustrates roughly the relative selectivity of various detector circuits. In A, a diode detector with no tuned circuit is shown. As can be seen from the diagram, it is virtually non-selective.

In B, a tank circuit consisting of an RF autotransformer and a variable capacitor has been added, and we begin to see a selectivity curve. However, the curve is rather broad. It may be improved by the injection of a local-oscillator signal as shown in C. The frequency of the local-oscillator signal is the same as that of the intercepted signal. It improves selectivity by increasing the "Q" of the tank circuit.

Loosely coupling the antenna, as shown in D, further increases the selectivity, while using an extremely low L/C ratio, as in E, increases selectivity. However, it should be pointed out that increasing the selectivity to the points shown in D and E will result in the

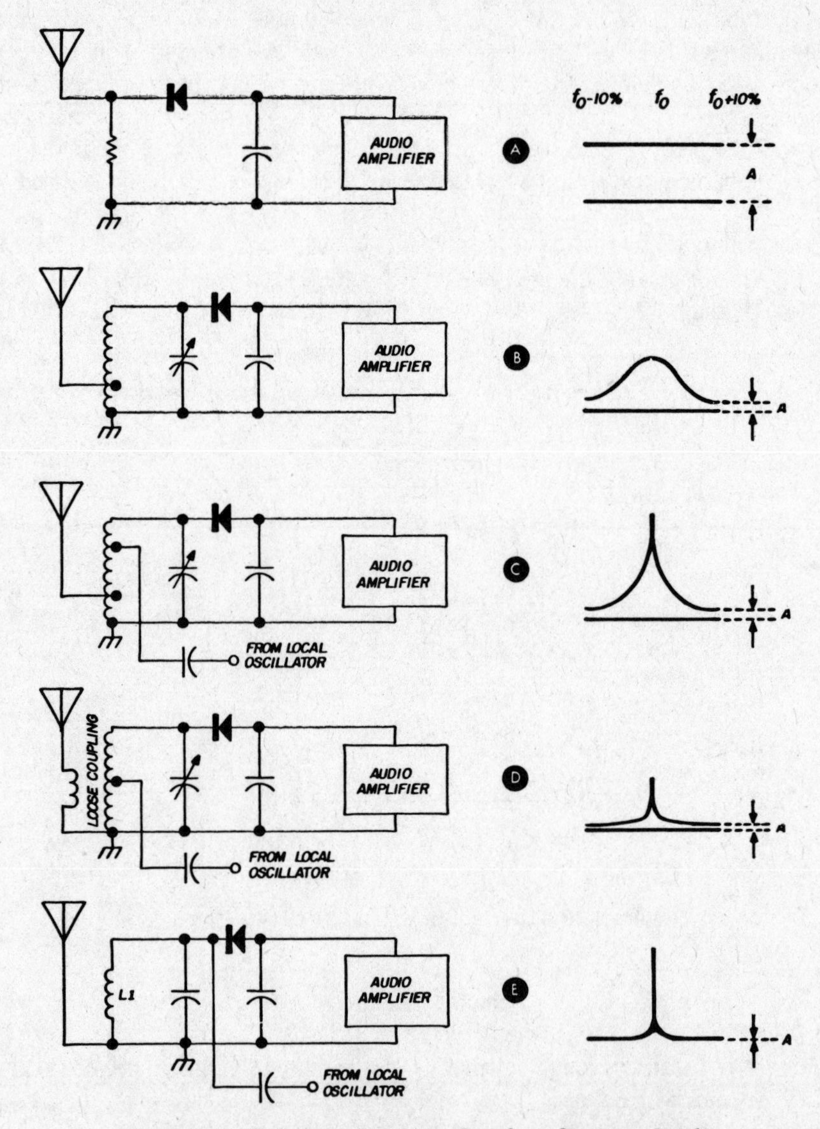

Figure 9-11. Relative selectivity of various detector circuits.

loss of much of the signal, and extremely high-gain audio stages will be needed to restore intelligibility.

When the local oscillator (not shown) is tuned to zero beat with the intercepted radio carrier signal, a "heterodyne" receiver is formed which will demodulate both AM and FM signals. (Illustration courtesy of *Ham Radio*.)

Diode Modulator

In the diode modulator circuit shown in Figure 9-12, the audio signal is applied through Cg to the grid of the control tube. Its plate-cathode current increases and decreases in step with the AF input signal. Since the current through the control tube also flows through the diode, the level of current flow through the diode also increases and decreases at the audio rate. This has the effect of increasing and decreasing the resistance and capacitive reactance of the diode's anode-cathode path. The variations in reactance injected into the oscillator tank circuit vary the frequency of oscillation. A change in diode reactance has the same effect as a change in capacitance across the tank circuit, and the result is a frequency-modulated signal.

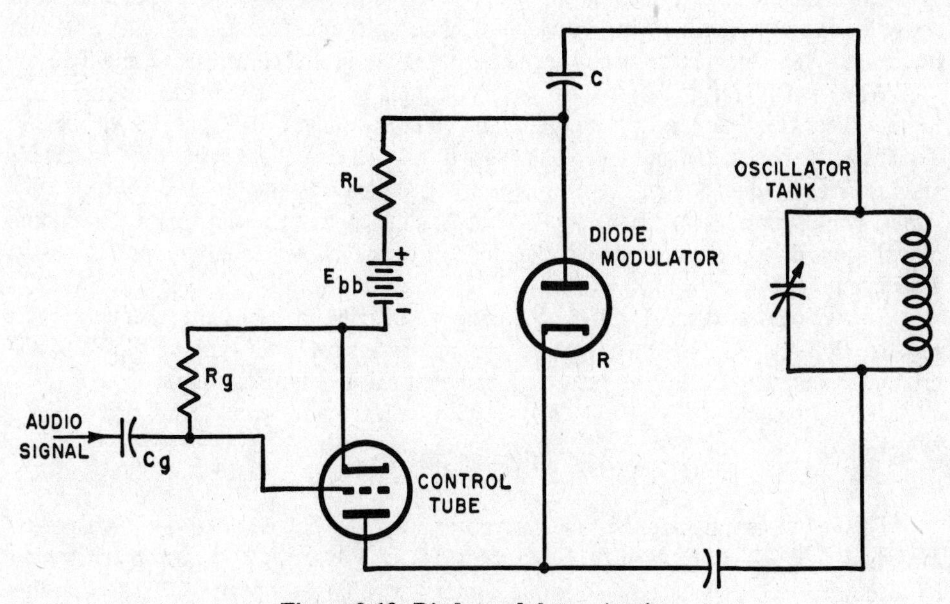

Figure 9-12. Diode modulator circuit.

Double-Tuned Discriminator

The double-tuned discriminator circuit, such as the one shown in Figure 9-13, was used in the early days of FM radio, although it has been largely replaced by more modern circuits. Designed to detect frequency and phase variations of a frequency-modulated

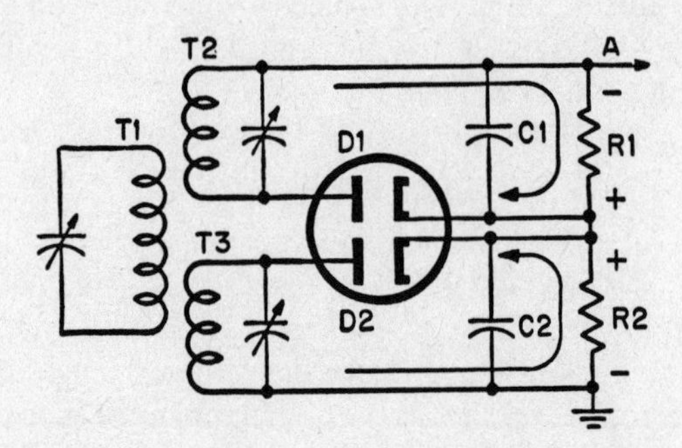

Figure 9-13. Double-tuned discriminator circuit.

signal, the detector consists of tuned circuits T1, T2, and T3, diodes D1 and D2, and networks C1-R1 and C2-R2.

The secondaries, T2 and T3, are set to resonate at two different frequencies—one above the carrier center frequency, and the other an equal distance below the carrier center frequency. At the center frequency, equal voltages are produced across T2 and T3.

When an RF voltage that is constant in amplitude but varying in frequency (about the center carrier frequency) is applied to T1, the voltages induced in T2 and T3 will be 180 degrees out of phase, and alternate voltage polarities will appear at the plates of diodes D1 and D2. These induced voltages will increase and decrease in amplitude as the frequency varies. For example, if the frequency is lower than the center frequency, a greater amount of voltage will appear across the circuit tuned lower than center, and hence across its associated diode.

The voltages associated with the changing frequencies appear as DC potentials across resistors R1 and R2. Frequency variations are translated into variations of the discriminator output voltage. The recovered audio signal is obtained at point A.

Dual Balanced Modulator

This circuit is shown for historical purposes only. As shown in Figure 9-14, an audio modulating signal is fed through T1 to the grids of V1 and V2 in push-pull. A low-frequency RF carrier signal is fed to the center tap of the secondary of T1 so that this signal will drive the triodes in push-push. The carrier signal is balanced out and does not exist at T2. However, when an audio modulating signal is present, the RF signal and audio signal heterodyne with each other, and upper and lower sideband RF signals are generated which are passed through T2 to filter 1 which removes one of the sidebands. The remaining sideband signal is fed in push-pull to the grids of V3 and V4 through transformer T3. Another RF signal at a higher frequency than the first is fed to the center tap of the secondary of T3 in push-push configuration and is balanced out in the plate circuit. Again,

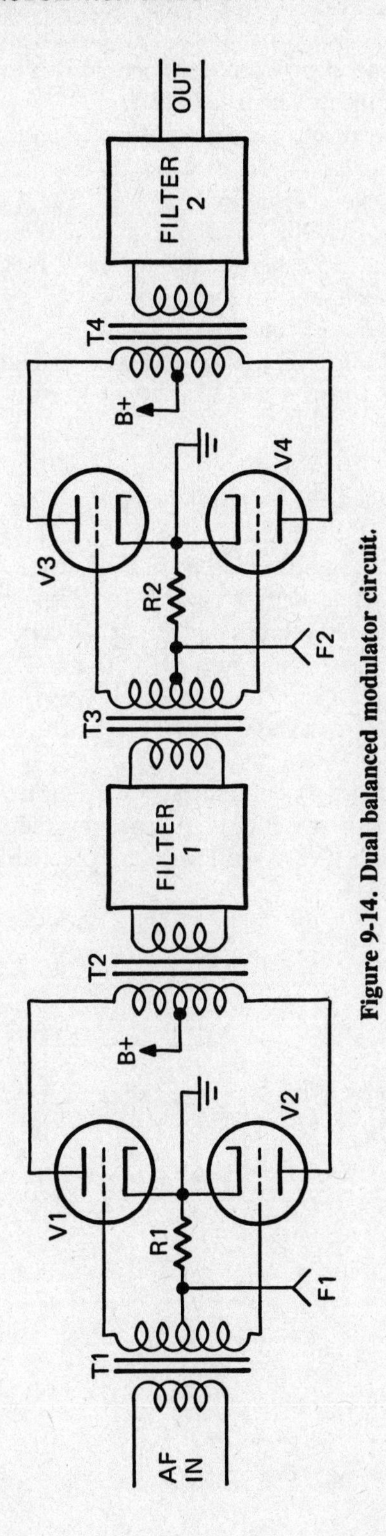

Figure 9-14. Dual balanced modulator circuit.

because of heterodyning action, upper and lower sidebands are formed which are fed out through T4. Filter 2 removes the unwanted sideband.

This is not a complete circuit and is shown only for the purpose of revealing an early technique for generating SSB signals. It was used in early international telephone systems. Assume that the AF signal contains frequencies between 300 and 3000 Hz and that the frequency of the RF signal fed in at Fl is 20 kHz. At T2, the sideband frequencies will be 17,000 to 19,700 Hz (lower sideband) and 20,3000 to 23,000 Hz (upper sideband). Then, if the upper sideband is passed through Filter 1 and the RF signal fed in at F2 has a frequency of 100 kHz, the sideband frequencies at T4 will be 77,000 to 79,700 Hz (lower sideband) and 120,300 to 123,000 Hz (upper sideband). If Filter F2 passes the upper sideband, the suppressed carrier frequency (100 kHz) will be removed 20,300 Hz from the transmitted sideband.

Dual-FET Balanced Modulator

Figure 9-15 shows the circuit of a balanced modulator using one of the dual field-effect transistors which are commercially available. There are two FET's with identical characteristics within the same case. In this circuit, the RF carrier signal from the oscillator is fed to the paralleled gates of the dual FET. The FET drains are connected in push-pull to obtain carrier cancellation. Provision has also been made here for the connection of a VFO (variable frequency oscillator) in place of the frequency determining crystal in the oscillator circuit.

The audio modulating signal is fed through a transformer to the FET sources in push-pull. The drain current of the FET's is varied by the constant-frequency RF carrier signal and the sidebands (upper and lower) are generated, but the RF carrier is

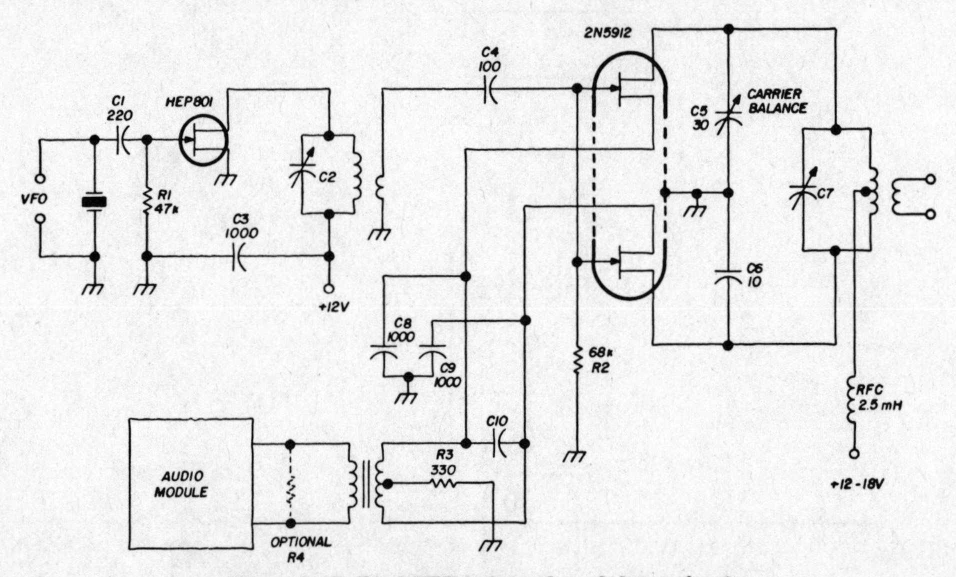

Figure 9-15. Dual-FET balanced modulator circuit.

balanced out. The DSB-SC (double sideband-suppressed carrier) output signal is fed through a bandpass filter when an SSB-SC (single sideband-suppressed carrier) signal is required. The filter passes only the desired sideband.

If the level of the modulating signal from the audio module is too great, R4 may be added to reduce the input level. Capacitor C10 provides a high-frequency audio roll-off characteristic. Capacitor C5 is used for carrier-balance adjustments. (Illustration courtesy of *Ham Radio*.)

FET Regenerative Detector

The simple receiver circuit shown in Figure 9-16 uses a field-effect transistor (FET) in a modified Hartley regenerative circuit. Incoming radio signals are amplified and demodulated by the FET. Regeneration is provided from the FET source back to a tap on L2. The tank circuit is tuned to the desired frequency by adjusting C1. The amount of regeneration is controlled by varying the drain voltage by adjusting potentiometer R2. Maximum sensitivity for AM signals is just below the point of oscillation.

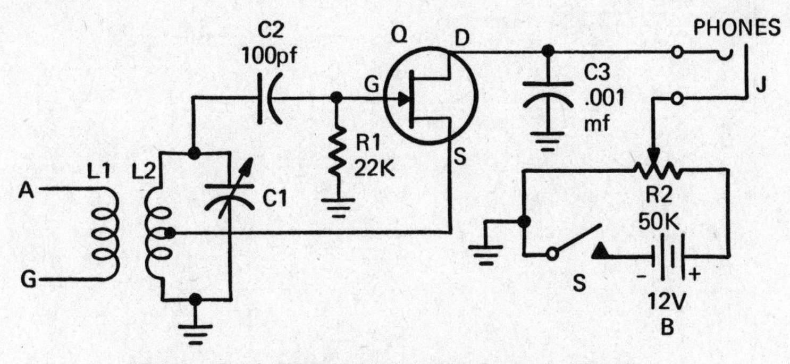

Figure 9-16. FET regenerative detector circuit.

FM IF System with Ratio Detector

The FM receiver IF and detector system circuit, shown in Figure 9-17, employs two integrated circuits and a two-diode ratio detector. The FM signal is amplified by the IC at the left and is fed to the second IC through a bandpass filter (T1 through T2). The FM output signal of the IC at the right is fed through T3 to the ratio detector diodes. The audio signal recovered by the ratio detector is fed to one of the amplifiers within the second IC for pre-amplification.

Grid-Leak Detector

Figure 9-18 shows a triode tube in a grid-leak AM detector circuit. The values of R1 and C2 are such that C2 charges during positive peaks of the incoming signal and

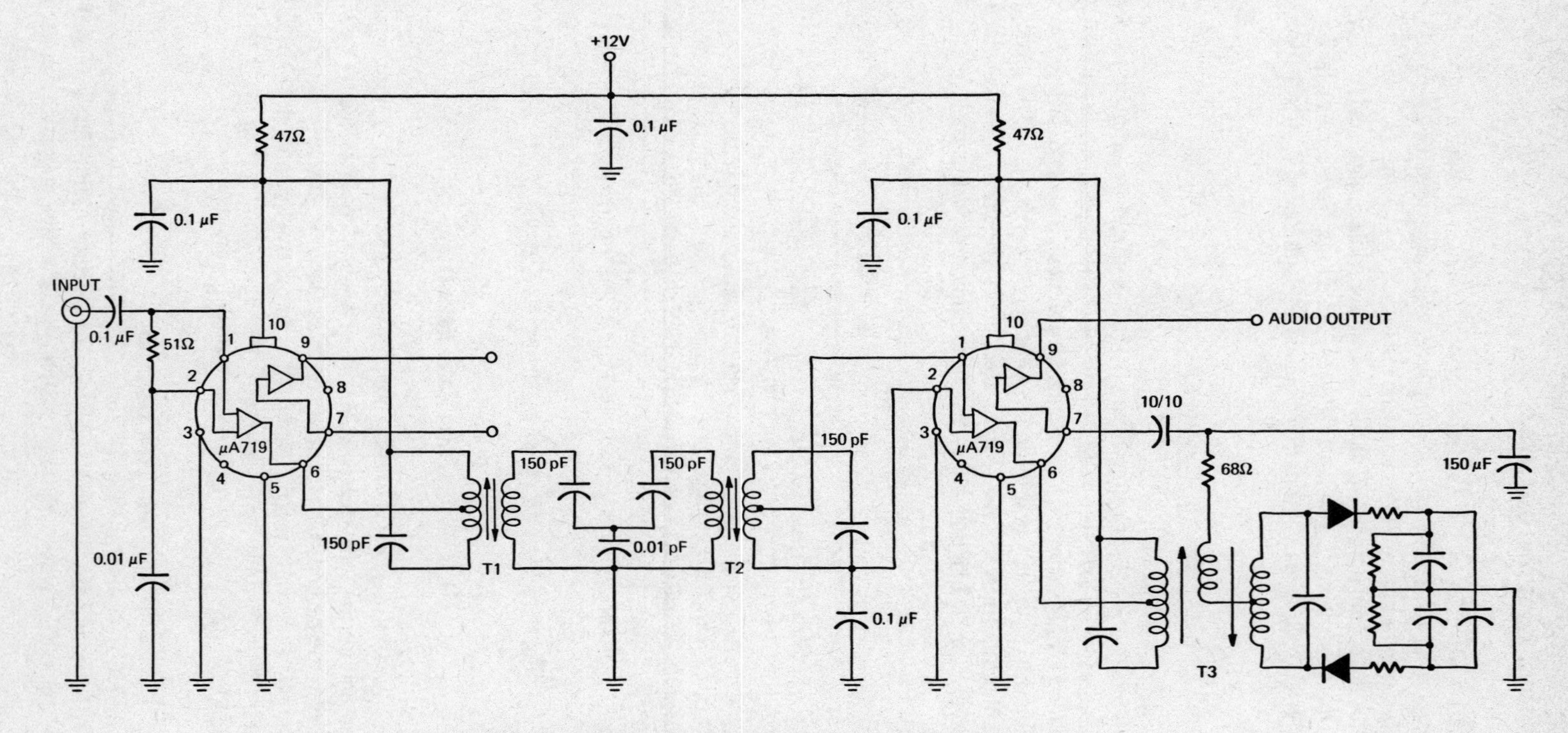

Figure 9-17. FM receiver IF and detector system circuit.

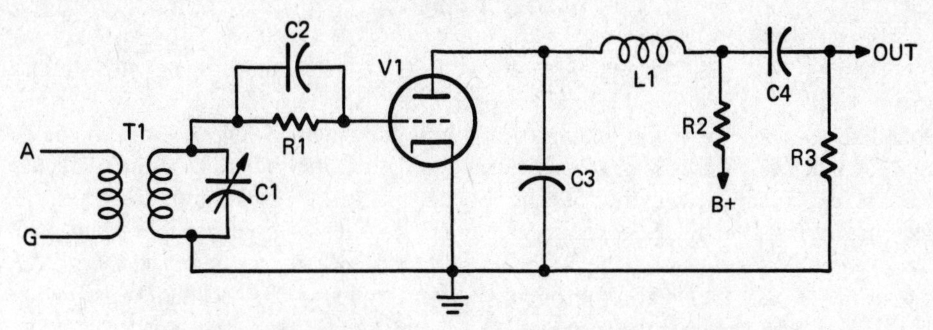

Figure 9-18. Triode tube in a grid-leak AM detector circuit.

discharges during negative peaks. During positive peaks of incoming signal, current glows from cathode to grid, making the grid act like the plate of a diode, and C2 charges. During negative-going peaks of the input signal, there is no grid current, and C2 discharges through R1. This results in recovery of the modulation contained in the RF envelope. Capacitor C3 bypasses the RF signal to ground. The recovered (and amplified) audio signal is fed through C4 and developed across R3.

Grid Modulation

The type of modulating circuit shown in Figure 9-19 is used in some AM transmitters. Considerably less audio power is required than when high-level plate modulation is used. The AF signal alternately bucks and boosts the fixed bias on the control grid of the tube, causing plate current to vary at both the RF and AF rates.

This type of modulator has somewhat poor plate efficiency since only about 25 percent as much output power can be obtained using grid modulation as compared with plate modulation.

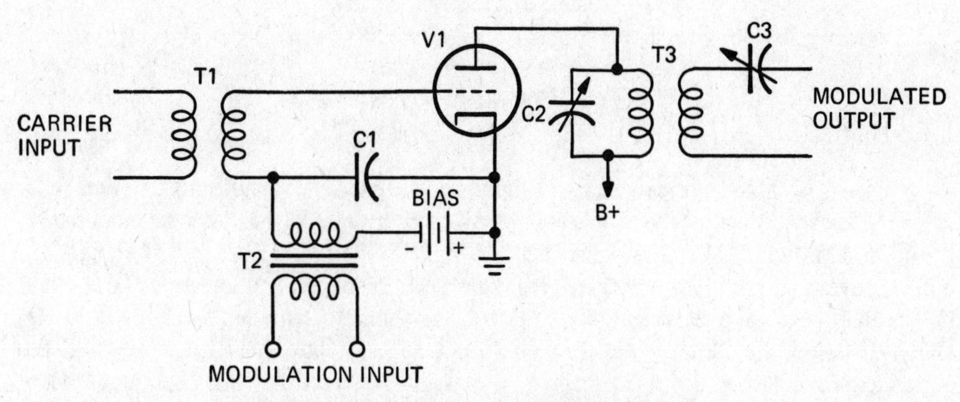

Figure 9-19. Grid-bias modulation circuit.

Homodyne Detector

The homodyne detector circuit shown in Figure 9-20 is shown only for historical purposes since its use would now be prohibited because of interference-causing capabilities and because it uses archaic components and circuitry. Triode V1 functions as a grid leak detector and V2 as a tunable oscillator and Q multiplier. With the oscillator (V2) not functioning, selectivity would be very poor because of the relatively low Q of the tank circuit consisting of C1 and the secondary of T1. With the oscillator functioning and tuned "exactly" (zero beat) to the frequency of an incoming radio signal, and C1 and C2 adjusted to resonate their respective circuits to that frequency, the oscillator signal introduces negative resistance to the secondary of T1, causing its Q and, hence, its selectivity, to increase dramatically.

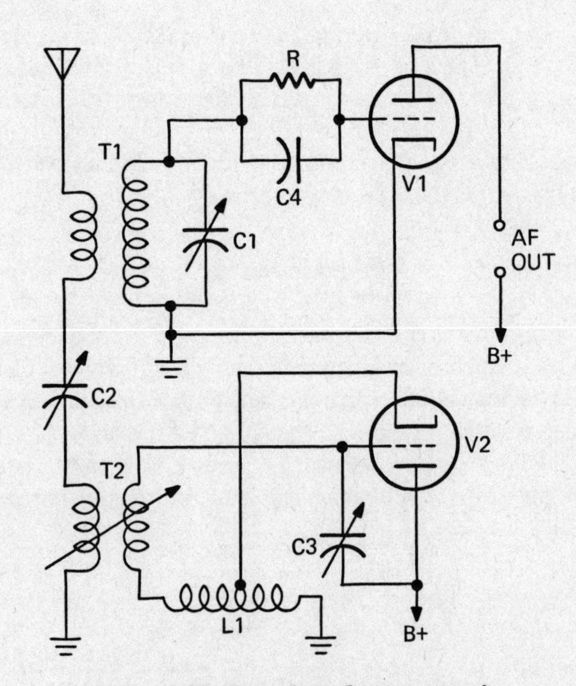

Figure 9-20. Homodyne detector circuit.

When C1 is tuned to receive an AM signal and the oscillator is tuned to zero beat with the signal's carrier, the oscillator increases the selectivity and V1 recovers the modulation from the AM signal. When receiving an FM signal, V1 will sense the FM signal frequency deviations (from its carrier frequency) and will recover the modulation. To receive CW (continuous wave radiotelegraph) signals, the oscillator is tuned to about 1000 Hz above or below the radio signal frequency so that an audible beat frequency will be generated every time a dot or dash is received. The receiver then functions as a "heterodyne" (not superheterodyne) receiver.

Inductive Modulator

The flow of electrons from the cathode to the plate of an electron tube can be modulated by winding a coil around the tube, which functions as an electromagnet, and by applying an AF signal or a steady frequency AC voltage to the coil. As shown in Figure 9-21, the coil is wound around the glass envelope of the tube. The tube may be used in either an oscillator or RF amplifier circuit. If 60 Hz AC is applied to the coil through the rheostat with which current level is adjusted, the electron stream will be modulated at the 60 Hz rate or, if an AF signal is applied to the coil, the electron stream will be modulated at the various audio frequencies present in the signal. To get a reasonably high percentage of modulation, fairly heavy current must flow through the coil.

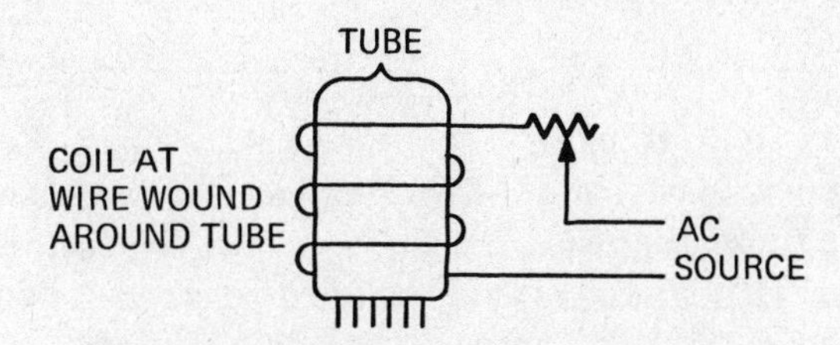

Figure 9-21. Inductive modulator circuit.

Infinite Impedance Detector

The circuit shown in Figure 9-22 closely resembles a cathode follower stage. However, the relatively large value cathode resistor R causes a high negative bias to be applied to the grid. The triode rectifies the RF input signal because the modulated positive half-cycles of input signal have a much greater effect on plate current than the negative half-cycles. This type of detector provides no gain. But there is virtually no loading of the input signal at the grid, thus improving selectivity. This circuit is known for its low-distortion characteristics.

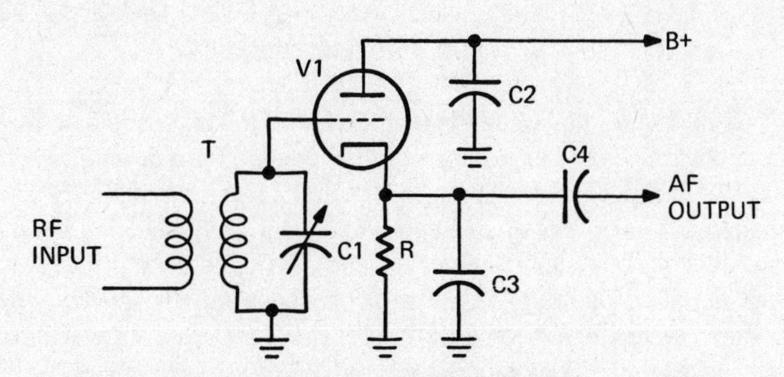

Figure 9-22. Infinite impedance detector.

Integrated Circuit Quadrature FM Demodulator

Most tube-type television receivers employ a gated beam or sharp cutoff pentode tube in a quadrature-type FM demodulator circuit for recovering the audio signal from the translated (IF) aural carrier signal because of its high AF output to FM deviation ratio. A solid state quadrature-type FM demodulator circuit is shown in Figure 9-23, a schematic of an IF amplifier system employing integrated circuits.

The IC at the left contains two IF amplifiers which are coupled through IF transformer T1. The output of this IC is coupled through IF transformer T2 to the IC at the right. As the intercepted FM signal frequency deviates with modulation, the signal voltage developed across the quadrature tank circuit (L) varies as the signal frequency deviates above and below its unmodulated carrier frequency. As a result, the audio modulation is recovered and amplified by one of the amplifiers within the IC at the right.

Link Phase Modulator

The Link phase modulator circuit (named after Fred M. Link) shown in Figure 9-24 uses the variations in transconductance of a tube with a varying audio signal as the basis for a phase modulator. The omission of the cathode bypass capacitor permits utilization of the variations in cathode current to provide degeneration. This reduces the transconductance of the tube and causes the phase of the amplified signal at the plate of the tube to vary with respect to the signal fed to the plate through the grid-to-plate interelectrode capacitance (Cgp) of the tube.

When an audio signal is applied through C4, the transconductance of the tube is varied at an audio rate, and the amplified component at the plate of the tube varies in amplitude. The signal coupled through the grid-to-plate capacitance of the tube, however, does not change in amplitude or phase. Since the amplified voltage changes and the capacitive voltage fed through Cgp does not, the amplitude and phase angle of the resultant total signal at the plate of the tube must change. The phase of the output signal varies in accordance with the amplitude of the input signal. The amplitude of the plate signal also varies, but the variations are not very great, and are eliminated by succeeding Class C amplifier stages.

Nonlinear Coil Modulator

Figure 9-25 illustrates how a nonlinear coil, L2, is connected to a conventional RF amplifier circuit to produce FM. The output of the RF amplifier is developed across a plate load network consisting of the resonant circuit C1, L1, and the nonlinear coil L2. The audio information is applied through RF choke coil L3 to the junction of L1 and L2. The resulting phase-modulated signal is coupled to the next stage through C2.

When the amplified signal from the crystal oscillator (not shown) flows through L2, the coil is saturated during part of each half-cycle. As the coil drops in and out of saturation, it changes the oscillator sine-wave signal to a series of spikes, or pulses. The audio signal applied to L2, by either bucking or boosting the oscillator signal, changes the

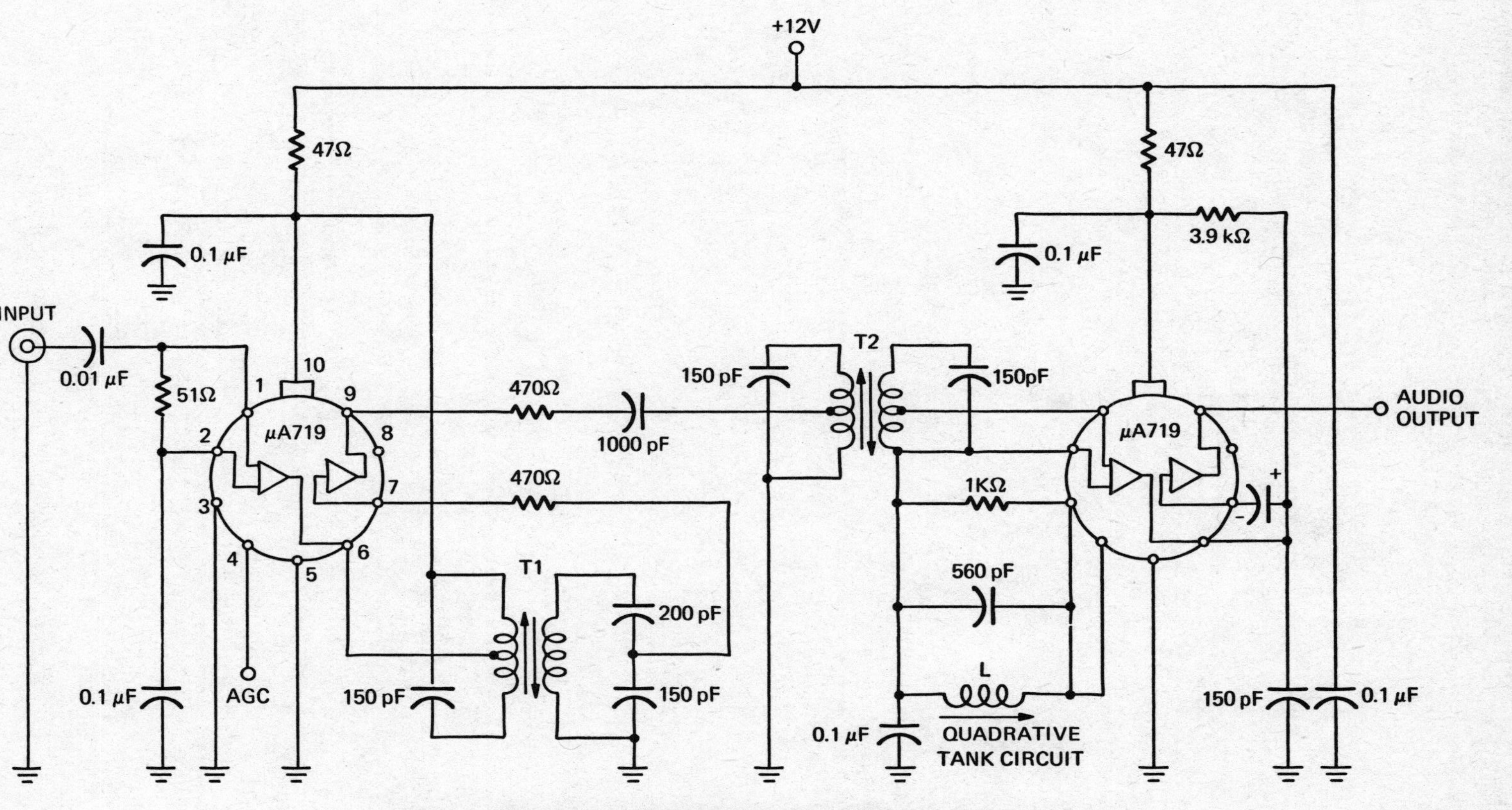

Figure 9-23. Solid stage quadrature-type FM demodulator circuit.

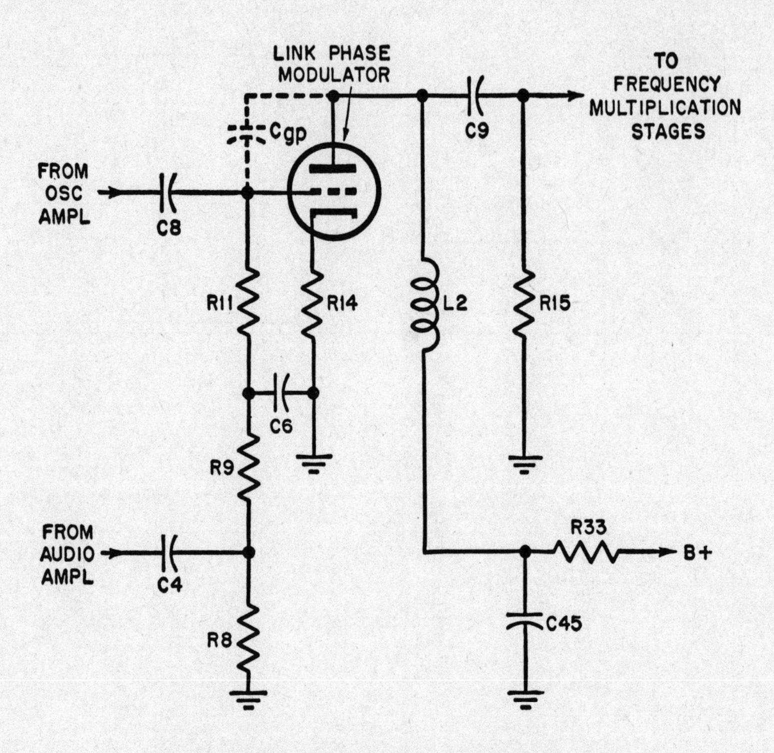

Figure 9-24. Link phase modulator circuit.

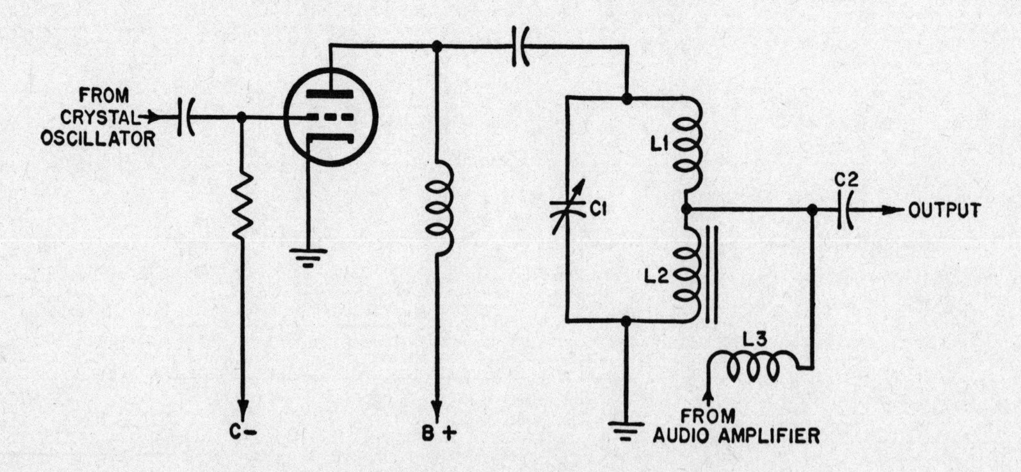

Figure 9-25. Non-linear coil modulator circuit.

saturation points of coil L2. Hence it changes the position of the pulses with respect to one another. The pulses are converted into a phase-modulated sine wave signal by the tank circuits of succeeding Class C amplifier stages.

Phase Discriminator

Figure 9-26 shows a phase discriminator circuit used in FM radio receivers to recover audio information from a frequency-modulated signal. Because of its popularity in FM circuits for a long period of time, this type of circuit is often merely called a discriminator.

L1 and L2 form the two windings of an IF transformer which has a center-tapped secondary. The RF voltage, El, across L1 is also induced across L2 and L3 and is fed to the plates of diodes D1 and D2. These induced voltages are identified as E2 and E3. The voltage E1 is also fed to inductor L through capacitor C, and the whole of E1 is also across inductor L. Thus, the secondary system receives its voltages in two ways—through inductive coupling and through capacitive coupling.

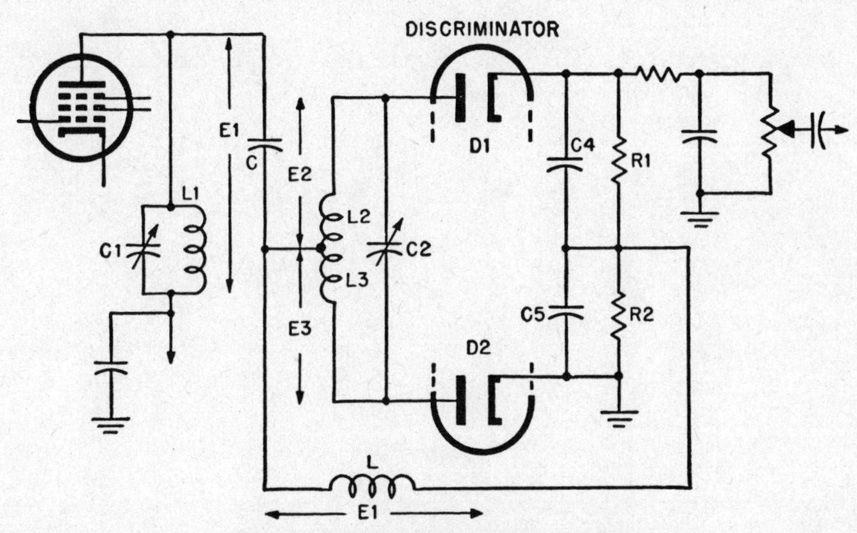

Figure 9-26. Phase discriminator circuit.

The voltage induced in L2-L3 is 90 degrees out of phase with the current in L1. The primary signal is also fed through C to the center tap of the secondary (between L2 and L3), and the secondary voltages combine on each side of the center tap so that one side leads the primary signal while the other side lags the primary signal by the same amount. When rectified by diodes D1 and D2, the two voltages are equal and of opposite polarity, resulting in a zero-voltage potential across R1 and R2.

When the frequency of the carrier departs from center, the phase relationships in the secondary of the transformer change, with the output of one-half of the secondary increasing, and the output of the other half decreasing. Since the rectified voltages appearing across R1 and R2 are now equal, an off-frequency condition at L1 results in a DC voltage at the output of the discriminator. It is this varying DC voltage, responding to variations frequency that is the recovered audio information.

Since the discriminator circuit cannot reject signals which are also amplitude-modulated, it is usually operated preceded by one or more amplitude limiter stages.

Phase-Locked Loop FM Detector

The phase-locked loop (PLL) circuit, a block diagram of which is shown in Figure 9-27, had its inception in the early 1930's, but the high number of discrete components necessary to assemble one kept it from being widely used. With the advent of integrated circuit techniques, however, the phase-locked is finding ever-increasing numbers of applications.

An incoming FM signal is applied to the phase comparator through pins 12 and 13 of the IC, while a reference signal from the built-in voltage-controlled oscillator (VCO) is applied internally. The VCO signal frequency is adjusted with a tuning capacitor (CO) applied externally between pins 2 and 3, and/or by means of a control voltage applied to pin 6.

If the frequency of the VCO is adjusted to be roughly equal to that of the input signal,

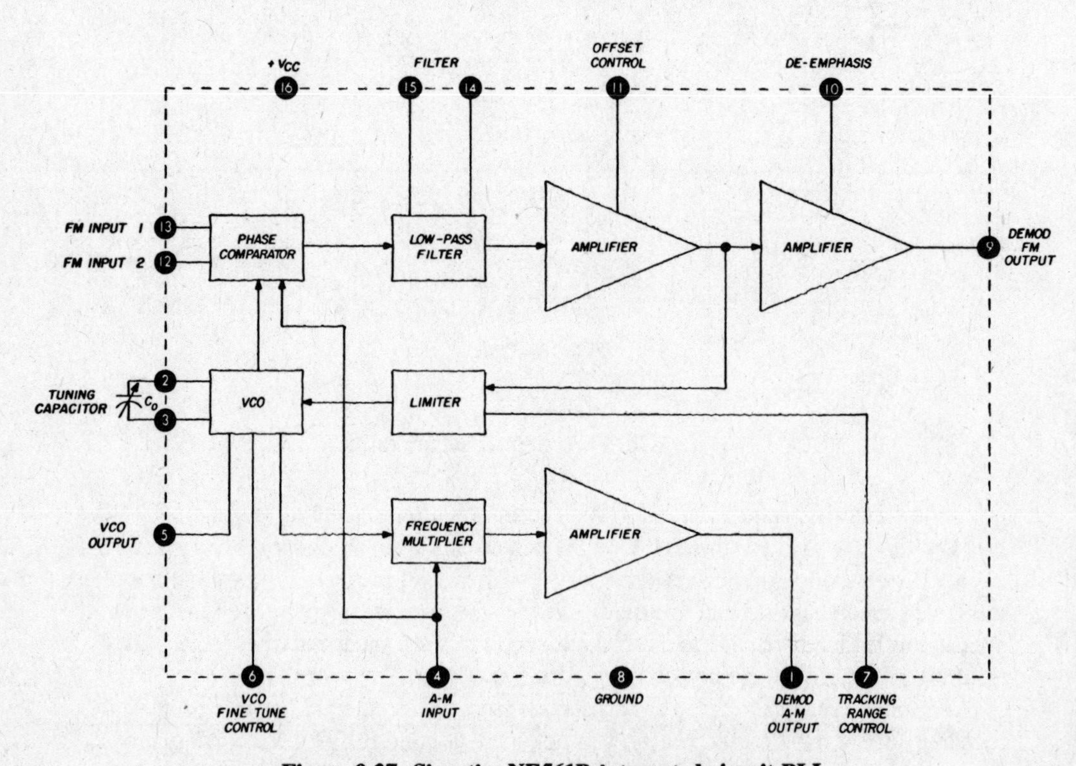

Figure 9-27. Signetics NE561B integrated circuit PLL.

the output of the phase comparator will be proportional to the difference in phase between the modulated input signal and the unmodulated signal from the VCO.

The output of the phase comparator is applied to a low-pass filter. The characteristics of this filter are controlled with an external network applied between pins 14 and 15 of the IC. Following the filter is a two-stage amplifier. Connection of a capacitor from pin 10 of the IC to ground controls the FM de-emphasis characteristic to offset the pre-emphasis applied at the transmitter. The recovered audio output is available at pin 9.

The DC error voltage at the output of the first amplifier is applied through a limiter to the voltage-controlled oscillator. This error voltage causes the VCO to phase-lock to the incoming signal. The VCO does not need to be tuned exactly to the frequency of the incoming signal since the error voltage pulls the VCO into frequency and phase synchronism with the carrier. When the incoming FM signal varies in frequency, the error voltage developed at the output of the first amplifier corresponds to the demodulated output.

For AM demodulation, the incoming AM signal is applied to a multiplier which functions as a mixing-type AM demodulator. The AM signal is also applied to the phase comparator, and ultimately to the VCO which locks onto the signal. The output of the VCO is at the same frequency as the input carrier, but without modulation.

The VCO signal is also applied to the multiplier. A low-pass filter at the output of the multiplier filters out the RF carrier and sideband components, and produces a difference signal corresponding to the AM modulation. This type of demodulation is called phase-lock AM detection. (The block diagram is that of a Signetics NE561B integrated circuit PLL and is reproduced through the courtesy of *Ham Radio*.)

Plate Detector

In the AM plate detector circuit shown in Figure 9-28, capacitor C1 and coil L1 are resonant at the intercepted carrier frequency and the modulated RF signal is applied directly to the grid of tube V. Cathode bias resistor R1 has such a high value that the triode tube is actually cut off during the negative swings of RF input signal. Cathode bypass capacitor C2 holds the DC voltage steady across R1. The positive swings of the RF input

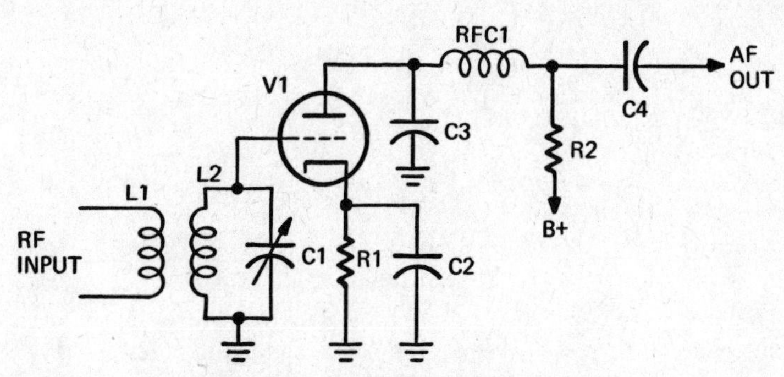

Figure 9-28. AM plate detector circuit.

signal reduce the net grid-cathode voltage so that plate current flows. It is the amplitude variation (due to modulation) of the positive swings of the RF signal that causes the plate current through R2 to vary at the audio frequency rate. The AF component is developed across R2 and capacitor C3, which is typically about .002 microfarad and which shunts RF to ground without significant effect on the recovered audio signal. Choke RFC allows the audio component to pass to the output while blocking RF fluctuations. The plate detector circuit is more sensitive than a conventional diode detector because the signal is amplified by the tube.

Plate Modulator

The circuit in Figure 9-29 is of a single-ended plate modulator used in some AM transmitters. The efficiency is approximately 65 percent. The carrier is applied to the grid of the RF power amplifier tube V1, and the modulating signal is applied to the grid of the other tube, V2. The modulating signal is passed through a modulation transformer through an RF choke coil which prevents the RF signal of the carrier from penetrating into the modulation transformer. When modulation is applied, the polarity of the AC voltage across the secondary of the modulation transformer alternately bucks and boosts the DC plate voltage fed to the RF power amplifier. The modulator must deliver AF power equal to half the carrier power for 100 percent modulation. The AF power is equal to the power in the sidebands when modulation level is 100 percent and the total RF output power is 50 percent greater than the unmodulated carrier power.

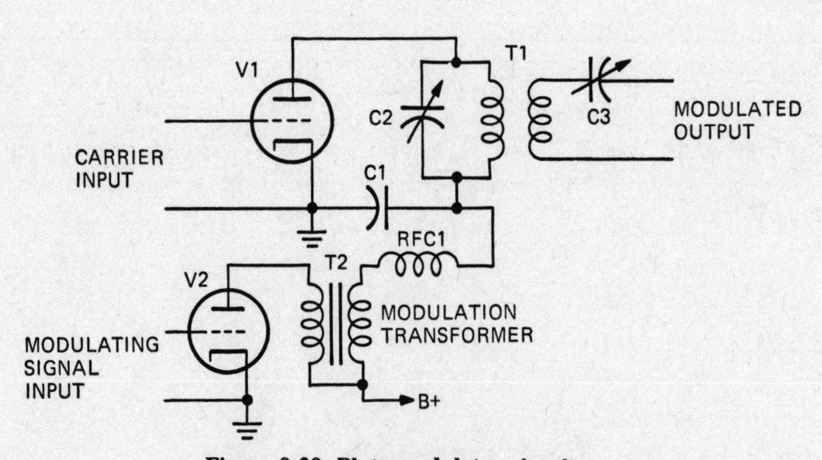

Figure 9-29. Plate modulator circuit.

Product Detector

The product detector circuit for single sideband signals, illustrated in Figure 9-30, is a cathode-coupled mixer circuit into which a BFO (beat frequency oscillator) signal is fed

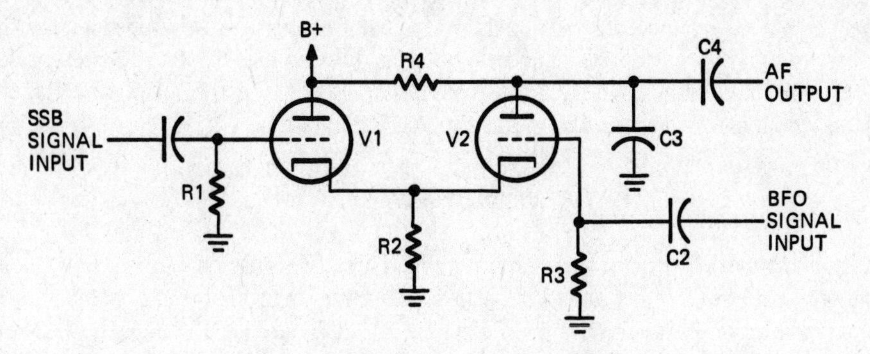

Figure 9-30. Product detector circuit.

to restore the carrier signal that is suppressed at the distant transmitter. The SSB signal from the IF amplifier which is applied through C1 is mixed with the BFO signal applied to the grid of V2 because V1 and V2 have a common cathode resistor (R2). The heterodyning of the constant-frequency BFO signal with the SSB signal causes the audio information contained in the SSB signal to be recovered and appear at the plate of V2, where it is developed across R4 and coupled to the audio amplifier through C4.

The BFO is tuned to the down-converted frequency (in a superheterodyne receiver) of the missing carrier.

Push-Pull Plate Modulator

Various techniques can be used to amplitude modulate a radio transmitter. One most widely used technique is illustrated in Figure 9-31. Here, the plate voltage fed to the RF amplifier output stage is fed through modulation transformer T whose primary is connected to the plates of modulator tubes V1 and V2 in push-pull. When there is no

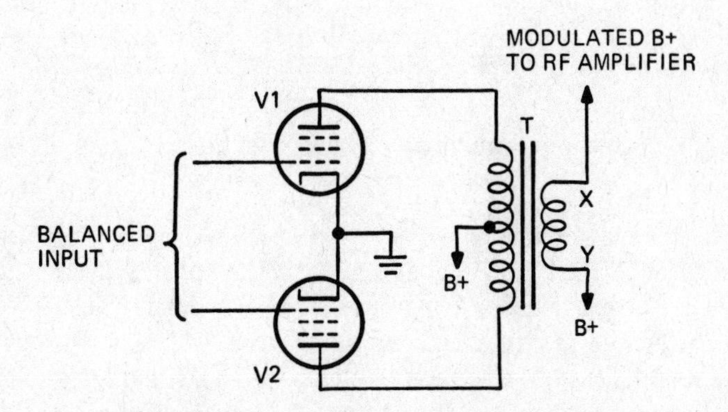

Figure 9-31. Push-pull plate modulation circuit.

modulating signal, RF amplifier plate voltage remains constant. When there is a modulating signal, an AC voltage at the modulating signal frequency is developed across the secondary of T. This AC voltage alternately bucks and boosts the DC plate voltage applied to the RF amplifier. At the 100 percent modulation level, RF amplifier plate voltage is alternately doubled and reduced to zero.

Quadrature FM Detector

The quadrature FM detector is very popular since it has only one tuning adjustment, the variable core of the quadrature coil with which its resonant frequency is changed. In the circuit shown in Figure 9-32, the FM IF signal is fed to the base of transistor Q1. From the emitter of Q1, the signal is fed to the base of Q2 through C2 and at the same time through C1 to the tap on the quadrature coil T, which is actually an autotransformer. The signal at the emitter of Q2 is fed to the emitters of Q3 and Q4. The quadrature tank circuit is connected to the bases of Q3 and Q4.

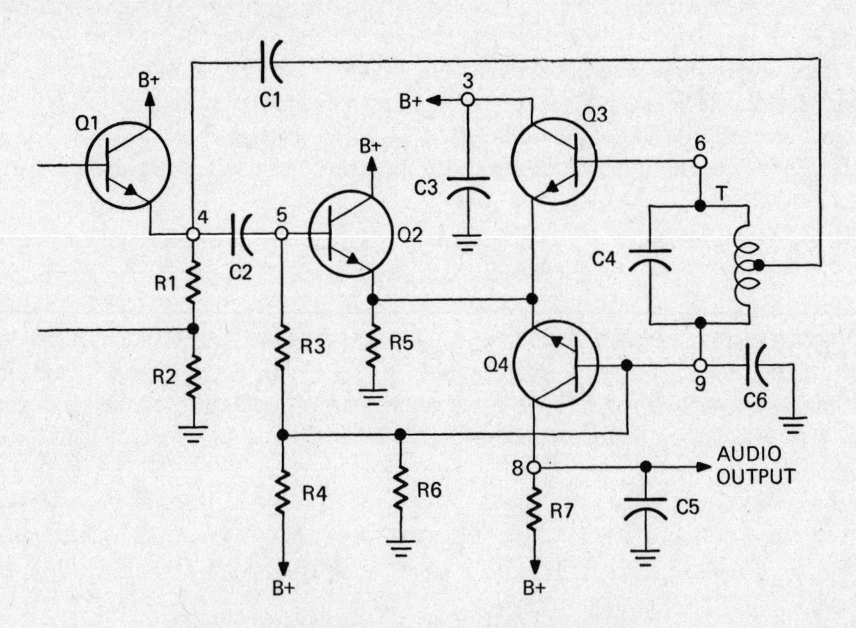

Figure 9-32. Quadrature FM detector circuit.

When the IF signal is not modulated, there is a steady DC voltage drop across R7 and no AF signal is present. When the IF signal is frequency modulated, current flow through R7 is varied by the voltages at the bases of Q3 and Q4. When the frequency deviates, the RF voltage across the quadrature tank circuit varies as the frequency swings above and below its resonant frequency. This causes a variation in Q3 and Q4 conduction and, hence, in the voltage drop across R7. The audio is recovered across R7 and fed out from the emitter of Q4 to the AF amplifier.

Ratio Detector

A ratio detector circuit, an example of which is shown in Figure 9-33, is widely used in FM radio receivers because of its ability to recover audio from a frequency-modulated RF signal without responding to amplitude variations of the signal.

Voltages E2 and E3 are each 90 degrees out of phase with El, and the same amount out of phase with the reference voltage which appears across L. The voltages rectified by diodes D1 and D2 are additive, and they appear across large capacitor C4. Capacitor C5 is an RF bypass which grounds one end of L at the radio signal frequency. Capacitors C2 and C3 function as radio capacitors.

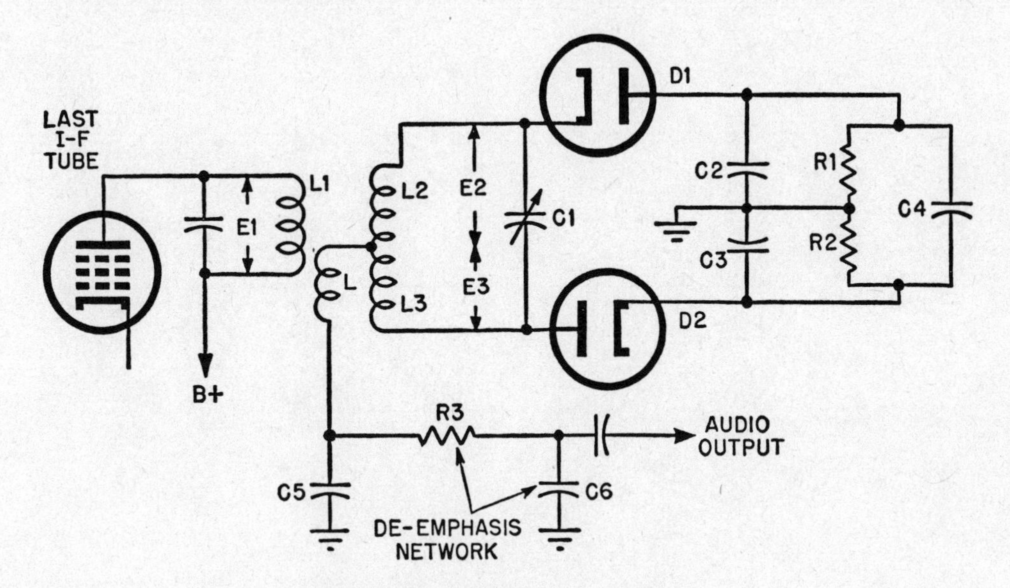

Figure 9-33. Ratio detector circuit.

When the carrier frequency of the signal applied to L1 changes, the voltages across C2 and C3 change, although their algebraic sum remains the same as the charge voltage in C4. When the input signal does not vary in frequency, the net current flow through L and C5 is zero when the diode currents are equal because the currents flow through L and C5 in opposite directions when the carrier is at the resonant frequency of the secondary. When the signal is frequency-modulated, however, the diode currents become unbalanced and the recovered audio voltage appears across C5 which has high reactance at audio frequencies.

Because the ratio of the rectified voltage is proportional to the level of the applied input signal voltage, which varies in frequency, the ratio detector does not respond to variations in the amplitude of the IF signal, and no limiter stage is required preceding it. The voltage stored in C4 varies slowly as intercepted radio signal voltage varies. It discharges slowly because of the long time constant of CR, R1, and R2.

Reactance Modulator

Figure 9-34 illustrates how a reactance modulator (V1) is used to vary the frequency of an oscillator (V2) to produce an FM signal. The instantaneous plate voltage of reactance tube V1 is applied across a load consisting of C3 and R2 which are also connected across the tank circuit (C4 and L1) of the Hartley oscillator.

With no audio input applied to the grid of V1, its plate voltage is steady. When an audio signal is applied at the grid of V1, it is also applied across R2. The combined effect of applying the audio signal to the grid of V1 and also across R2 is to change the combined reactance of C3 and R2 as the resistance of the plate-cathode path of V1 varies in accordance with the audio signal.

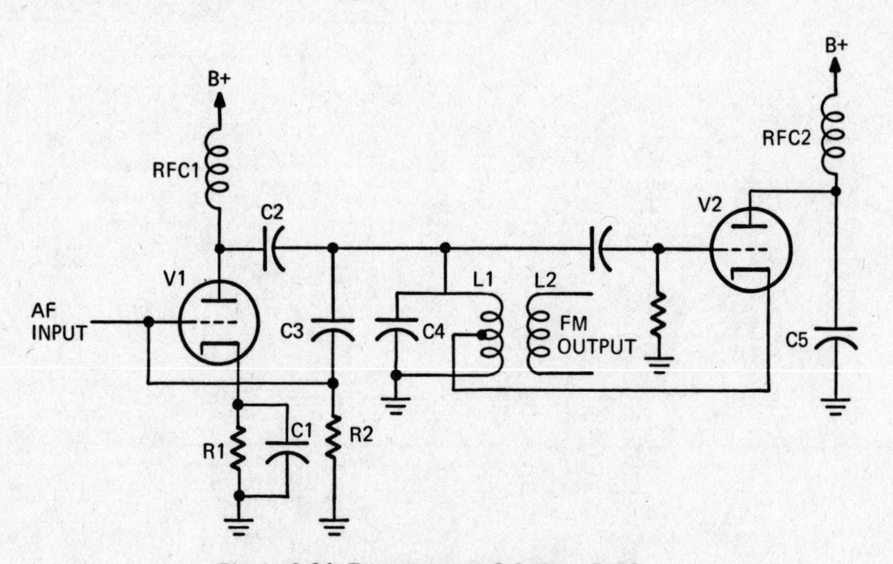

Figure 9-34. Reactance modulation circuit.

When an audio signal appears at the input of V1, the reactance across the oscillator is varied above and below the zero-signal value. This varies the frequency of oscillation of V2 above and below the center frequency. The variation in reactance, as well as the frequency deviation, depends upon the amplitude of the modulating signal. The result is a signal whose frequency is modulated by the audio input signal.

Regenerative Pentode Detector

A filament type pentode tube is used in the regenerative detector circuit shown in Figure 9-35. Only one battery (B) is used, and it supplies filament, control grid, screen grid, and plate voltage. Since the positive terminal of the battery is connected to ground, as is the control grid, the screen grid through R, and the plate through the AF output load,

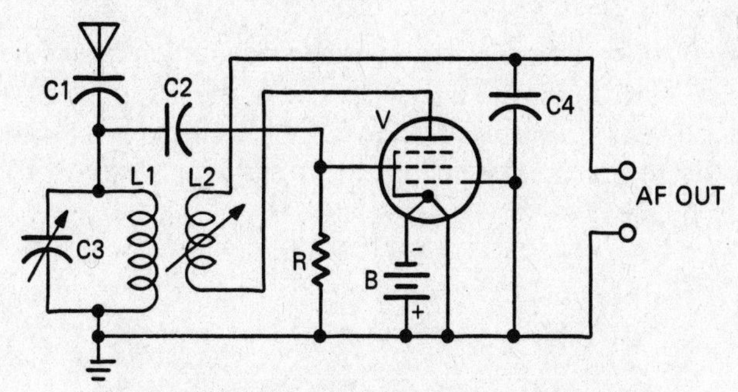

Figure 9-35. Regenerative detector circuit.

the control grid, screen grid, and plate are slightly positive with respect to the filament (cathode). These electrodes draw current. The intercepted radio signal is coupled through C2 to the screen grid which modulates electron flow from the filament to the plate. C3 and L1 are tuned to the receiving frequency. Regenerative feedback is obtained by coupling L2, the tickler coil, to L1. Regeneration is controlled by varying the degree of coupling between L1 and L2. Although greater sensitivity could be obtained by using a separate B battery for supplying plate voltage, this circuit offers economic advantages since it requires only one battery.

Reinartz Detector

The regenerative receiver circuit (simplified) shown in Figure 9-36 was conceived by John L. Reinartz. It employs a modification of the Hartley oscillator circuit. Since tuning capacitor C1 is connected across only a part of the secondary of T, "bandspread" tuning is achieved. Regeneration is controlled with variable capacitor C2 which is in the feedback path. The plate is isolated (for RF) by RF choke coil L.

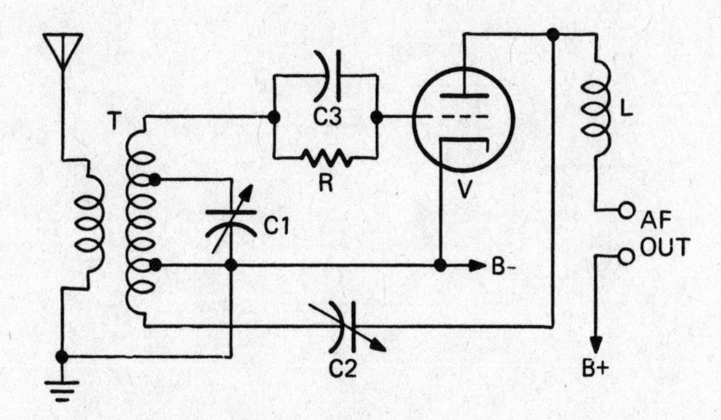

Figure 9-36. Reinartz regenerative detector circuit.

Reinartz Detector Variation

The antenna is connected to a low-impedance point tap on the tank coil L1 in the circuit shown in Figure 9-37. The upper tap can be moved to cut in as much of L1 as required for the band of frequencies to be covered by tuning of C1. Regeneration feedback is inductively coupled from tickler coil L2 to L1 and is controlled with C2. The plate is isolated (for RF) by RF choke coil L3.

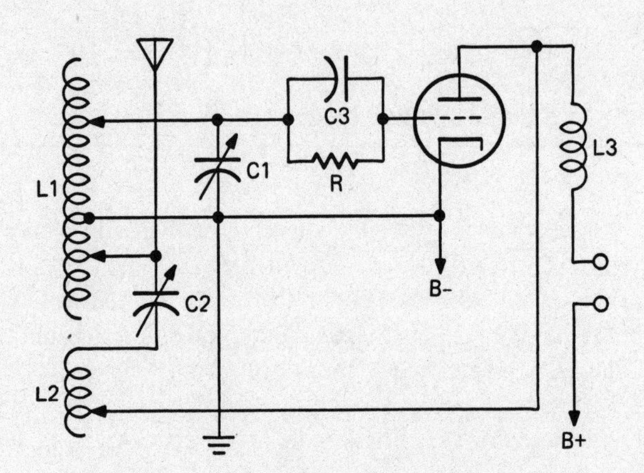

Figure 9-37. Variation of the Reinartz circuit.

Ring Modulator

Figure 9-38 is a circuit of a modulator in which four diodes function as an electronic double pole double switch operating at the speed of the carrier frequency. The carrier is applied to the center taps of the input and output transformers, and the modulated signal is fed across the input transformer. Diodes CR1 and CR2 rectify the signal during each half-cycle and CR3 and CR4 rectify the signal during the alternate half-cycles. The carrier

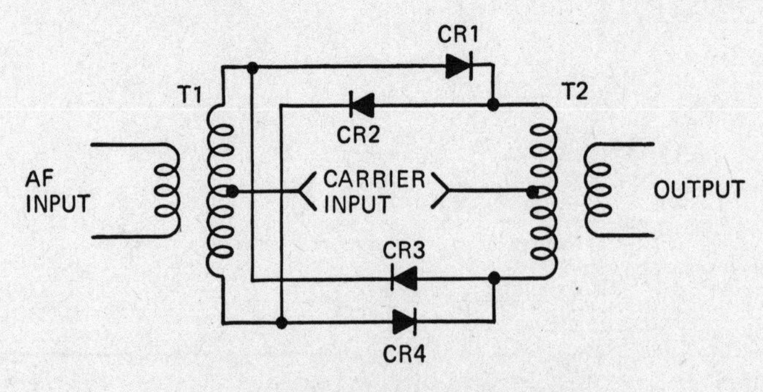

Figure 9-38. Ring modulator circuit.

signal is balanced out except when modulation is applied. Then, upper and lower sideband beat frequencies are generated. The output signal consists of the upper and lower sidebands with no carrier frequency being present.

Series-Resonant Crystal Detector

Early "crystal sets" employed a galena, silicon, or other semiconductor crystal as a detector. The crystal was held secure in a metal cup by a set screw. The cup was one terminal of the detector. The other terminal was a springy wire known as a "cat whisker," one end of which touched a sensitive spot on the surface of the crystal. This kind of crystal detector has long since been made obsolete by germanium and silicon diodes which have no adjustable cat whisker.

The simple receiver circuit shown in Figure 9-39 has a series-resonant tuning circuit (L and C1), a crystal diode (CR), and an RF bypass capacitor (C2). The unseen capacitance Cag between the antenna and ground is effectively in series with C1. The receiver is most sensitive at the frequency at which the inductive reactance of L is equal to the combined capacitive reactance of C1 and Cag. The RF voltage across C1 is maximum when C1 is tuned for series resonance at the frequency of an intercepted radio signal. The RF voltage across C1 is rectified by CR and the recovered audio modulation is developed across C2 and the load (headphones or AF amplifier input).

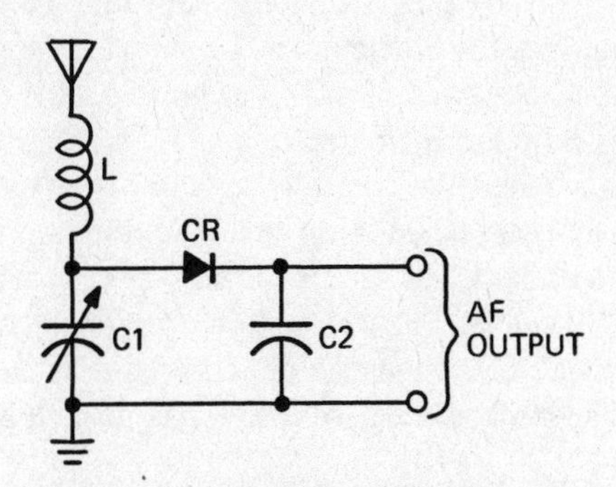

Figure 9-39. Simple receiver circuit.

Series-Tuned Regenerative Detector

In the early regenerative detector circuit (of historical value only) shown in Figure 9-40, a variocoupler (L1-L2), two adjustable inductors (L3 and L4), and a variable capacitor (C1) are used for tunning. The arrow through L2 indicates that it is rotatable inside of L1 so that the inductive coupling between them can be varied. The arrow through L1 indicates that it has taps that can be selected to obtain the required amount of inductance. The same is true of L3 and L4. L3 and C1 are adjusted to obtain resonance at the receiving frequency, L1 inductance is adjusted for maximum energy pickup by the an-

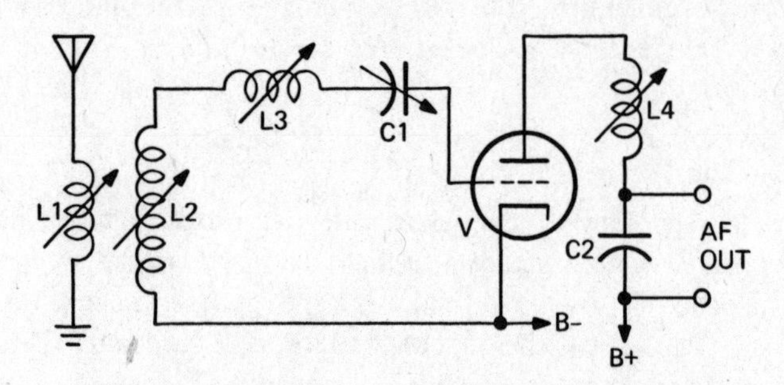

Figure 9-40. An early regenerative detector circuit.

tenna system and L2 is rotated to obtain adequate energy of selectivity. The inductance of L4 is adjusted to obtain the required amount of regeneration.

Single-Circuit Receiver

A regenerative detector circuit, which was popular in the 1920's, is shown in Figure 9-41. The receiver is tuned to the frequency of an incoming signal by variable capacitor C1 which is in series with L1. Coil L1 and C1 form the primary of a series-resonant circuit. Since current through L1 is maximum at the resonant frequency, maximum signal voltage is fed to the grid of the triode tube V. Grid-leak bias is developed by C2 and R. Regenerative feedback is introduced by tickler coil L2 which is inductively coupled to L1. (L1 is the fixed coil of an RF transformer known as a "variocoupler" and L2 is a rotatable coil within L1.) Regeneration is controlled by rotating L2 to vary the degree of coupling between the two coils. C3 is an RF bypass capacitor which grounds one end of L2 at the radio signal frequencies but which does not ground the plate voltage nor the recovered audio signal which flows through the AF output load.

A long antenna and an earth ground are required to make it possible for L1 and C1 to

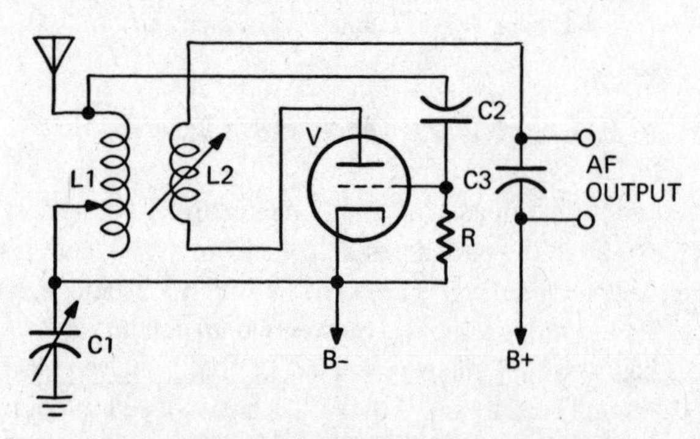

Figure 9-41. Regenerative detector circuit.

be series-resonant at low and medium frequencies, since the capacitor they form is a part of the tuning circuit. Since the regenerative circuit is connected directly to the antenna system, harmful interference can be caused if L2 is adjusted so that the detector oscillates. This circuit was popular because of its high sensitivity and quite good selectivity. In early receivers employing this circuit, the tube used was a type WD-11, WX-12, UV-199, VT-2, or 201A, all of which have a directly heated cathode (filament).

Single IC Quadrature FM Detector

A single IC is used as an IF amplifier and quadrature FM detector in the circuit shown in Figure 9-42. The FM input signal is fed through IF transformer T1 to terminals 11 and 12 of the General Electric PA189 integrated circuit. T2 is an autotransformer which functions as the quadrature coil. It senses variations in the signal frequency from its center carrier frequency and causes a voltage to be developed from which the audio signal can be recovered. The recovered audio signal is fed to the AF amplifier from terminal 8 of the IC.

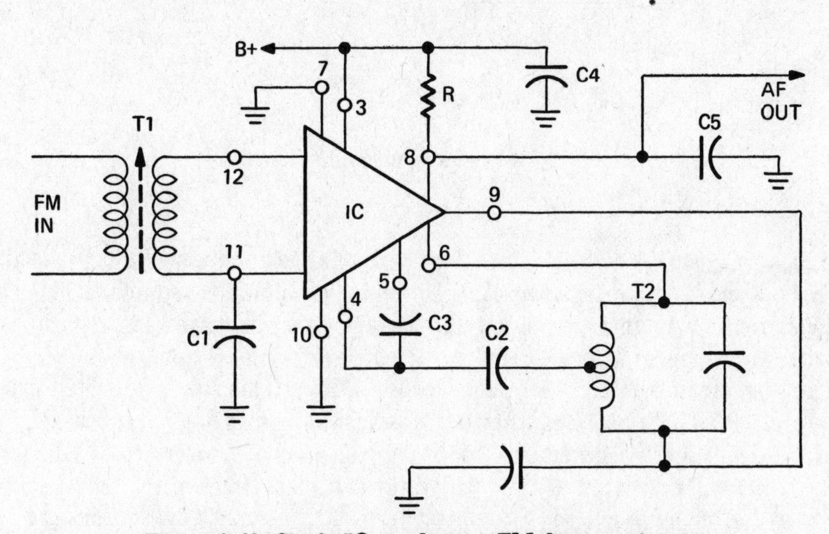

Figure 9-42. Single IC quadrature FM detector circuit.

Solid State Balanced Modulator

The balanced modulator circuit shown in Figure 9-43 employs two bipolar transistors and is used in SSB (single sideband) and suppressed-carrier DSB (double sideband) transmitters. The AF modulating signal is fed through transformer T to the bases of transistors Q1 and Q2, connected in push-pull. The collectors of Q1 and Q2 are capacity-coupled to the push-pull tank circuit (L1, C8A, and C8B). Since the inductance of L1 is extremely low at audio frequencies, the AF modulating signal does not appear across L1.

The unmodulated RF carrier signal is fed to the emitters of Q1 and Q2. Since the RF

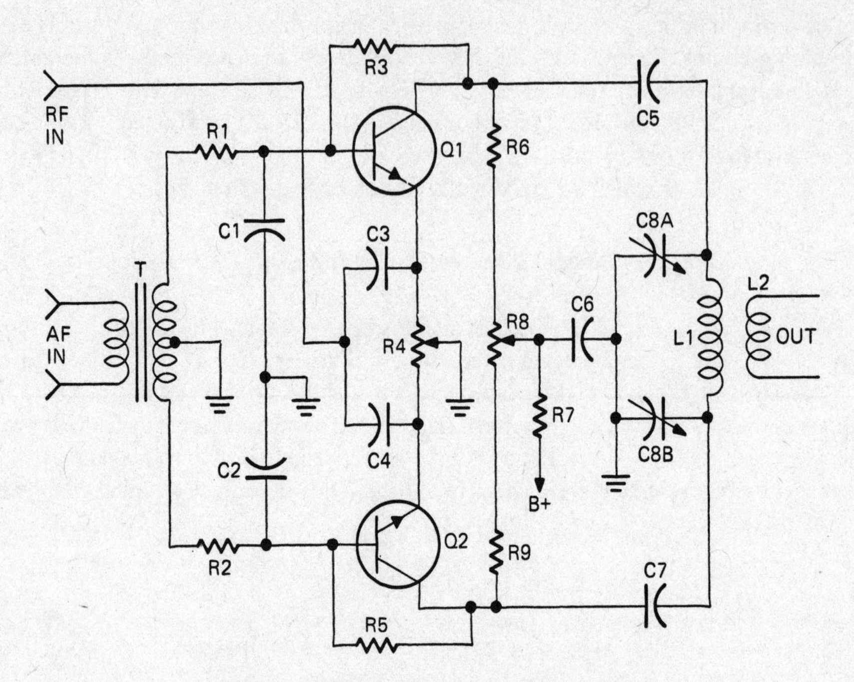

Figure 9-43. Balanced modulator circuit.

signal causes Q1 and Q2 collector current to rise and fall in unison, the RF carrier signal is balanced out and is not present across L1. However, when both the AF modulating signal and RF carrier input signal are present, the two signals interact and beat frequencies are generated which appear across L1. R4 and R8 are balancing controls.

For example, if the RF carrier frequency is 9 MHz and the AF modulating frequency is 1000 Hz, a 9001-kHz beat frequency (upper sideband) and an 8999-kHz beat frequency (lower sideband) will be generated. These two signals will appear across L1, but the 9-MHz carrier will not. If the AF modulating signal contains frequencies between 300 Hz and 3kHz, the upper sideband will extend from 9000.3 to 9003 kHz and the lower sideband from 8997 to 8999.7 kHz. Either sideband can be eliminated by a filter (not shown) when an SSB signal is required.

SSB Exciter

A beam deflection tube is used as the balanced modulator in the SSB (single sideband) exciter circuit shown in Figure 9-44 (Courtesy International Crystal Manufacturing Co., Inc.) The carrier signal is generated by the 6BH6 pentode tube shown at the left. This tube functions as a crystal controlled oscillator. Its output is fed to the control grid of the type 7360 beam deflection tube shown at the right. With no audio modulating signal applied, the amplified signals at the plates of this tube are fed in push-pull config-

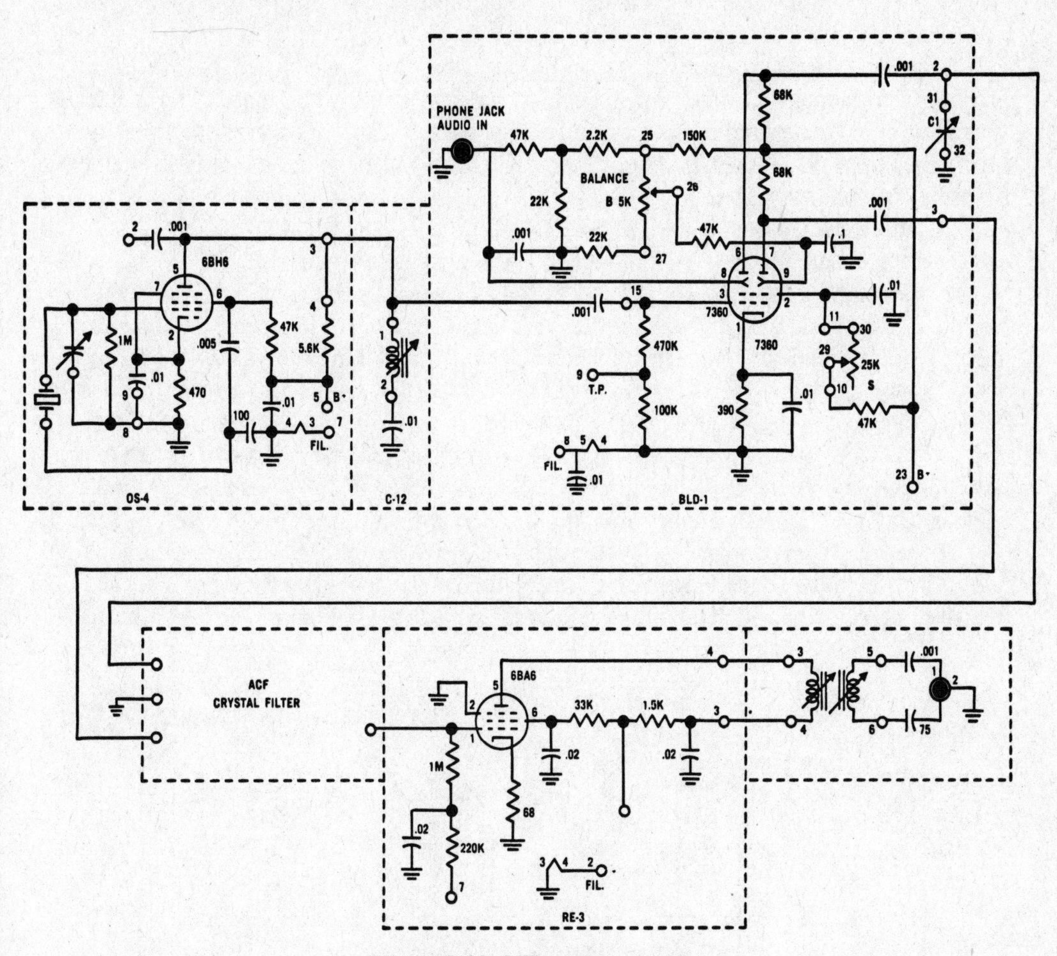

Figure 9-44. SSB exciter circuit.

uration to the input of the crystal filter. Since these two output signals are in phase opposition with respect to each other, they cancel each other out and there is no carrier signal fed to the 6BA6 pentode amplifier tubes. However, when an audio modulating signal is fed to the deflection electrodes of the 7360 tube, the carrier signal is alternately switched from one plate to the other. Because of the heterodyning action of the carrier signal and the audio modulating signal, upper and lower sideband RF signals are generated which are fed to the input of the crystal filter. However, the original carrier frequency signal does not exist because it has been balanced out. The filter removes one of the sidebands and the remaining sideband is amplified by the 6BA6 tube whose output is fed to a frequency converter (not shown). This particular SSB exciter generates a 9 MHz carrier signal. Its sideband output is then translated by a frequency converter to the required transmitting channel.

Superheterodyne Receiver

Figure 9-45 shows an example of an AM broadcast band superheterodyne receiver circuit. Radio signals are intercepted by loop antenna L which is tuned by variable capacitor C1 (C2 is a trimmer capacitor shunted across C1.) The signals are amplified by the remote cutoff 6BA6 pentagrid converter tube which functions as both the mixer and the oscillator and whose frequency is determined by C7, C8, C9, and T2. (C8 is the variable tuning capacitor which is ganged to C1 and C5 to enable single-dial tuning. C7 and C9 are trimmer capacitors. T2 is a two-winding oscillator transformer.)

As the tuning dial is turned, the local oscillator generates a signal 455 kHz higher than the desired radio signal. For example, if L and T1 are tuned to receive a 1000-kHz signal, T2 is automatically tuned to 1455 kHz. Heterodyning of the 1000-kHz radio signal at the locally-generated 1455-kHz signal causes a 2455-kHz sum beat frequency and a 455-kHz difference beat frequency to be produced. Since T3 is fixed-tuned to 455 kHz, only the 455-kHz difference beat frequency (intermediate frequency) signal is fed to the 6BA6 remote cutoff pentode IF amplifier tube. The amplified IF signal is fed to the diode detector section of the 6AV6 tube through T4 which is fixed-tuned to 455-kHz.

The recovered audio signal is developed across volume control R8 and is fed through C18 to the grid of the triode section of the 6AV6 tube which amplifies the AF signal. The amplified AF signal is coupled through C23 to the control grid of one of the 6AQ5 beam

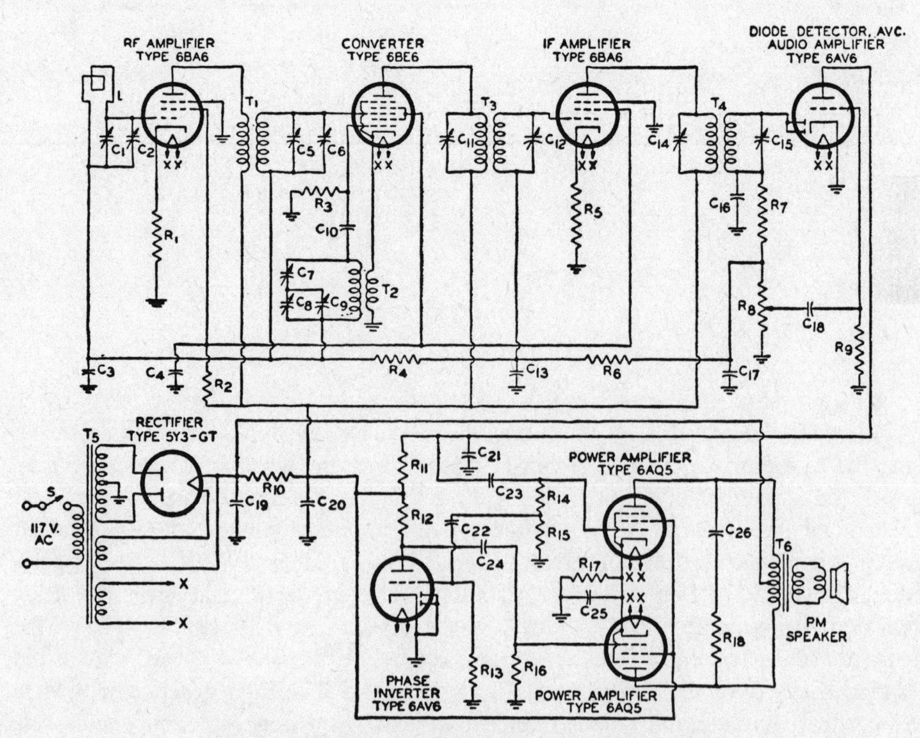

Figure 9-45. AM superheterodyne receiver circuit.

power tubes through C22; and also to the grid of the 6AV6 phase-inverter tube. The two 6AQ5 beam power tubes function as a push-pull AF power amplifier whose output is developed across the primary of output transformer T6. The low-impedance secondary of T6 is connected to the loudspeaker.

Plate and screen grid voltages are provided by a full-wave rectifier power supply employing a 5Y3 tube. The DC output of the rectifier is fed through a pi-section RC ripple filter (R10, C19, C20). Power transformer T5 furnishes 6.3 volts AC to the tube heaters and high voltage AC to the 5Y3 dual-plate rectifier tube. (Illustration courtesy of RCA Corporation.)

Superregenerative Detector

The superregenerative detector, which was invented by Major Edwin H. Armstrong, who also invented FM and the superheterodyne receiver, has unique characteristics. It will demodulate both AM and FM signals. It is extremely sensitive and has inherent limiting action (weak signals are reproduced almost as well as strong signals). Its selectivity is poor, and because it radiates a signal that can cause serious radio interference, it is virtually obsolete. The circuit shown in Figure 9-46 is shown only for historical purposes.

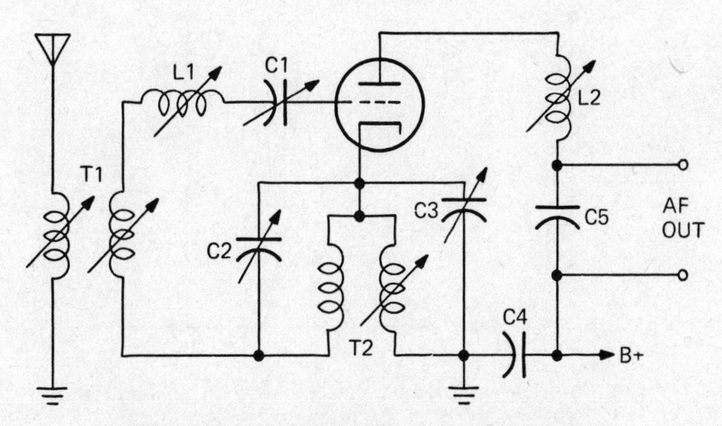

Figure 9-46. Superregenerative dectector circuit.

The triode tube functions as both a detector and quench oscillator. T1, L1, and C1 are tuned to the incoming signal frequency and L2 is adjusted to control regeneration. C2, C3, and T2 form the quench oscillator tuning circuit. The quench frequency is much lower than the receiving frequency. For example, the receiving frequency could be 27.255 MHz and the quench frequency could be 50 kHz. The quench signal alternately drives the 27.255-MHz part of the circuit in and out of oscillation 50,000 times per second. Since maximum sensitivity is obtained at the point just before which oscillation takes place, the circuit is automatically swung past this point at the quench frequency rate.

Superregenerative Detector with Quench Oscillator

A super regenerative detector utilizing only one tube or transistor is known as the self-quenching type since a separate quench oscillator is not used. In the circuit shown in Figure 9-47, two triode tubes are used. V1 is the detector and V2 is the separate quench oscillator. V1 employs the Hartley oscillator circuit so that it will function as a regenerative detector. V2 employs the Armstrong tuned grid oscillator circuit which operates at a much lower frequency than the receiving frequency. Plate voltage for V1 is routed through R5, regeneration control R4, the feedback coil of transformer T, R2, RF choke coil L3, and part of tank coil L2. The low frequency RF signal generated by L2 directly modulates V1 plate current and indirectly its control grid voltage. This causes V1 to swing in and out of oscillation at a rapid rate, causing it to operate as a very sensitive detector. The recovered audio signal is developed across R2 and fed out through C4 to a headset or an AF amplifier.

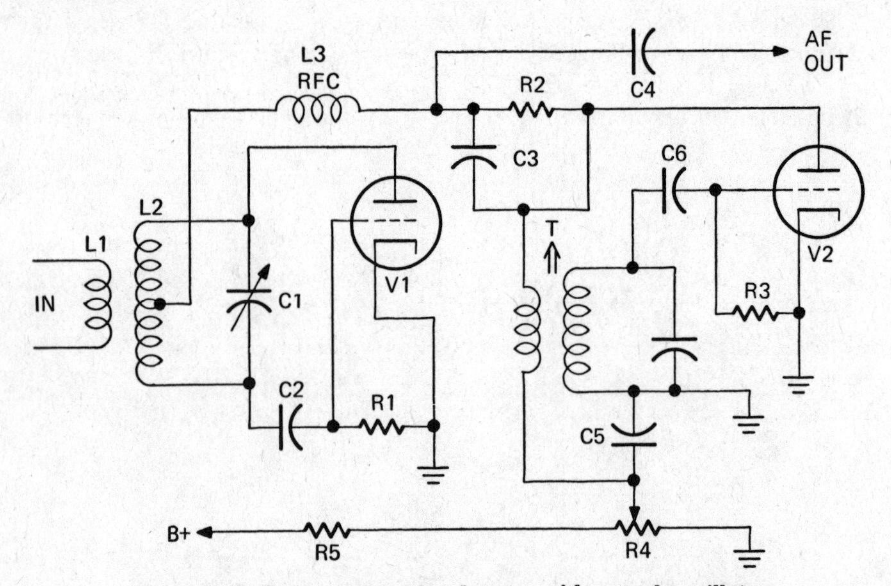

Figure 9-47. Superregenerative detector with quench oscillator.

Superregenerative Detector with Squegging Control

The self-quenching superregenerative detector circuit shown in Figure 9-48 includes a "squegging" controls (C3 and R3) with which quench frequency and amplitude can be varied. Triode V is connected in a Colpitts circuit whose tank circuit (L2, C1A, and C1B) is tuned to the receiving frequency. Quench oscillations are produced because the grid is returned (through R1) to a positive voltage point instead of ground, and because of blocking oscillator action whose frequency is determined by the time constant of C2-R1 and can be varied by adjusting C3 and R3. The hash-type noise produced by quench action is attenuated by a low-pass filter (L3, L4, and C4).

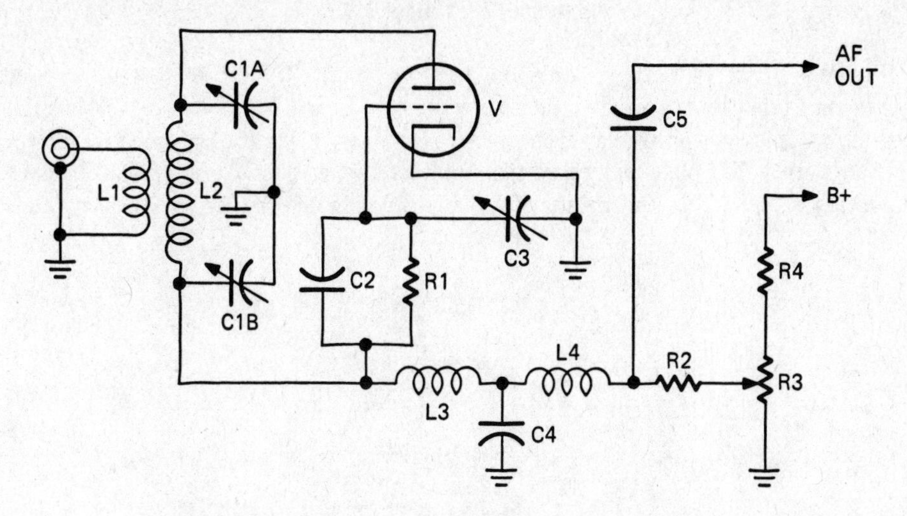

Figure 9-48. Self-quenching superregenerative detector circuit.

Super-Selective Crystal Detector

The crystal detector circuit shown in Figure 9-49 is not very selective although at the time of its conception it was considered so. It is a "Rube Goldberg" circuit conceived by an unknown radio pioneer who designed circuits experimentally in much the same manner as Thomas A. Edison did his inventing. In this circuit, C2 is tuned to make L2-C2

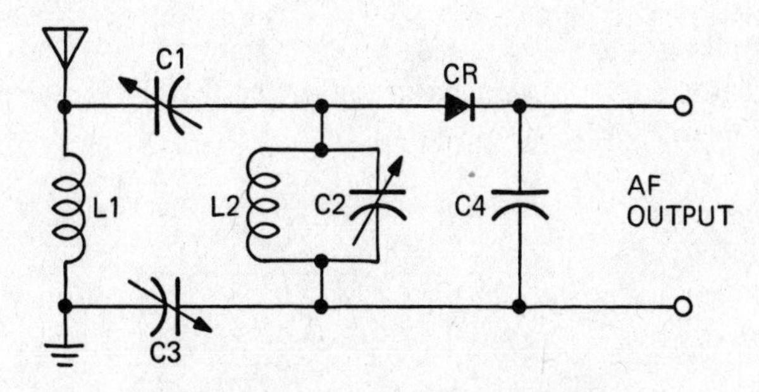

Figure 9-49. Crystal detector circuit.

parallel-resonant at the receiving frequency. C1 is used for varying antenna coupling. By reducing the coupling, the selectivity of L2-C2 is improved, but sensitivity is made less. C3 is adjusted to make L1-C3 series-resonant at the frequency of an interfering signal so that signal will be short circuited. Diode CR rectifies the signal and the recovered audio modulating signal appears across RF bypass capacitor C4.

Suppressor Grid Modulator

Figure 9-50 shows a suppressor grid modulator circuit. The carrier signal is applied to the control grid of the pentode tube and modulation is applied to the suppressor grid. When the modulation voltage varies the suppressor voltage, plate current and RF power output are varied. The plate tank tuned transformer removes the DC and audio frequency components of the signal and passes only the sidebands and carrier and, perhaps, unwanted harmonics.

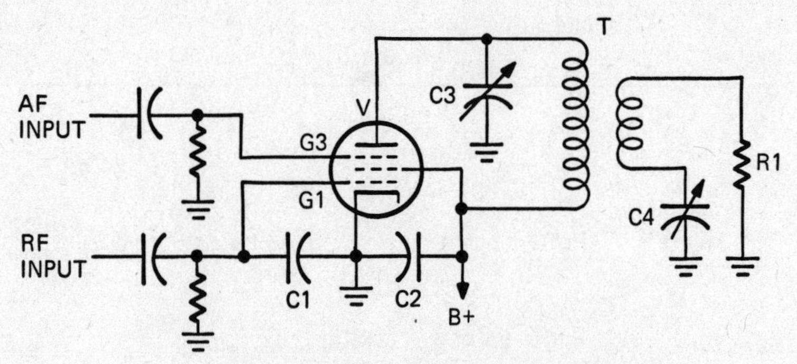

Figure 9-50. Suppressor grid modulator circuit.

Switch-Type Balanced Modulator

The circuit of a Cowan bridge balanced modulator employing four diodes is shown in Figure 9-51. When only the RF carrier signal is applied (to C and D), the bridge is balanced and there is no RF signal across A and B. But when an AF modulating signal is applied to A and B, the diodes are switched on and off at the AF rate, causing the bridge to be unbalanced. Then the resulting sidebands, generated by the heterodyning of the two signals, appear at the output.

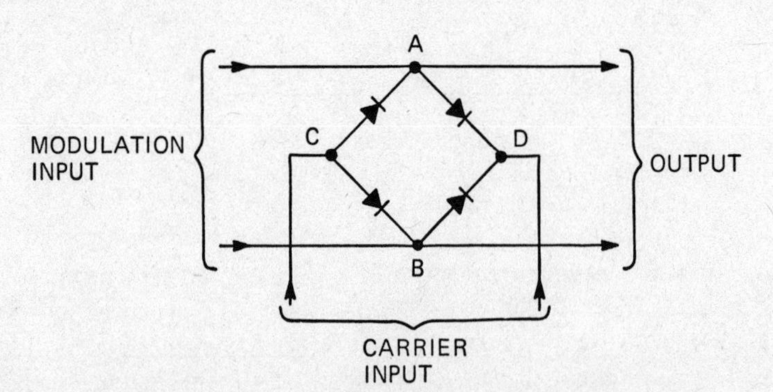

Figure 9-51. Cowan bridge balanced modulator circuit.

Three-Circuit Regenerative Detector

The regenerative detector circuit shown in Figure 9-52 employs grid-leak bias. The triode tube draws grid current during the positive half-cycles of input signal. The time constant of R and C2 is such that they respond to the audio-frequency variations of the input signal, but not to the RF variations. The audio variations, together with an RF component, are amplified and appear in the plate circuit. Tickler coil L2 is coupled to grid coil L3, and the RF component in the plate circuit is coupled back in phase to the grid tank circuit. Thus the voltage gain of the detector is increased. The amount of regeneration is controlled with C3. The regenerative detector is most sensitive to AM signals and is most selective when C3 is set just below the point at which positive feedback causes oscillation. To receive CW (continuous wave) radiotelegraph signals, C3 is adjusted just beyond the point where oscillation occurs.

An RF amplifier should be used ahead of a regenerative detector to minimize radiation of a potentially interfering signal that is generated when the detector oscillates.

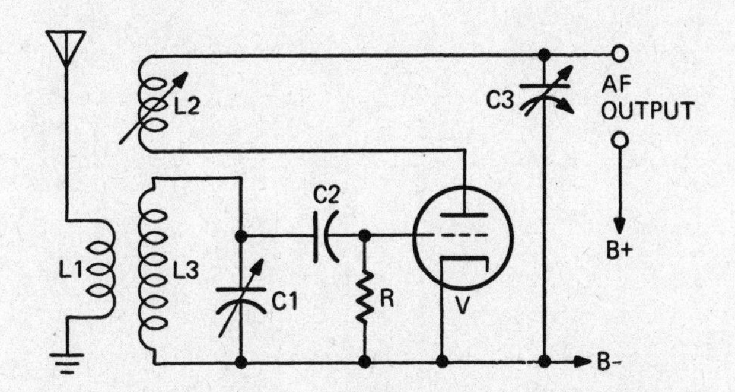

Figure 9-52. Three-circuit regenerative detector circuit.

Transistor-Amplified Crystal Detector

The recovered audio signal at the output of a crystal diode detector (D) is amplified by a transistor (Q) in the circuit shown in Figure 9-53. For AM broadcast band reception,

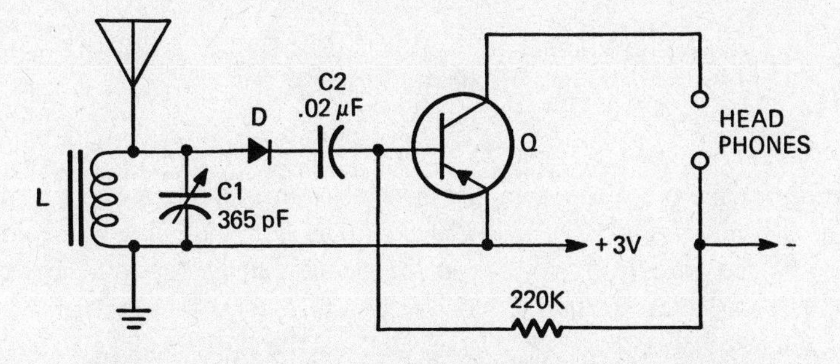

Figure 9-53. Transistor-amplified crystal detector circuit.

L can be a ferrite antenna (loopstick) and C1 a 365-pf variable capacitor. The RF signal is rectified by D and the recovered audio signal is fed to the base of Q which is forward-biased through a 220,000-ohm resistor. The AF signal is amplified by Q sufficiently to drive headphones at a relatively high sound level.

Triode Tube Balanced Modulator

The balanced modulator, whose circuit is shown in Figure 9-54, generates a DSB signal and is used in SSB transmitters (see Solid State Balanced Modulator). In this circuit, the AF modulating signal is fed in push-pull to the cathodes of triode tubes V1 and V2. The RF carrier signal is fed to the parallel-connected grids of V1 and V2 and is balanced out in the push-pull plate circuit. R3 is the balancing control. The upper and lower sidebands generated by heterodyning of the AF and RF signals appear across the output tank circuit (L1, C6A, and C6B). To derive an SSB signal, a filter (not shown) is connected between L2 and the next stage to eliminate one of the sidebands.

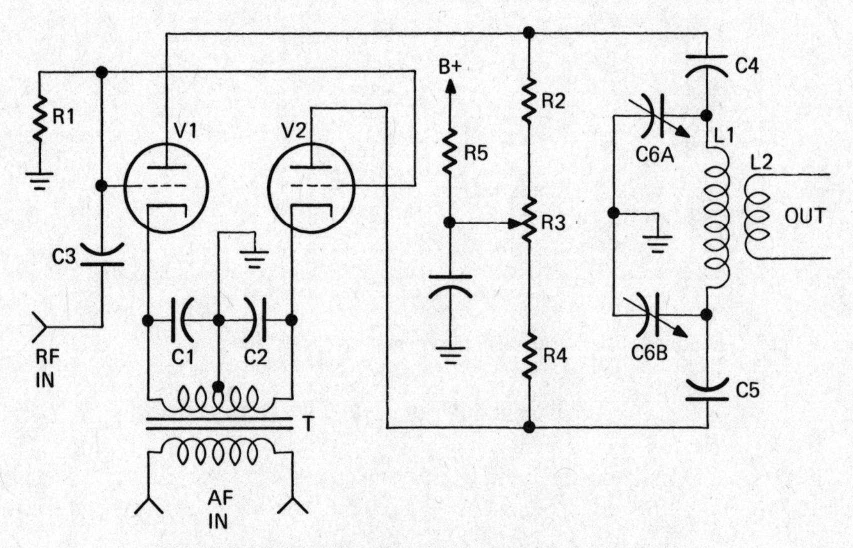

Figure 9-54. Triode tube balanced modulator circuit.

Ultraudion Detector

The ultraudion detector circuit (Figure 9-55) was devised by Dr. Lee De Forest, inventor of the triode tube. It is a regenerative detector utilizing a series-resonant tuning circuit. The antenna is coupled to inductance L1 and the plate of the triode through series capacitor C whose primary function is to isolate the antenna from the DC plate supply voltage. When variable capacitor C1 is adjusted to receive a radio signal at a specific frequency, the reactances of L1 and C1 are equal and the signal voltage across C1, applied to the grid of the tube, is at a maximum. The amplified signal at the plate of the tube is fed

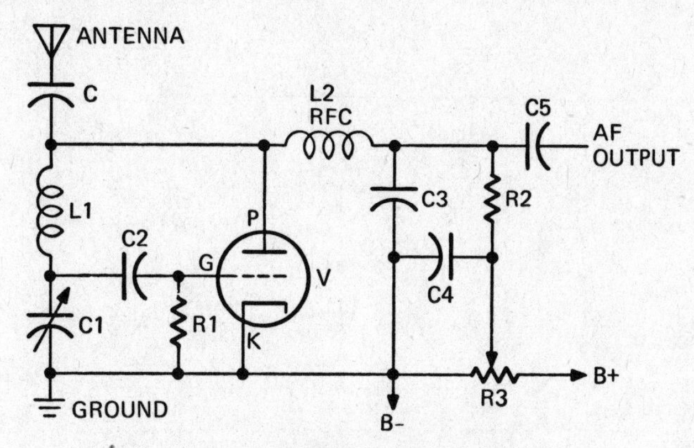

Figure 9-55. Ultraudion detector circuit.

back to L1 at a phase angle that supports oscillation. To prevent the detector from oscillating when receiving AM signals, the regeneration control R3 is adjusted to lower the plate voltage to the point just below which oscillation would result. Grid leak bias is furnished by C2 and R1. The RF choke coil L2 permits tapping off the audio modulation from the plate without loading the plate circuit by the AF output circuitry. C3 bypasses any remaining RF signal to ground. The AF signal is developed across R2 and coupled to an AF amplifier through C5. Capacitor C4 bypasses the AF signal to ground so that no AF signal will be developed across R3, and so that only R2 will function as the plate load.

Varactor Frequency Modulator

A voltage variable capacitance diode (Varactor, Varicap, Capistor, etc.) is used in the frequency-modulated Colpitts oscillator circuit shown in Figure 9-56. With no AF modulation applied, the oscillator frequency is held constant by crystal Y. When an AF signal is applied to the anode of the varactor through C2 and R3, the capacitance of the varactor varies. Since the varactor is in series with the crystal, these capacitance variations cause the oscillator frequency to deviate up and down. Thus, direct frequency modulation of a crystal controlled oscillator is achieved.

Variometer-Tuned Receiver

The variometer and variocoupler were very popular radio receiver tuning components in the early 1920's, but have been superseded by fixed coils and transformers, and variable capacitors which were first manufactured by Hammarlund. Figure 9-57 shows the circuit of a regenerative receiver employing two variometers and a tapped inductor (L1). Intercepted radio signals are coupled through variometer L2 and C1 to the grid of the triode tube. L1 has taps which are selected by means of rotary switch S to vary the amount of inductance.

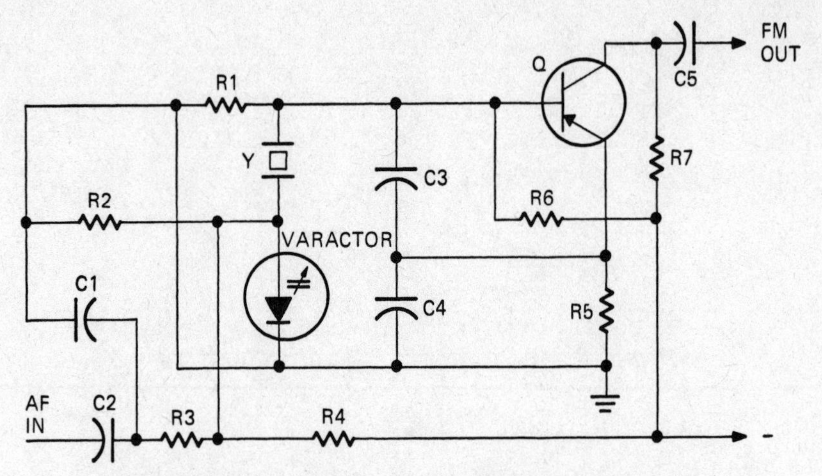

Figure 9-56. Frequency-modulated Colpitts oscillator circuit.

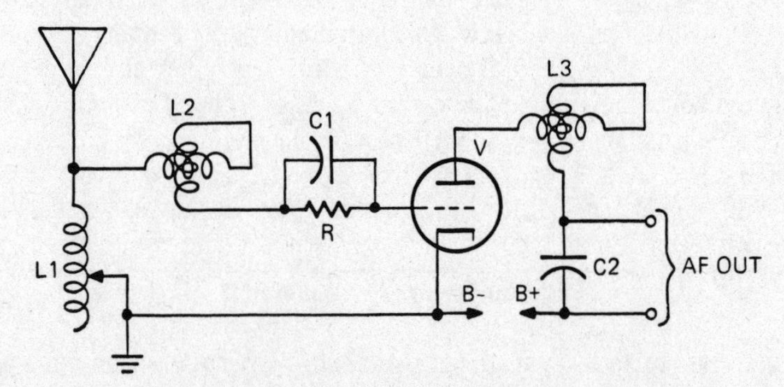

Figure 9-57. Variometer-tuned receiver circuit.

Tuning is accomplished by setting S to insert the required amount of inductance in series with variometer L2 which is rotated (by means of a tuning dial) to make L1 and L2 resonant at the receiving frequency. Variometer L3 is the regeneration control. When resonant at close to the same frequency as the input circuit, oscillations will occur because of positive feedback through the internal grid-plate capacitance of tube V. For maximum sensitivity to AM signals, L2 is adjusted to the point just before which oscillations occur. As in most other regenerative detector circuits using a tube, grid-leak bias is developed by C1 and R.

A variometer is a variable inductor consisting of a rotatable coil within a fixed coil, connected in series with each other. Rotating the inner coil causes the inductance of the coils to add to or subtract from their combined inductance, depending upon their orientation with respect to each other. The modern counterparts of the variometer are the variable core coils and RF transformers used in AM auto radios.

Voltage Doubler Detector

The voltage doubler AM detector circuit shown in Figure 9-58 provides 6 dB of voltage gain over a conventional diode detector. When the input signal at the secondary of T makes the anode of diode CR1 positive, capacitor C1 is charged. CR1 does not conduct since it is reverse-biased. During the half-cycles when the signal makes the cathode of diode CR2 negative, C2 is charged. CR2 does not conduct because it is reverse-biased.

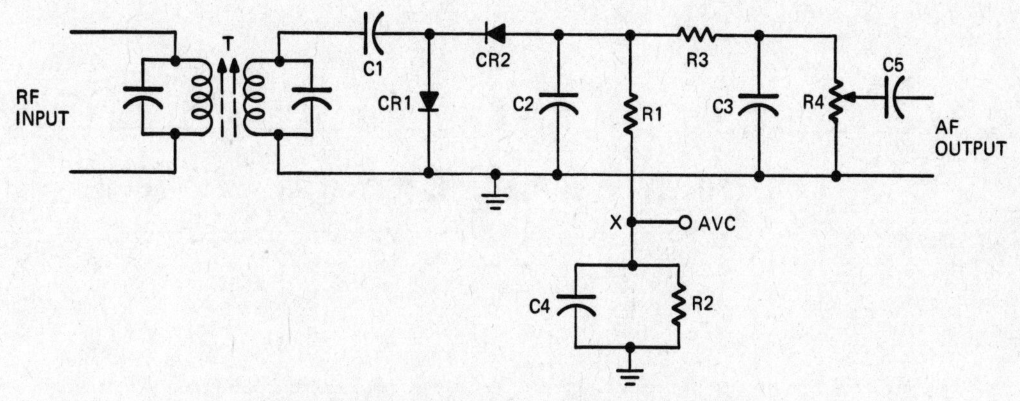

Figure 9-58. Voltage doubler AM detector circuit.

During the negative half-cycles, the charge voltages of C1 is in series-aiding with the signal voltage, causing the C2 charge voltage to be twice as great as the C1 charge voltage. The capacitances of C1 and C2 are small enough to allow the recovered audio signal to be developed across C2. The voltage across C2 is DC which varies at the audio rate. To use this DC voltage for AVC purposes, it is fed through R1 to R2 and C4. The AVC voltage is prevented from varying at the audio rate by C4 whose reactance at audio frequencies is very low.

10

Power Supply Circuits

Early electron tube radio receivers were powered by external batteries. When so-called "battery eliminators" were developed in the late 1920's, enabling the powering of radio receivers from the AC powerline, the demand for radio batteries declined sharply. Later, the demand fell almost to zero when radio receivers werre introduced which included an AC-to-DC power supply. The circle was almost closed when transistor radios, powered by an integral battery, were produced by the millions. However, the need for power transmogrifiers did not disappear. Although batteries could be used to power most solid state electronic devices, many depend on power supplies so that it will not be necessary to replace batteries as they wear out.

Described and illustrated in this chapter are examples of power supply circuits. As in the case of logic circuits, available space does not permit covering all of the possible combinations. Indeed, it would require several volumes.

AC-DC Power Supply

Figure 10-1 shows how many of the earlier table model radios using tubes were wired to enable operation from either an AC or DC powerline. When AC is supplied to the line plug and the switch S1 turned on, the AC flows through ballast resistor R13 and is applied to the series-connected heaters of the tubes and the pilot light. The sum of the voltage drops across the components in this string is approximately 115 volts, the full value of the AC line voltage.

The AC is also supplied to the plates of the dual-diode rectifier, the 25Z5 tube. After half-wave rectification by the parallel diodes, the resultant pulsating DC is filtered by the pi-section filter consisting of C17, C16, and L2. Instead of a permanent magnet, the

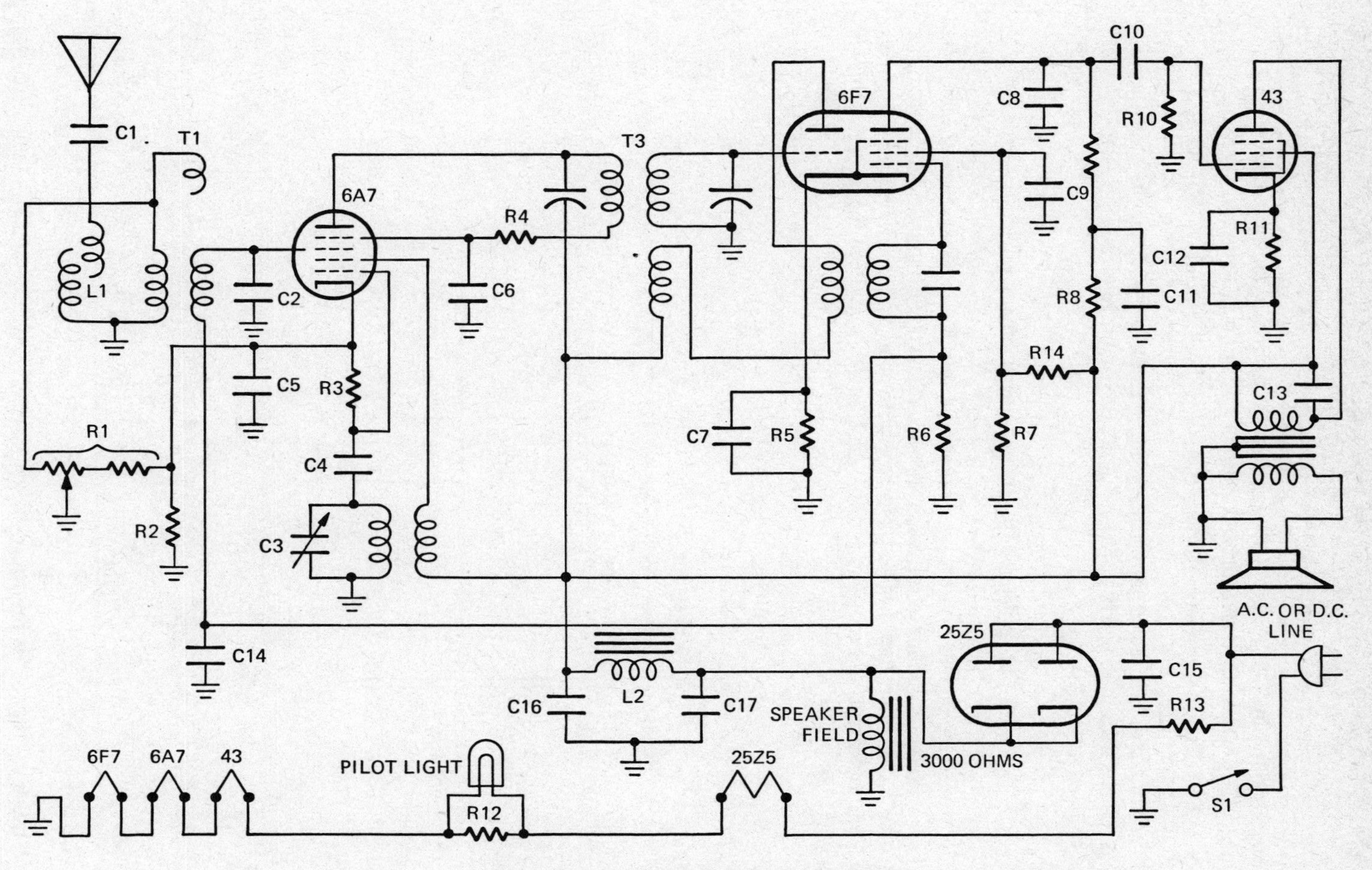

Figure 10-1. AC-DC receiver power circuit.

magnetic field for the loudspeaker is provided by an electromagnet, current for the coil of this magnet is obtained by connecting it to the cathodes of the diodes and to ground. The B+ voltage is applied to the receiver circuitry from the junction of C16 and L2.

In the case of DC operation, the positive side of the powerline is connected to the plates of the 25Z5 dual diode through the plug, and the negative side is connected to ground. The rectifier then merely passes the direct current for application through the filter to the receiver circuitry.

Amplifier or Transmitter Power Supply

Four different output voltages are provided by the power supply whose circuit is given in Figure 10-2. AC for tube heaters at 6.3 or 12.6 volts is provided directly by one of the transformer secondaries. Two different positive DC voltages (B++ and B+) are provided for tube plates, and a negative DC voltage (C-) is provided for power amplifier stage grid bias.

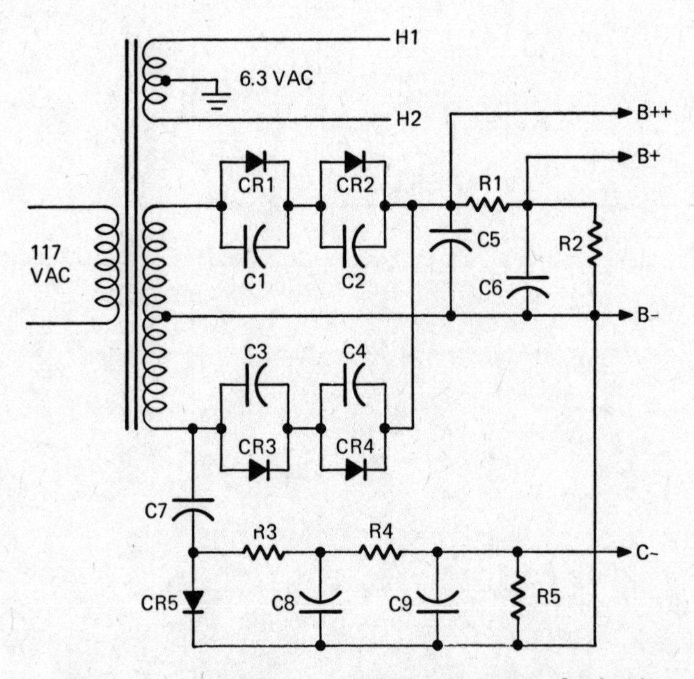

Figure 10-2. Amplifier or transmitter power supply circuit.

A full-wave rectifier is formed by diodes CR1, CR2, CR3, and CR4, which are fed by the center-tapped high voltage secondary of the transformer. Each diode is shunted by a capacitor (0.01 mf) which protects it from over-voltage surges. The DC output of the rectifier is filtered by C5 (20 mf or higher) and C6 (40 mf or higher) and by R1. Maximum high DC voltage (B++) is obtained at the junction of C5 and R1. Reduced DC voltage

(B+) is obtained at the junction of R1 and R2 which form a voltage divider. Their values depend upon voltage across C5 and load current. Actually, a voltage divider is usually not needed. But for protection reasons, R2 is used as a bleeder and as a by-product becomes one leg of a voltage divider.

The unique feature of this circuit is the manner in which negative DC grid bias voltage is derived. Diode CR5 functions as a *shunt* rectifier fed through C7. Its negative DC output voltage is filtered by R3, C8, R4 and C9. Since this part of the power supply delivers no current (only voltage), the DC voltage at the junction of R4 and R5 is equal to the voltage across C8 minus the voltage drop across R4 which is determined by the bleeder current of R5. The voltage across C8 depends upon the values of C7 and C8 and the transformer voltage.

Balanced Power Supply

Figure 10-3 shows the circuit of a balanced power supply that provides positive and negative regulated DC output voltages. The zener diodes D1 and D2 used in the circuit act as a voltage divider and as voltage regulators. They should be capable of dissipating more power than is normally fed to the loads. Because of the series ballast resistors, the input voltage from the rectifier should be higher than the combined avalanche voltages of the zener diodes.

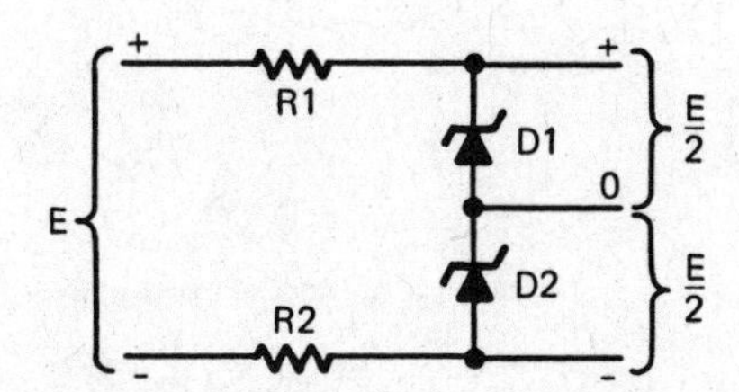

Figure 10-3. Balanced power supply circuit.

Battery Eliminator

Figure 10-4 shows the circuit of a power supply which provides 9 volts DC to power a transistor radio or other battery-operated device. The power supply consists of 12.6-volt filament transformer, rectifier diode, filter capacitor, a pass transistor voltage regulator, and a zener diode voltage reference for the transistor. The voltage regulator offsets DC output voltage variations which would otherwise result because of AC input voltage variations and load current variations. (Illustration courtesy of *Ham Radio*.)

Bridge Rectifier

Figure 10-5 shows how four diodes are connected in a full-wave bridge rectifier circuit. During the AC half-cycle when point X is positive with respect to point Y, electron current flow is from point Y through diode CR4 load resistor RL, diode CR1 to

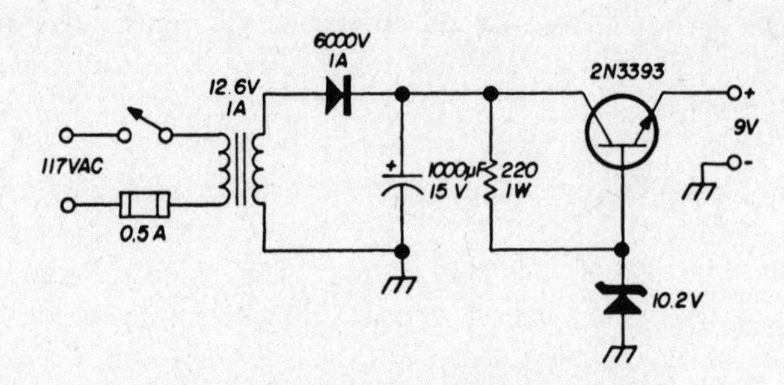

Figure 10-4. Battery eliminator circuit.

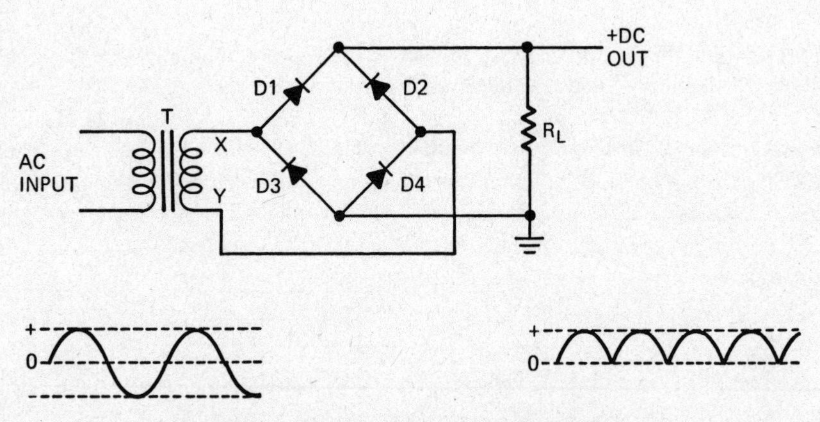

Figure 10-5. Full-wave bridge rectifier circuit.

point X. During the opposite half-cycles, current flows from point X through CR3, RL, CR2 to the transformer secondary at point Y. The bridge rectifier delivers a positive DC pulse during each AC half-cycle, regardless of polarity. The DC output can be smoothed by connecting a large value capacitor across RL. It has the advantage over a conventional full-wave rectifier in that a center-tapped transformer is not required. Also, for a given transformer, the bridge rectifier will provide twice the output voltage as the conventional full-wave rectifier. Bridge rectifiers may also employ diode tubes instead of semiconductor diodes.

Capacitance-Type Filter

Figure 10-6 shows a simple capacitance filter circuit. When the pulsating DC voltage from a rectifier is applied to the capacitor, it charges to the peak value of the supplied voltage. When the output from the rectifier falls to zero, the capacitor partially discharges through the load (RL). This gradual discharge prevents the DC output voltage from falling

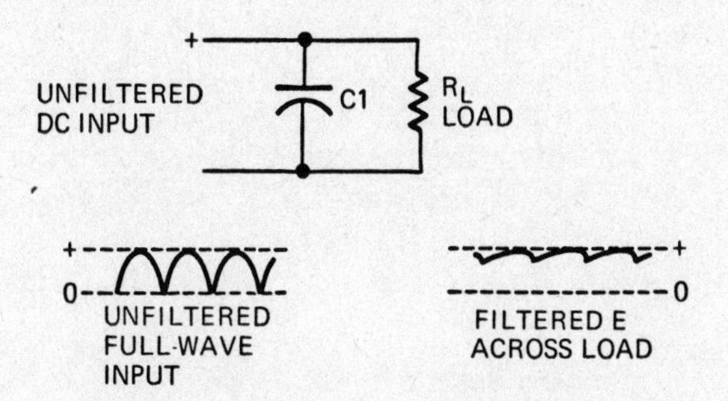

Figure 10-6. Simple capacitance filter circuit.

to zero and reduces the ripple riding on the DC voltage. The average DC output voltage depends upon whether a half-wave or full-wave rectifier is used, load resistance and the value of the capacitor. If load current is zero, the capacitor charges to the peak value of the pulsating DC input voltage. The capacitor, in addition to acting as a storage device, serves as a reactance shunted across the ripple voltage. If C has a value of 100 uf, for example, its reactance at 60 Hz (half-wave rectifier ripple frequency) is approximately 27 ohms and 13.5 ohms at 120 Hz (full-wave rectifier ripple frequency).

DC Heater Voltage Source

In high-gain AF amplifiers employing tubes, hum can be a problem. It is customary to apply AC to the heaters of all tubes. Hum can often be reduced by applying DC to the heaters of preamplifier stage tubes. This DC can be obtained by rectifying and filtering the AC voltage obtained from the heater winding of the power transformer. An alternative way is illustrated in Figure 10-7. Tubes V1, V2, V3, and V4 are power amplifiers connected in push-pull parallel. (The other tubes used in the amplifier are not shown.) Tube V5 is a dual-triode used as a preamplifier. Its series-connected heaters are used as the 84-ohm cathode bias resistor for V1, V2, V3, and V4. Capacitor C bypasses the AF

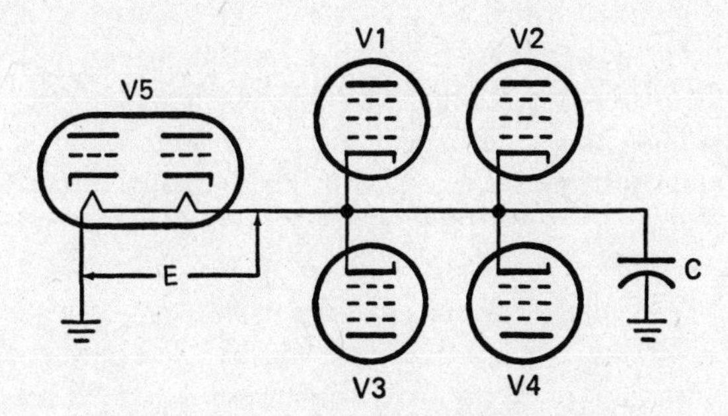

Figure 10-7. DC heater voltage source.

signal to ground and insures the integrity of the DC flowing through the heaters of V5.

Assume that the series-connected heaters of V5 draw 0.15 ampere when 12.6 volts is applied. To obtain a 12.6-volt drop across the heaters of V5, the combined cathode current of the power amplifiers must be 150 mA (approximately 37 mA each). If more than 12.6 volts of bias is required by the power amplifiers, a resistor can be connected in series with the heaters of V5. But, it is necessary that the current through the heaters of V5 be maintained as close as possible to 150 mA. In some cases, however, less hum, ample gain, and longer tube life can be obtained when heater current is lower than its rated value.

Dual Voltage Power Supply

This power supply delivers approximately 9 or 18 volts DC depending upon the setting of switch S1 which is wired as shown in Figure 10-8. Transformer T is a 6.3 volt filament transformer and diodes CR1 and CR2 are conventional rectifier diodes. Capacitors C1 and C2 are 15 volt electrolytic capacitors with a value between 100 and 1000 microfarads depending upon ripple and low current requirements. Switch S is an SPDT toggle switch and is used for selecting either low voltage or higher voltage. (Illustration courtesy of *Ham Radio*.)

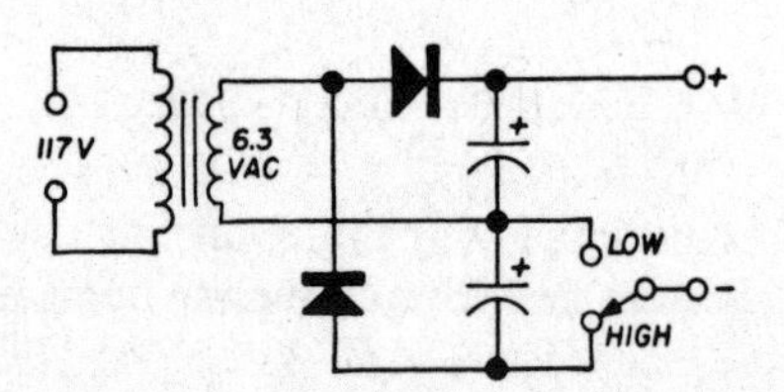

Figure 10-8. Dual voltage power supply.

Electron Tube Voltage Regulator

Figure 10-9 shows a circuit of a voltage regulator using electron tubes. The output voltage of the regulator is developed across R3, R4, and R5, in parallel with the resistance of the load. All of the load current must flow through the cathode-to-plate resistance of triode V1; this resistance is controlled by the other components in the circuit.

V3 is a gas-filled VR (voltage regulator) tube which holds the cathode of pentode V2 at a constant positive potential with respect to ground. R4 is set so that the control grid of V3 is biased so that the tube passes a certain amount of plate current. The plate current of V2, flowing through R1, determines the bias on the grid of V1 which, in turn, affects plate-to-cathode resistance of V1.

If the load voltage tends to rise because of an increase in supply voltage or a decrease in load current, the voltage between the wiper arm of R4 and ground will also increase. The difference in this voltage and the fixed voltage across V3 decreases, thus decreasing V2 grid bias. This increases the flow of current through V2, increases the IR drop across

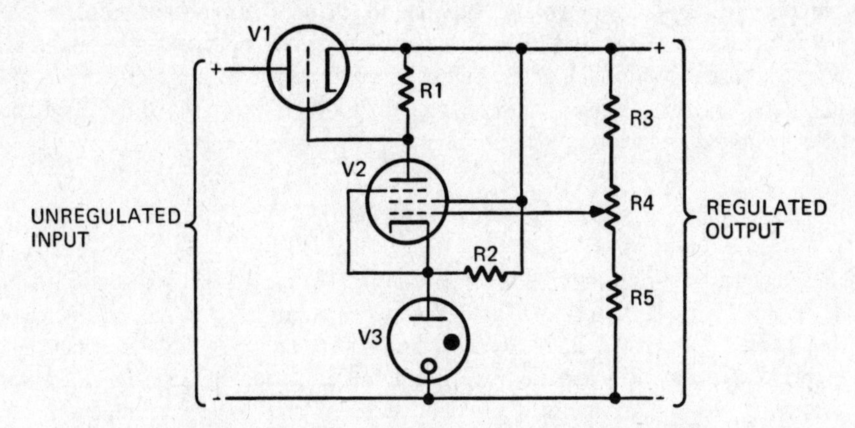

Figure 10-9. Electron tube voltage regulator circuit.

R1, and increases the negative grid bias on V1. This increase in V1 bias causes the cathode-to-plate resistance of V1 to rise, increasing the voltage drop across V1, and effectively keeping the output voltage constant. A drop in supply voltage or an increase in load current will have the opposite effect.

The anode of the VR tube, V3, is connected to the positive terminal of the regulated output voltage through resistor R2 in order to cause the gas in the tube to ionize when the power supply is first turned on. Since all of the load current must pass through V1, this tube must have a heavy current-carrying capability. In many cases, several identical tubes are connected in parallel.

Five-Volt Regulated Power Supply

The power supply whose circuit is shown in Figure 10-10 is suitable for digital equipment applications where a low-noise, well-regulated, 5-volt DC source is required. The MOV across the transformer primary is a metal-oxide varistor which provides tran-

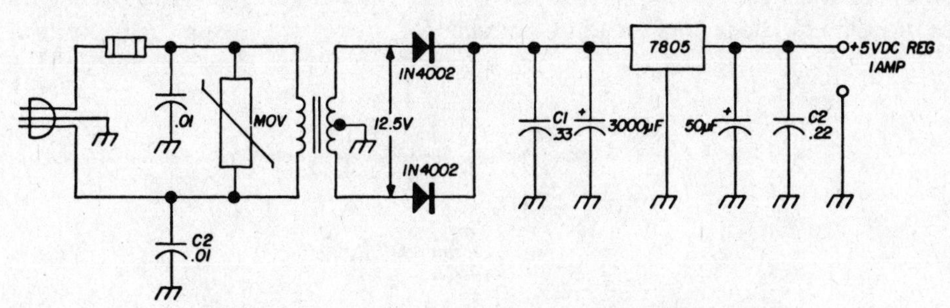

Figure 10-10. Five-volt regulated power supply circuit.

sient protection. The IC in the output line is a commercially available voltage regulator, which reduces and regulates the voltage supplied to it from the full-wave rectifier. Note that shunted across the 50 uf and 3000 uf electrolytic filter capacitors are lower value non-electrolytic capacitors which shunt high-frequency transients more effectively than the electrolytic capacitors which are more effective at low frequencies. (Illustration courtesy of *Ham Radio*.)

Full-Wave Rectifier

The full-wave rectifier circuit shown in Figure 10-11 employs two semiconductor diodes (CR1 and CR2). During the AC half-cycle when point X is positive, no current can flow from point Z through CR2, but current flows from point Y, the center tap of the transformer, load resistor RL, and diode CR1 to point X. During the opposite half-cycles, when point Z is positive and point X is negative, CR1 won't pass current, but current flows from point Y through R1, and diode CR2 to point Z. Thus, the voltage across load RL will be a series of pulses, one pulse for each positive half-cycle and one pulse for each negative half-cycle of input voltage. The voltage at the end of RL connected to the junction of CR1 and CR2 will be positive with respect to its other end. The ripple frequency is twice the AC supply voltage frequency.

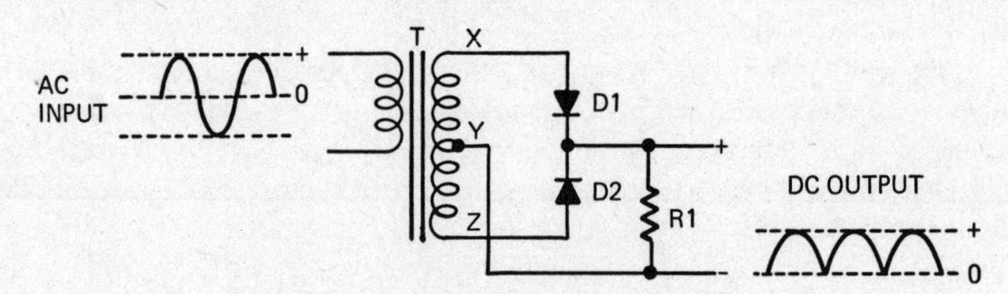

Figure 10-11. Full-wave rectifier circuit.

Full-Wave Voltage Doubler

Figure 10-12 shows a full-wave voltage doubler rectifier circuit. During the AC half-cycles when point X at the secondary of transformer T is positive, capacitor C1 charged through diode CR1. During the other half-cycle, when point Y is positive,

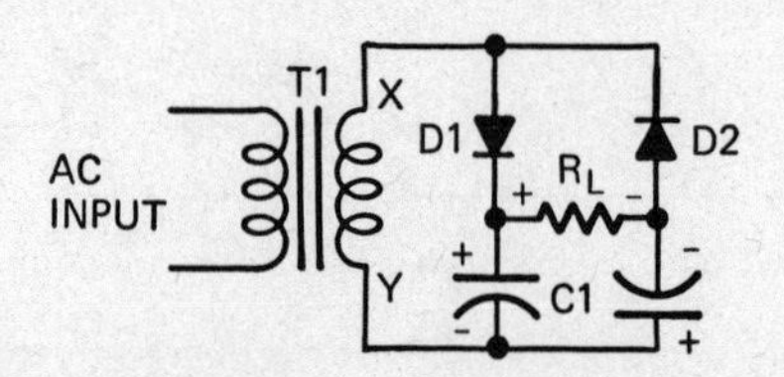

Figure 10-12. Full-wave voltage doubler rectifier circuit.

capacitor C2 charges diode CR2. The voltage charges in C1 and C2 are in series-aiding and the voltage across load RL is twice as high as the charge in either C1 or C2. Since CR1 and CR2 alternately conduct once during each AC input cycle, they function as a full-wave rectifier.

Fuse Circuits

Figure 10-13 shows overload-protector fusing circuits. In A is shown how both a primary and a secondary fuse are used to protect electronic equipment, as well as the transformer. One fuse is connected in series with the transformer primary and the other in series with the center tap of the secondary (when the transformer feeds a full-wave rectifier.) If a bridge rectifier is used, fusing can be provided as shown in B. In case of excessive load current, the fuse should blow and thus protect the power supply and/or the load. (Illustrations courtesy of *Ham Radio*.)

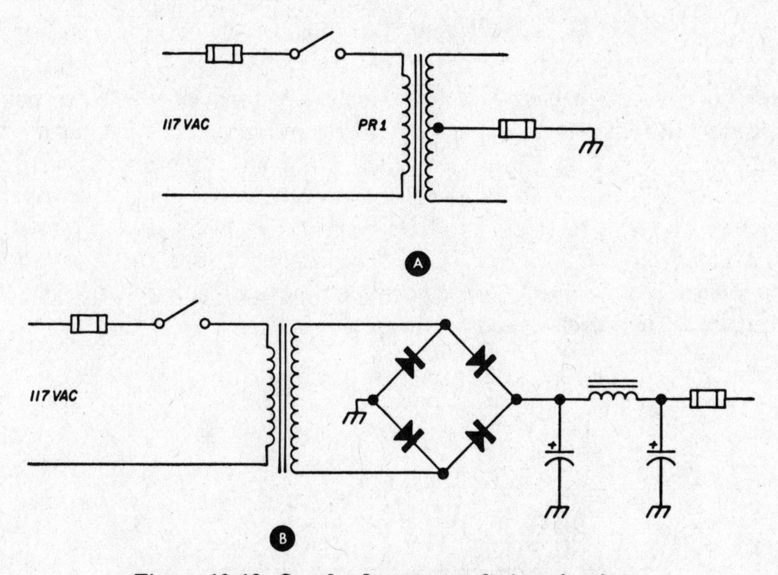

Figure 10-13. Overload-protector fusing circuits.

Half-Wave Rectifier

Figure 10-14 shows a half-wave rectifier circuit employing a single diode. AC line voltage is fed through transformer T and stepped up or down by the transformer to the desired level. During the AC half-cycles when point X is positive with respect to point Y, current will flow through load resistor R, and diode CR (from cathode to anode). However, when point X is negative with respect to point Y, during the opposite AC half-cycles, no current will flow because the diode will be reverse biased. The output of a half-wave rectifier consists of a series of DC pulses, occurring once during each cycle,

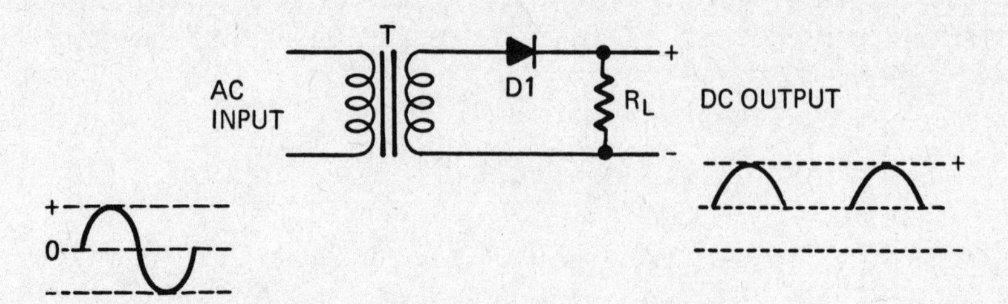

Figure 10-14. Half-wave rectifier circuit.

spaced by periods of zero voltage. Output voltage can be obtained at all times by connecting a high value capacitor across R.

Half-Wave Voltage Doubler

Figure 10-15 shows half-wave voltage doubler rectifier circuit. When point Y is negative, current flow is from Y, through diode CR1 and capacitor C1, to point X and C1 is charged. When the AC polarity is reversed and point X becomes negative, CR2 conducts and C1 adds its charge to the voltage at point X. The voltage across the load resistor (R) and the charge stored in C2 is equal to twice the voltage stored in C1. Capacitor C2 holds its charge when CR2 is nonconducting. However, C2 cannot hold its full charge if there is any appreciable load current. Since current flows through CR2 only during alternate AC half-cycles, the ripple frequency is the same as the AC supply voltage frequency.

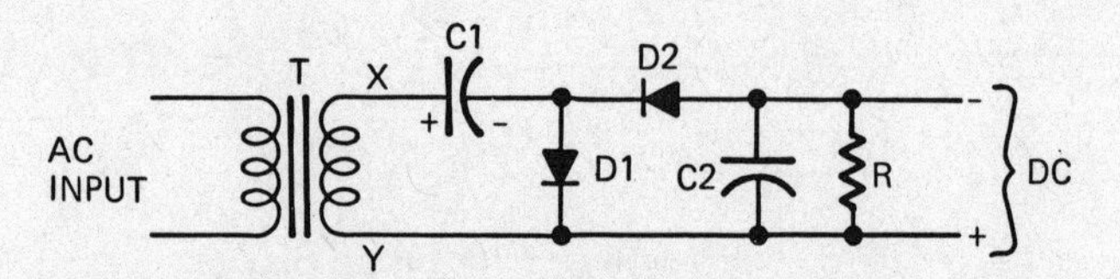

Figure 10-15. Half-wave voltage doubler rectifier circuit.

Inductive Filter

Figure 10-16 shows how an inductor is connected in series with the unfiltered voltage from a rectifier to provide ripple filtering. The inductor prevents the current from building up or dying down quickly, smoothing out the ripple that is present in the unfiltered DC. When only an inductor is used as the ripple filter, the DC output voltage is not as high as when a capacitive filter is used, partially because of the DC voltage drop across the inductor. An inductive ripple filter will tend to offset sudden load current variations.

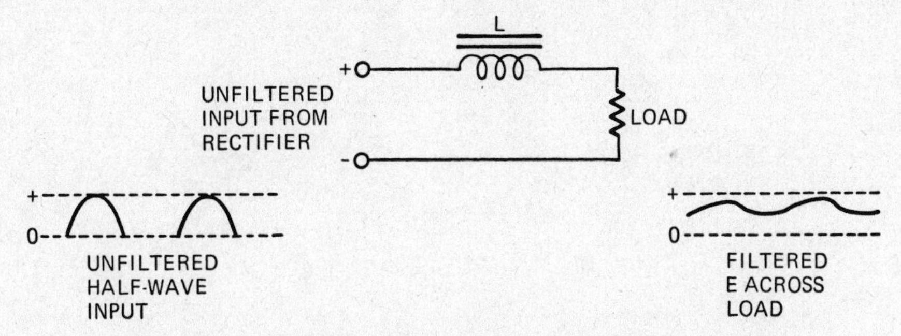

Figure 10-16. Inductor connected in series with the unfiltered voltage from a rectifier.

Line Voltage Adjuster

Powerline voltage can be raised or lowered with a step-down transformer (T) having a center-tapped secondary, when connected as shown in Figure 10-17. The voltage fed to the electric outlet receptacle J depends upon the setting of the three-position switch. When set to position Y (as shown), the voltage at J is the same as at the power plug. If T has a 10-volt center-tapped secondary and a 115-volt primary, the voltage at J can be raised or lowered by approximately 5 volts. Assuming that the line voltage is 115 volts, the voltage at J can be made 120 volts by setting the switch to position X or 110 volts by setting the switch to position Z. The 5 volts between the center tap of the transformer secondary and either end of the winding can be used to buck (series-opposing) or boost (series-aiding) the line voltage.

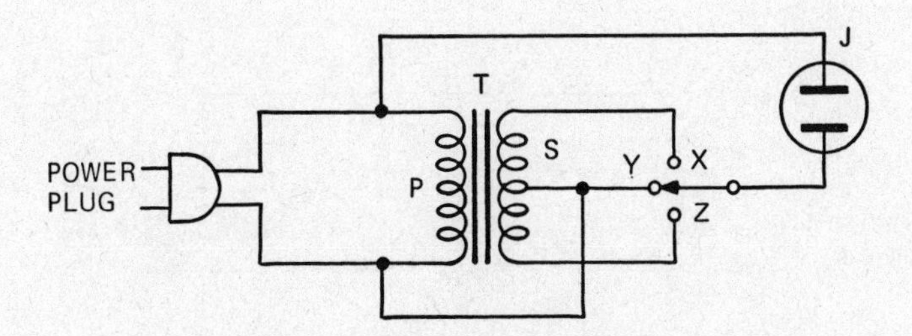

Figure 10-17. Line voltage adjuster.

L-Section Filter

The L-section filter circuit shown in Figure 10-18 derives its name from its resemblance to an inverted "L." It is also called a choke input filter. The input choke (inductor L) limits the current the rectifiers must pass when capacitor C is initially charging. Because of the current stabilizing effect of the inductor, the L-section filter is often used in

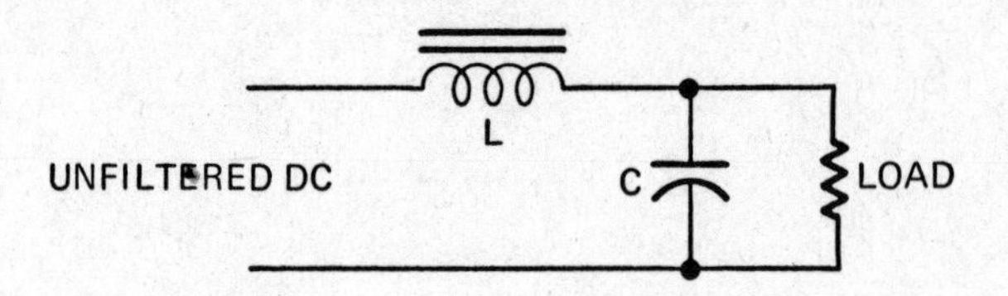

Figure 10-18. L-section filter circuit.

high-power applications where voltage regulation is important. Two L-section filters are often used in series to provide better ripple filtering.

Mercury Arc Rectifier

The rather large and cumbersome mercury arc rectifier tube is used where relatively high voltage and high current are necessary such as for furnishing DC to motion picture projector arc lamps and spotlights in theaters. An example of a typical circuit is shown in Figure 10-19. Transformer T1 supplies the AC which is to be full-wave rectified by the mercury arc tube. Transformer T2 supplies AC to a full-wave rectifier employing diodes CR1 and CR2. The DC is smoothed by reactor L. The DC voltage supplied by this rectifier is used to heat the pool of mercury at the bottom of the tube so that it will vaporize and cause the tube to provide a low resistance path between the plates and the pool of mercury. As can be seen in the diagram, the positive output is taken from the pool of mercury which acts as a cathode. The negative output is taken from the center tap of transformer T1.

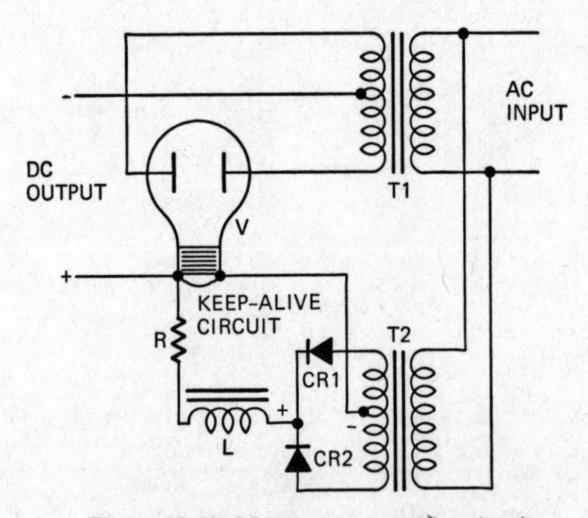

Figure 10-19. Mercury arc rectifier circuit.

Multivoltage Power Supply

Figure 10-20 shows the circuit of a power supply which provides AC voltage and several different DC voltages as required by a complex device such as a transmitter. The 900-volt DC section of the power supply employs a full-wave rectifier using series-connected diodes, each shunted by a resistor and a capacitor to equally divide the peak inverse voltage across them. Ripple filtering is provided by a choke-input filter. Four 100,000-ohm resistors, connected in series across the high-voltage output, form a network which equalizes the charge voltages across the series-connected filter capacitor network. This permits the use of capacitors rated at 450 volts DC.

Another full-wave rectifier, followed by a pi-network filter, provides three DC voltages, 350, 210, and 105 volts. The 210-volt output is regulated by a pair of OB2 (105-volt) VR (voltage regulator) tubes connected in series. The 105-volt output is also regulated by the VR tubes.

A half-wave rectifier, fed from the same transformer winding, provides two grid bias voltages, regulated 18 volts DC and another DC voltage that can be varied from -40 to -70 volts DC by means of a potentiometer. One winding of the transformer supplies 6.3 volts AC for a pilot light and tube heaters voltage. Two other windings (5 volts and 6.3 volts) are connected in series to furnish 11.3 volts to a half-wave rectifier which supplies 12 volts DC. (Illustration courtesy of *Ham Radio*.)

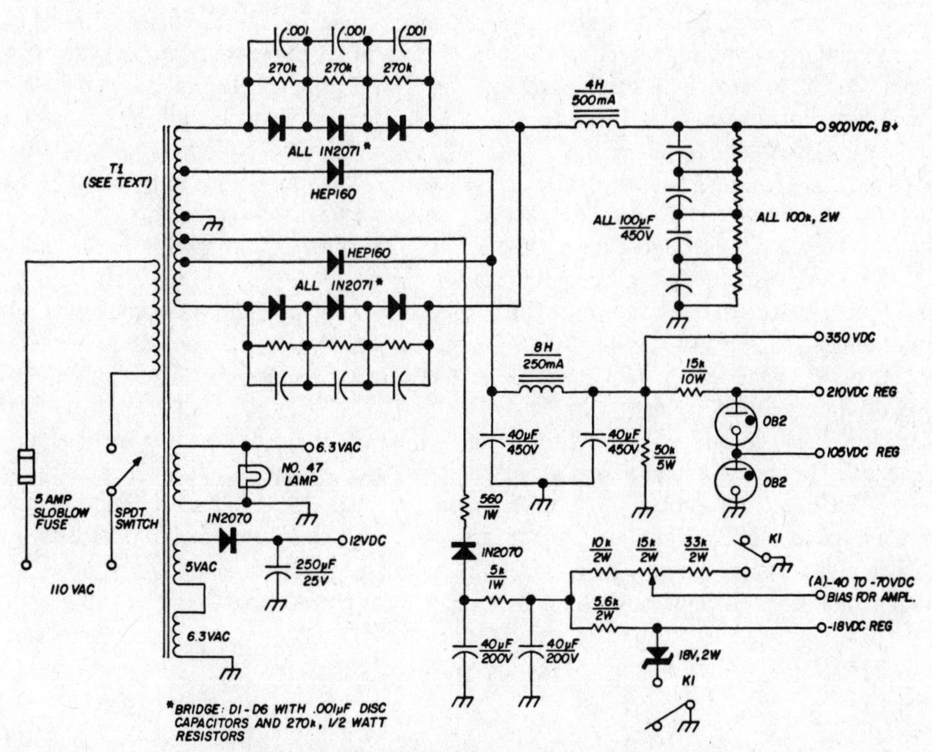

Figure 10-20. Multivoltage power supply circuit.

Non-Synchronous Vibrator Power Supply

Figure 10-21 is a schematic diagram of a power supply employing a non-synchronous vibrator to convert low-voltage DC (from say, a car battery) to a relatively high AC voltage which is subsequently rectified to provide relatively high DC voltage. In this circuit, G101 represents the electromechanical vibrator which contains an electromagnet, a vibrating contact, and three stationary contacts. The vibrating contact is connected to terminal 1 of the plug-in vibrator, two of the stationary contacts are connected to terminals 2 and 3, and one end of the electromagnet's coil to terminal 4. The other stationary contact is connected internally to the other end of the electromagnet's coil. It is this contact that causes the vibrator action. In its quiescent state, this contact touches the vibrating contact. When current flows through the coil, the vibrating contact is pulled away from the stationary contact. This disconnects the vibrator coil. Then the vibrating contact swings back and again mates with the stationary contact and the vibrator coil exerts a magnetic pull on the vibrating contact, and again the contacts open, cutting off current flow through the coil. This action reaches a constant speed of about 150-250 times per second.

At the same time, the vibrating contact alternately mates with stationary contacts 2 and 3. When the vibrating contact touches contact 2, current flows from ground (negative side of battery), through the vibrating contact, contact 2, and the 5-6 section of the primary of the power transformer to the positive end of the battery. When the vibrating contact touches contact 3, current flows through the 6-7 section of the transformer primary. As the vibrating contact alternately touches contacts 2 and 3, current alternately flows through in opposing directions through sections of the transformer primary, simulating alternating current. This causes a relatively high AC voltage to be induced into the 3-4 secondary winding.

This AC voltage is buffered by C168 and fed to a full-wave, voltage doubler rectifier system (CR101, CR102, C144, and C171A) whose DC output is filtered by an RC filter (R145, C171B) which reduces the ripple. The highest DC voltage is obtained at the junction of R145 and R146. Lower DC voltage is obtained at the junction of R146 and C171C which also function as a ripple filter.

This particular power supply circuit is used in a two-way radio transceiver operable from either a 12-volt DC or 115-volt AC source. When operated from an AC source, winding 1-2 of the transformer functions as the primary and winding 7-8 supplies 12.6 volts AC to the tube heaters.

It was the invention of the vibrator by William P. Lear that made the auto radio practical. Before the vibrator was developed, auto radios required bulky B batteries to supply DC plate and screen voltages. The vibrator made it possible to step up the auto battery voltage to 200 volts or more. In tube-type auto radios and mobile transceivers, switching transistors are now more widely used than the vibrator. And in solid state auto radios, neither is required since high DC voltages are not required.

OP-AMP Regulator

Figure 10-22 shows a circuit employing a commercially-available operational amplifier in a voltage regulator circuit. The non-inverting (+) input of the op-amp is held at a

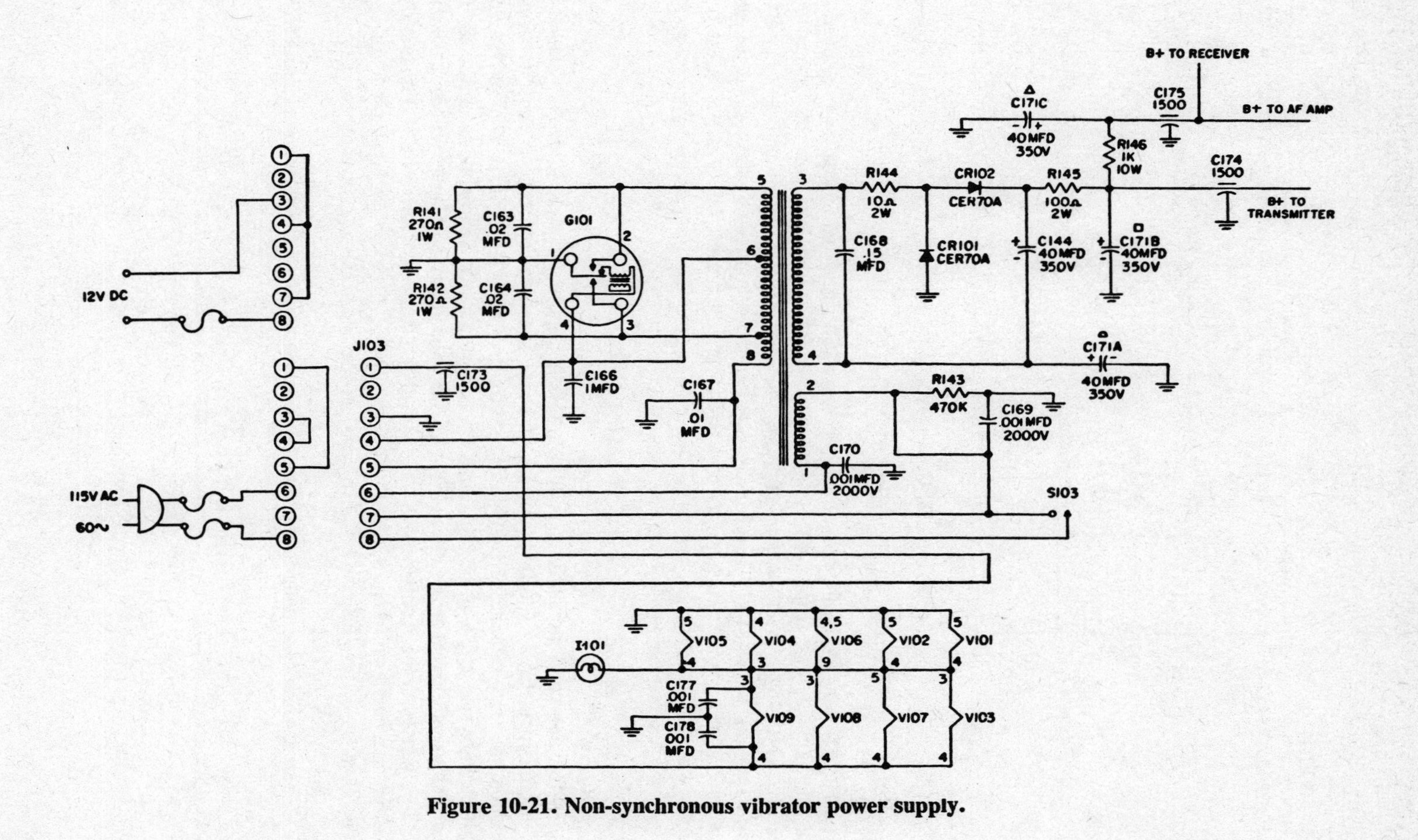

Figure 10-21. Non-synchronous vibrator power supply.

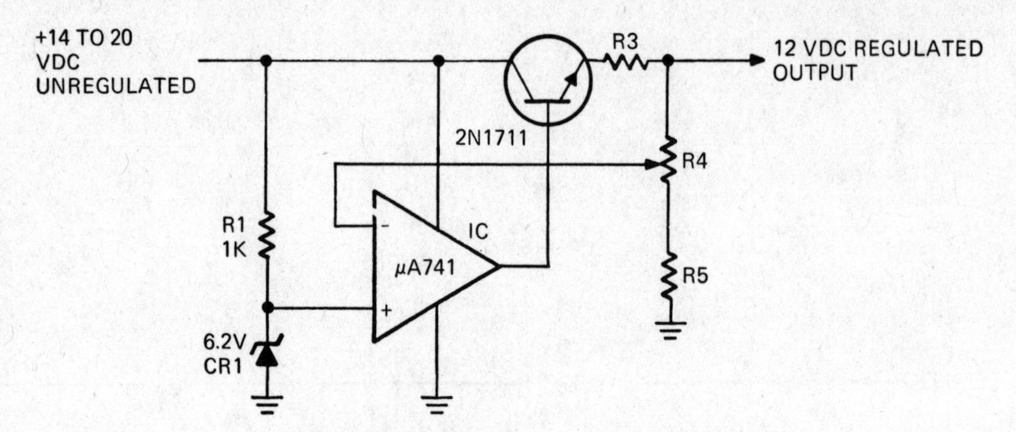

Figure 10-22. Op-amp regulator circuit.

constant 6.2 volts DC by the zener diode CR1, while the inverted (-) input is fed a portion of the output voltage tapped off by potentiometer R4. Error voltage is reflected in the feedback loop to the inverting input of the IC, while the output controls the base current of the series pass transistor Q1. Since the resistance of base-emitter path through the transistor increases as input voltage rises or load current decreases, and vice versa, output voltage is held constant.

Pi-Section Filter

The pi-section ripple filter circuit shown in Figure 10-23 is sometimes called a capacitor input filter. The filter derives its name from the fact that the capacitors and inductor in the schematic diagram seem to form the Greek letter, π, or pi. This type of filter can reduce ripple so that the output waveform closely approximates a pure direct current waveform. C1 acts to bypass the greatest portion of the ripple component of the DC voltage from the rectifier to ground. The choke coil (inductor L) helps maintain the current at a nearly constant level during the charge and discharge cycles of capacitor C1. The final capacitor, C2, serves to bypass any residual fluctuations that remain after filtering by C1 and L1.

The capacitors provide low reactance shunt paths and the inductor provides a high

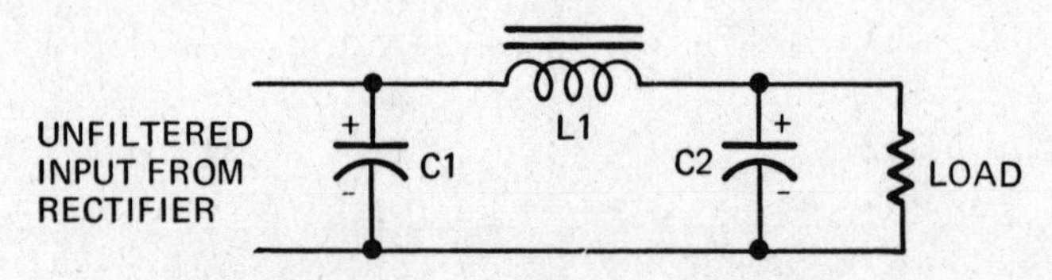

Figure 10-23. Pi-section ripple filter circuit.

reactance series path at the ripple frequency. After charging, the capacitors do not shunt the DC voltage and the DC voltage drop across the inductor is very low because of the low resistance of its winding. Actually, this circuit is of a low-pass filter, passing DC with little opposition but offering great opposition to passage of the ripple.

Powerline Filters

A powerline filter at the primary of a power transformer helps prevent interception of interference from the powerline and feeding of signals into the powerline. A very commonly used powerline filter consists of two capacitors (usually 0.1-0.5 mf) connected in series across the powerline, as shown in Figure 10-24A. The junction of C1 and C2 is grounded to the chassis. Simpler, but not always as effective, is a filter consisting of a single capacitor connected across the powerline, as shown in Figure 10-24B. Both types provide some immunity from short-duration powerline transients which they bypass. More complex power line filters use series-connected RF choke coils in addition to shunt-connected capacitors.

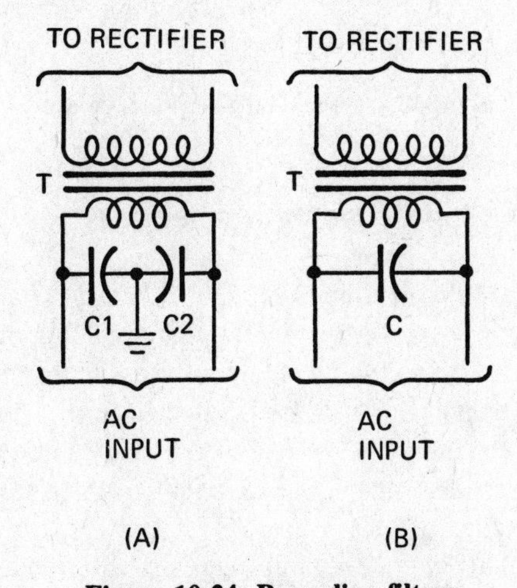

Figure 10-24. Powerline filters.

Precision Voltage Regulator

Figure 10-25 shows a schematic diagram for a precision voltage regulator, the Fairchild MA723, all components of which are contained within a single IC chip. The zener diode D1 changes with temperature to compensate the reference amplifier for changes in the rest of the circuitry. Pin 4 is the reference voltage connection, and typically is coupled through a resistor to the non-inverting input of the second op-amp, the error amplifier. The inverting input of the error amplifier and the current sense terminal, pin 1, are connected externally to the load to sense and respond to variations in current or voltage.

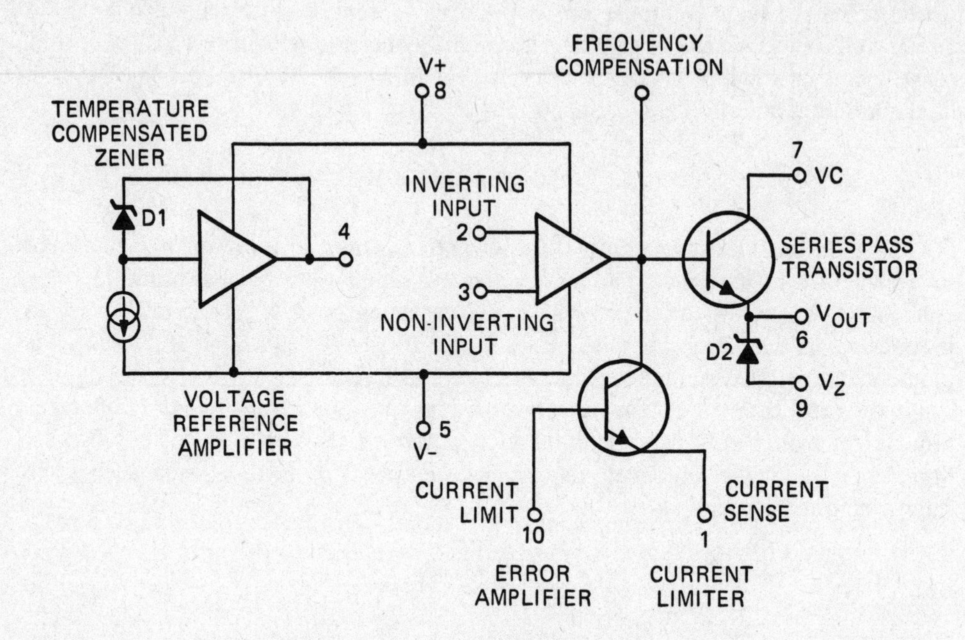

Figure 10-25. Precision voltage regulator.

Series-Type Voltage Regulator

In this voltage regulator circuit, transistors Q3, Q5, and Q7 are connected in parallel and act as a variable resistance between the unregulated DC input and the regulated DC output. As shown in Figure 10-26, the collectors of these three transistors are connected together as are their bases. Their emitters are connected in parallel through R2, R4, and R6, 0.1-ohm resistors which cause the transistors to equally share the load current. Zener diode CR provides the reference voltage. As the output voltage tends to rise, the increase in voltage is distributed across R8, R9, and R10. This tendency to change voltage is sensed by the base of Q6. This voltage change is then transferred to Q4. Since Q4 and Q6 share R5 as the emitter resistor, any change in voltage causes a change in the base-emitter voltage of Q4. As shown, the zener diode and R3 are connected in series. The voltage drop across zener diode remains constant, but any variation in voltage across R3 changes the bias on Q1. A change in current through Q1 causes a change in the current through Q2. The conduction of Q3, Q5, and Q7 is controlled by the voltage fed to their bases from the emitter of Q2.

A rise in input voltage or a decrease in load current decreases the conduction of the three parallel connected transistors. Conversely, a decrease in input voltage or an increase in load current increases their conduction. Essentially, these three transistors act as an automatically controlled variable resistor which maintains output voltage constant. (Illustration courtesy of RCA Corporation.)

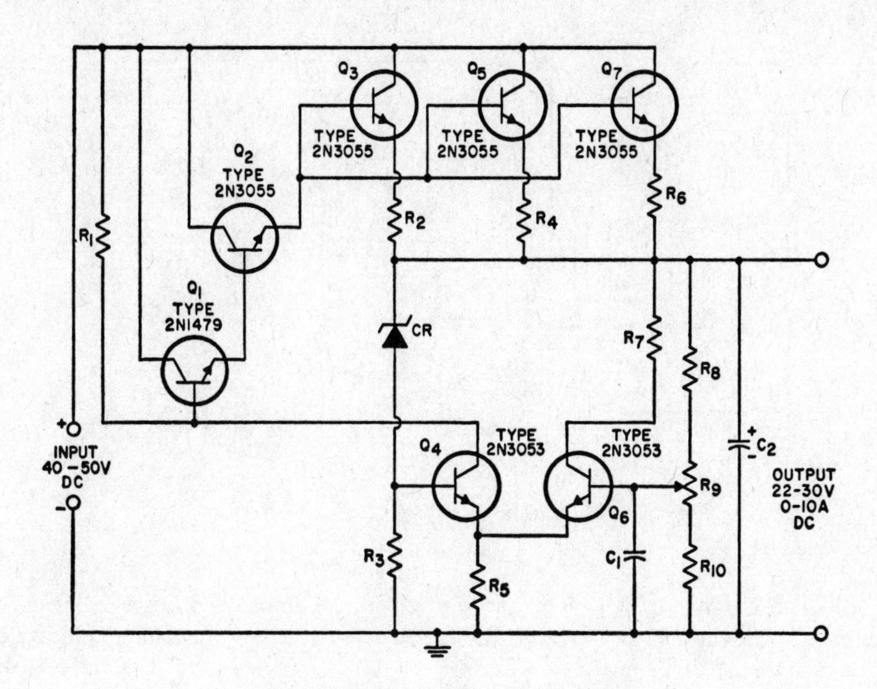

Figure 10-26. Voltage regulator circuit.

Stabilized Voltage Regulator

Two dual triodes and a VR (voltage regulator) tube are used in the stabilized voltage regulator circuit shown in Figure 10-27. Unregulated DC voltage from a rectifier (not shown) is fed through R1 to the plates of the type 6080 dual triode tube whose grids and cathodes are connected together. The plates are also essentially connected in parallel through R1 which is used to adjust plate current balance. The two triode sections of the type 6080 tube are in series between the input voltage and the regulated output voltage. The conduction of this tube is controlled by the type 5751 dual triode. The type 5651A VR tube supplies a fixed reference voltage to one of the grids (socket terminal 7) of the 5751 tube. Variations in input voltage are amplified by the 5751 tube which is utilized as a two-stage DC amplifier. As input voltage rises, the voltage at one of the plates of the 5751 tube (socket terminal 1) becomes less positive and reduces the conduction of the 6080 tube. Conversely, a decrease in input voltage causes the 5751 tube to increase the conduction of the 6080 tube. This regulator circuit also compensates for changes in load current. By means of potentiometer R9, the output voltage can be varied from 0 to 250 volts. Load current may be as high as 225 milliamperes provided the rectifier can furnish that much current.

When AC line voltage remains constant, the DC output voltage of the regulator, when R9 is set so that it delivers 250 volts, will vary less than 0.2 volt at any load current

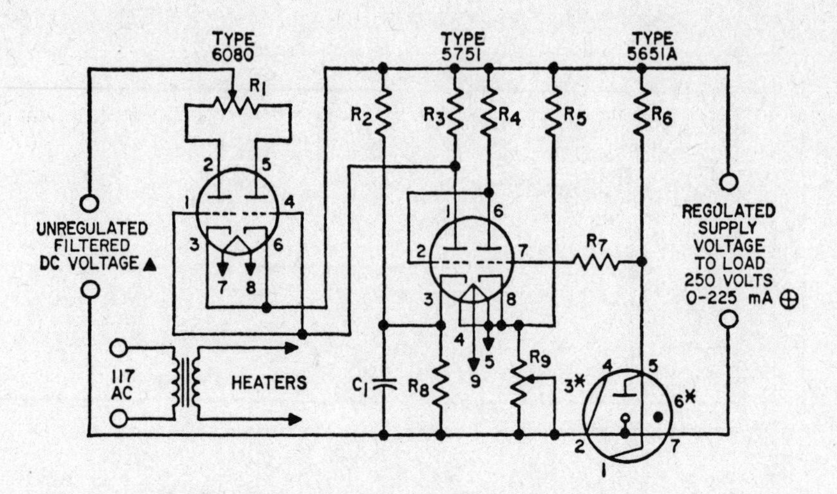

Figure 10-27. Stabilized voltage regulator circuit.

between 0 and 225 milliamperes. Only a 0.1 volt variation in output voltage will result at full current when line voltage varies as much as +10 percent. (Illustration courtesy of RCA Corporation.)

Shunt Transistor Voltage Regulator

Figure 10-28 shows the circuit of a shunt type regulator which is used to provide a relatively constant voltage where the load current is relatively constant. If the input voltage increases, the voltage drop across R2 increases, since the voltage drop across zener diode CR1 remains constant. Increasing the voltage drop across R2 increases the forward bias on transistor Q, causing it to conduct more heavily. The increase in collector current increases the voltage drop across resistor R1, keeping the output voltage constant. Conversely, if the input voltage decreases, the Q forward bias decreases. This decreases the Q collector current, decreasing the voltage drop across R1 and holding the output voltage constant.

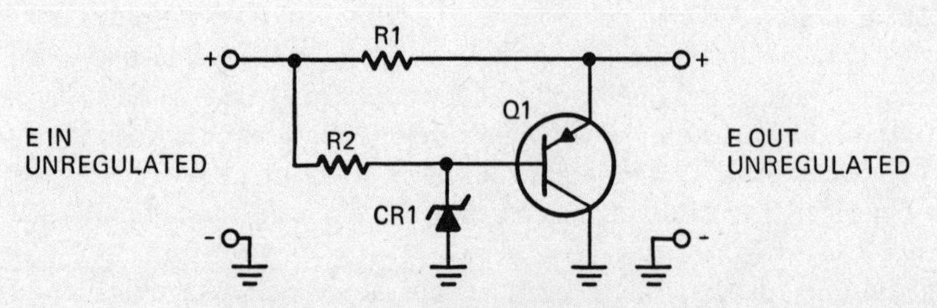

Figure 10-28. Shunt type regulator circuit.

Synchronous Vibrator Power Supply

Figure 10-29 shows the circuit of a DC-to-DC power supply employing a synchronous vibrator. A separate rectifier for converting AC into DC is not required. When switch S is closed, current from battery B flows through the vibrator coil L and the top half of the primary winding of transformer T1. The magnetic field of the vibrator coil causes the vibrating arm to swing up and mate with the two upper contacts. Mating of these contacts shorts the vibrator coil, releasing the vibrating arm to spring downward and mate with the bottom pair of contacts. This allows battery current through the bottom half of the primary of the transformer.

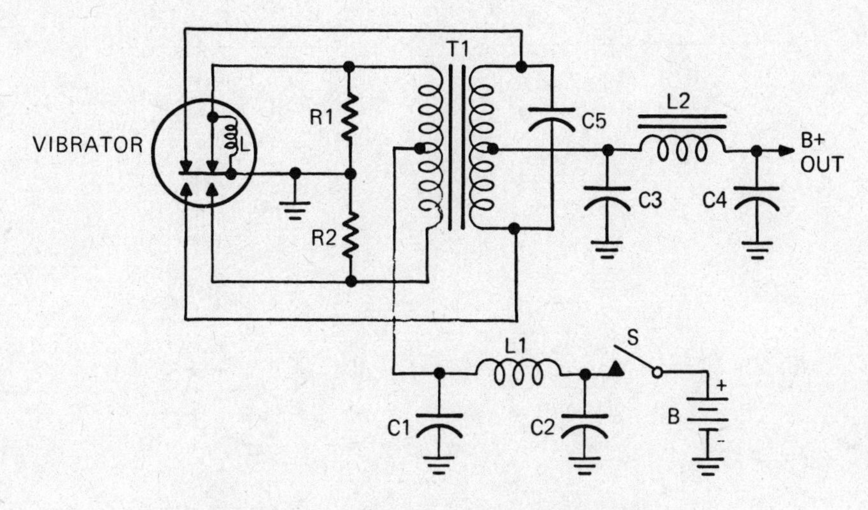

Figure 10-29. DC-to-DC power supply circuit.

As the vibrating arm oscillates back and forth between the top and bottom sets of contacts, typically at about 135 Hz, an alternating current is set up in the primary and the secondary of the transformer. However, between the center tap of the secondary and ground, only pulsating DC is present because of the action of the vibrating arm. This pulsating DC voltage is filtered by the pi-section filter C3, L2, and C4, and delivered to the load as an almost ripple-free DC voltage. Buffer capacitor C5 bypasses inductive kick spikes. Its capacitance must be carefully matched to the characteristics of the transformer and vibrator.

Three-Phase Rectification

Many heavy-duty electronic devices are powered by a three-phase AC power source instead of a single-phase AC power source. When the power source is three-phase AC, DC can be obtained by half-wave rectifying each phase in sequence or by taking advan-

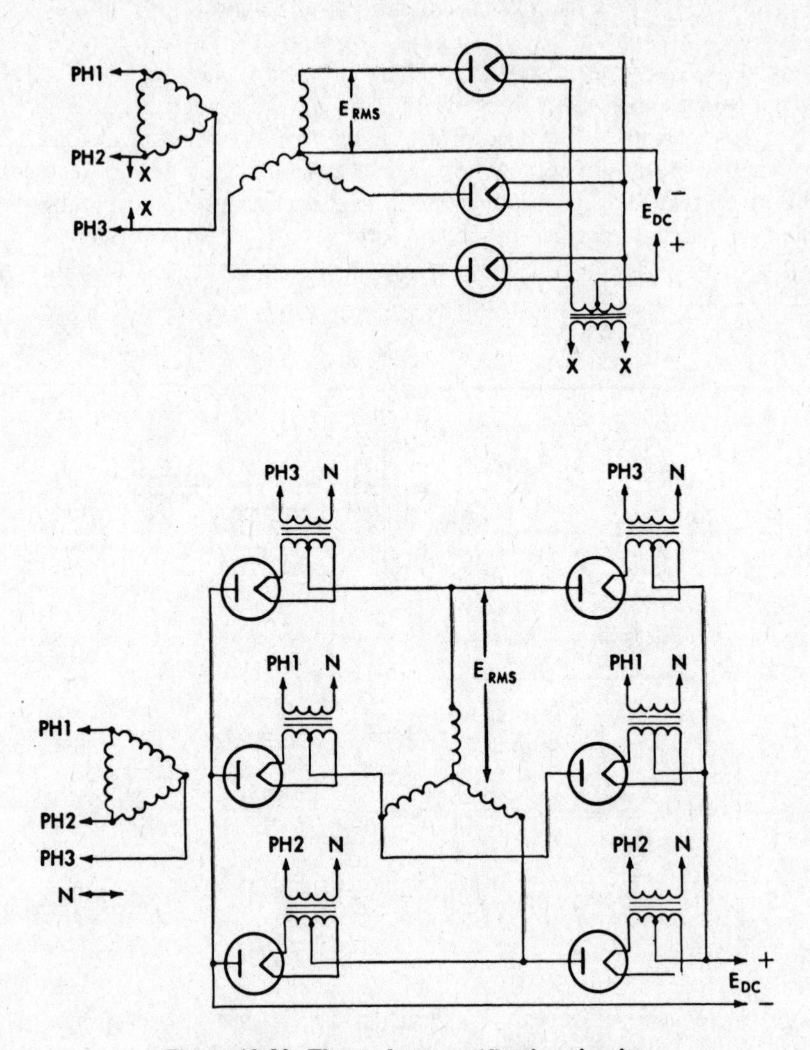

Figure 10-30. Three-phase rectification circuits.

tage of the three available phases and utilizing full-wave rectification, rectifier life can be increased and the percentage of ripple in the DC output can be significantly reduced. Both types of circuits are shown in Figure 10-30.

The circuit of a half-wave, three-phase rectification system using a delta-wye transformer and three thermionic diode tubes is shown in A. Because of the utilization of the three available phases, the ripple frequency is three times the AC supply voltage frequency. Without filtration, the ripple level of the DC output voltage is only 18 percent RMS, compared to 47 percent for a single-phase bridge or full-wave rectifier and 109 percent for a single-phase half-wave rectifier.

When a delta-wye transformer and six rectifiers are used in a three-phase, full-wave rectifier circuit, as shown in B, an RMS ripple is only 4 percent of the DC output voltage and ripple frequency is six times the AC supply voltage frequency. By adding a ripple filter, ripple can be reduced to an insignificant value.

Transistor Rectifier

Figure 10-31 shows how a power transistor may be connected for use as a rectifier. The rectifying diode is formed by the collector-base junction; the emitter is not used. Power transistors may be used as half-wave rectifiers (as shown), or in any of the other rectifier configurations. The collector serves as the anode of the rectifier, and the base serves as the cathode.

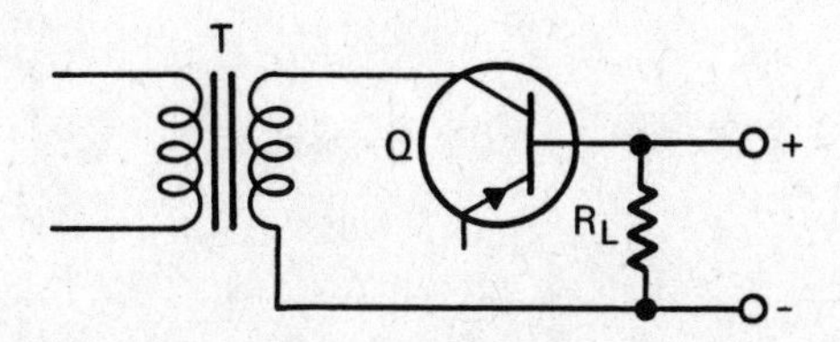

Figure 10-31. Power transistor connected for use as a rectifier.

Tungar Rectifier

The Tungar bulb identified as V in Figure 10-32 is a filament type diode that was widely used in battery chargers. This type of diode is particularly suitable for high current low-voltage application. As can be seen from the diagram, the filament is energized from a tap on the secondary of transformer T. In the configuration shown, the positive pulsating DC output is from the transformer secondary and the negative lead is connected to the anode of the Tungar bulb.

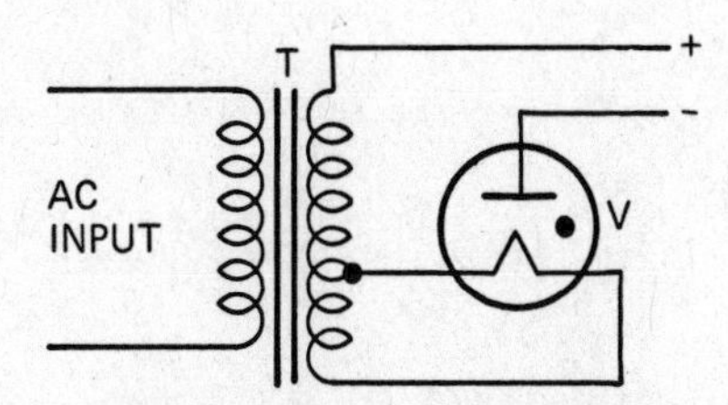

Figure 10-32. Tungar rectifier.

Varistor Transient Suppressors

Figure 10-33 shows two transient suppression circuits employing varistors. A shows the use of a varistor across the power transformer primary to control high-peak transient

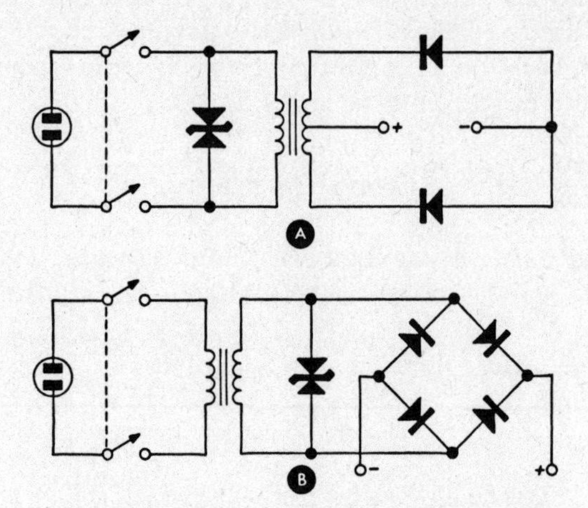

Figure 10-33. Transient suppression circuits.

surges that may occur. The varistor offers a very high resistance at up to 2-½ times the RMS voltage of the input, but at voltages above that, it breaks down and presents a very low resistance, thus bypassing the transformer primary when high-voltage transients are present. B shows the connection of a varistor transient suppressor across the secondary of a transformer. (Illustrations courtesy of *Ham Radio*.)

VR Tube Regulator

A VR tube is a gas-filled diode with a cold cathode. It conducts only when its plate is made sufficiently positive with respect to its cathode. An example of a voltage regulator circuit employing a VR tube is shown in Figure 10-34. The unregulated DC voltage (generally from a rectifier) is fed through a series resistor to the load. Shunted across the

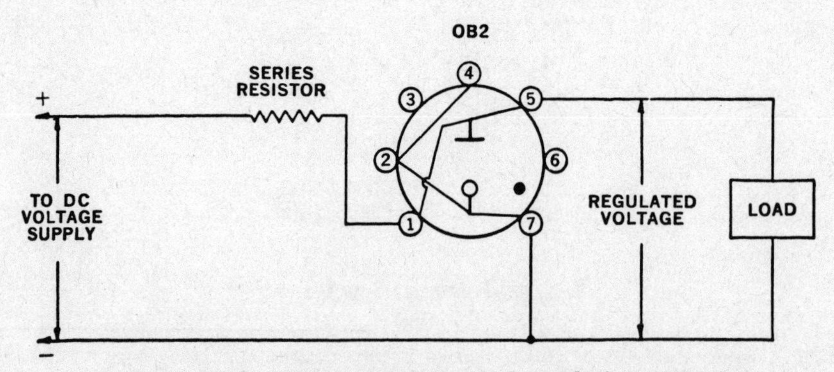

Figure 10-34. Voltage regulator circuit employing a VR tube.

load is the plate-cathode path of the VR tube. VR tubes are available in various voltage ratings. The type OB2 has a maintaining voltage rating of approximately 108 volts. Maximum current flow through the OB2 tube should not exceed 30 milliamperes. Assume that the current drawn by the load is normally 15 milliamperes, and that the value of the series resistor is such that 15 milliamperes of current will flow through the load and 15 milliamperes will flow through the VR tube. When the supply voltage increases or load current decreases, more current will flow through the VR tube. Conversely, when the input voltage falls or load current increases, less current flows through the VR tube. Therefore, the VR tube acts as an electronic resistance shunted across the load.

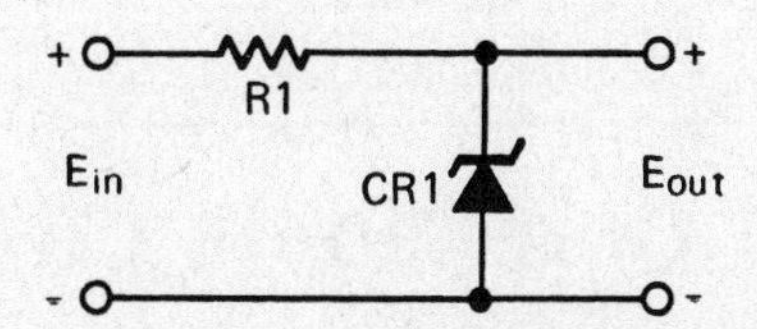

Figure 10-35. Zener diode for voltage regulation.

Zener Diode Voltage Regulation

Figure 10-35 shows a circuit using a zener diode for voltage regulation. The zener diode CR is designed to have a specific breakdown voltage and to operate over a small voltage range. The selected zener diode should have a breakdown voltage equal to the required regulated output voltage. The voltage drop across limiting resistor R must be equal to the voltage difference between E in and E out. It must pass both in the zener diode current and the load current. When the input voltage rises or the load current decreases, current through the diode and the voltage drop across R1 both increase, so that output voltage will remain constant. If the input voltage decreases or the load current increases, the zener diode current decreases, and thus the voltage drop across R1 also decreases to maintain output voltage constant.

11

Signal Conditioning Circuits

The term ''signal conditioning'' came into popular use with the dawn of the Space Age. Signal conditioning can cover many things. For example, the transducers of the prototype of the Saturn missile generated analog signals which were ''conditioned'' and then converted into digital signals which were required for entry into a digital computer. Signal conditioning generally means changing the level or the shape of a particular waveform. Included in this chapter are rudimentary signal conditioning circuits.

Amplitude Limiter

In FM receivers and in other circuits where the amplitude of a signal is to be prevented from rising above a specified level, an amplitude limiter is often used. Figure 11-1A is a schematic diagram of a simple amplitude limiter employing two diodes. If these diodes have a barrier voltage of 0.2 volt, the amplitude of the voltage across X and Y cannot exceed 0.2 volt. When the input voltage is less than 0.2 volt, the diodes do not conduct and have no effect on the signal. If the input voltage rises to 1 volt, for example, the diodes will conduct. During one half-cycle of the signal, CR1 is forward-biased and conducts, and CR2 is reverse-biased and does not conduct. During the other half-cycle, CR1 does not conduct and CR2 conducts. The voltage drop across each diode, when conducting, cannot exceed 0.2 volt. When the diodes are conducting, the input voltage will be equal to the output voltage plus the voltage drop across series resistance R.

Another type of amplitude limiter circuit is shown in Figure 11-1B. It employs two transistors, Q1 connected as an emitter-follower and Q2 as a common-base amplifier. When the input signal increases the forward bias on Q1, the forward bias on Q2 is reduced and the output voltage rises. When the signal reduces the forward bias on Q1, Q2 forward

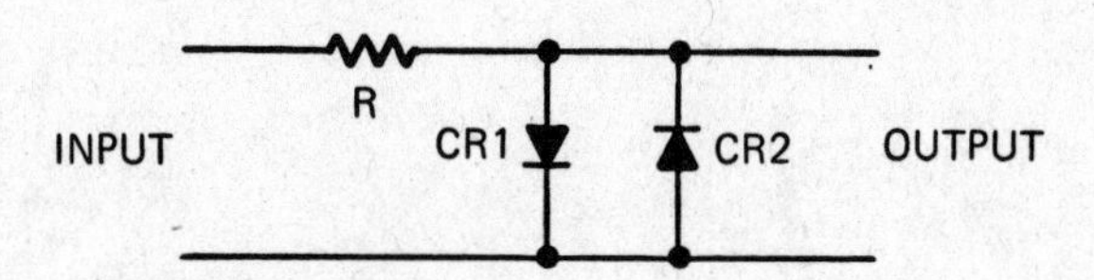

Figure 11-1A. Simple amplitude limiter circuit employing two diodes.

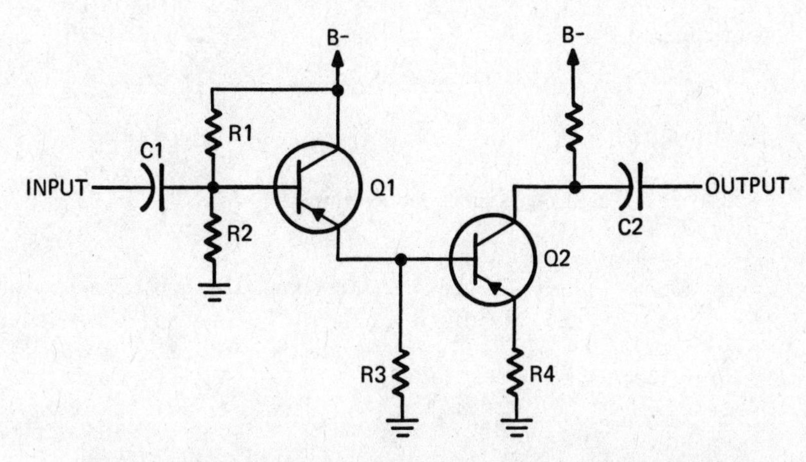

Figure 11-1B. Amplitude limiter circuit employing two transistors.

bias is increased and the output voltage falls. However, as the input signal excursions exceed a certain level, Q1 is cut off when the input signal reduces its forward bias below the transistor conduction level. During the other half-cycles of the input signal, Q1 conducts, but Q2 is cut off when its forward bias is reduced below the transistor conduction point. Thus, this circuit acts as an amplifier when the signal level is such that both Q1 and Q2 can conduct. When the input signal level exceeds a certain value, both transistors are alternately cut off. The level of the output signal cannot exceed the supply voltage level.

Clamping Circuit

Figures 11-2A and B show a simple diode clamping circuit that clamps both output levels of the transistor Q. This is the direct method. The collector level is fixed to ground and -VL (load voltage) is less than the supply voltage -VC. When the transistor is on as in Figure 11-2A, conducting collector current is high, but is limited by resistor R3 which also passes the current flowing through diode CR1. The clamping current is chosen by considering expected loading. Clamping reduces the effects of transistor variations and

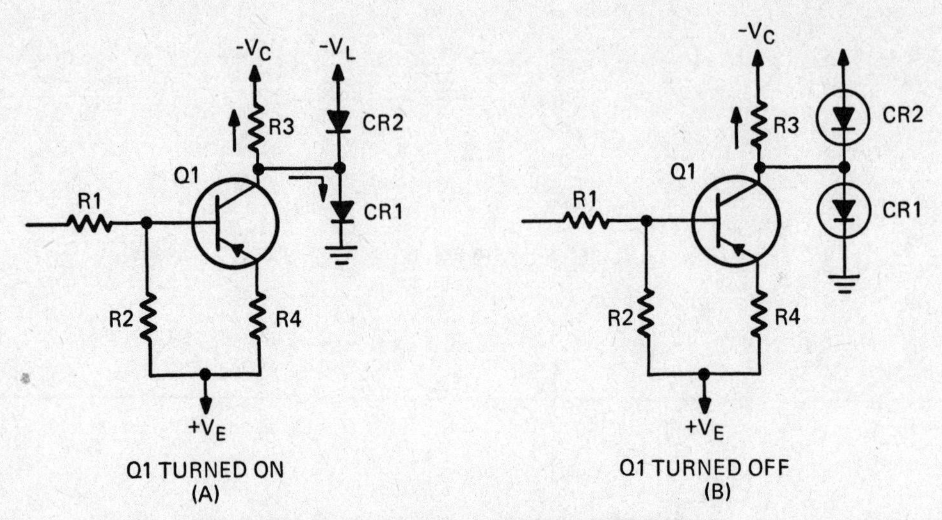

Figure 11-2. Simple diode clamping circuits.

ensures nonsaturation. Clamping increases circuit complexity and requires additional supply voltages. The "off" case is shown in Figure 11-2B. Clamping is used to improve response time.

Diode Noise Limiter

Figure 11-3 shows a noise limiter circuit employing a diode. AM signals are demodulated by diode CR1 and the C1-R1 network. The recovered audio signal is through C2 to the noise limiter. Normally, the anode of diode CR2 is positive with respect to the cathode, and the diode is forward-biased, allowing audio signals to pass through unchecked. Reception of a noise pulse develops a large negative pulse to be coupled through C2 to the anode of CR2. When this negative pulse exceeds the value permitted by R4, it cuts off CR2, preventing the noise pulse from being transmitted through it to the audio amplifier. When the noise pulse disappears, CR2 again is forward-biased and passes the audio signals.

Double Shunt Diode Clipper

Connecting a battery (or other DC voltage source) in series with a clipping diode sets the clipping level above or below the forward-voltage drop characteristic of the diode depending upon the battery voltage and polarity. In the circuit shown in Figure 11-4 diodes CR1 and CR2 conduct when anode voltage is positive with respect to cathode voltage. In this circuit, CR1 and CR2 are connected complementarily—one conducts while the other is quiescent. As a result of clipping, the output signal has a flat topped positive and negative waveforms.

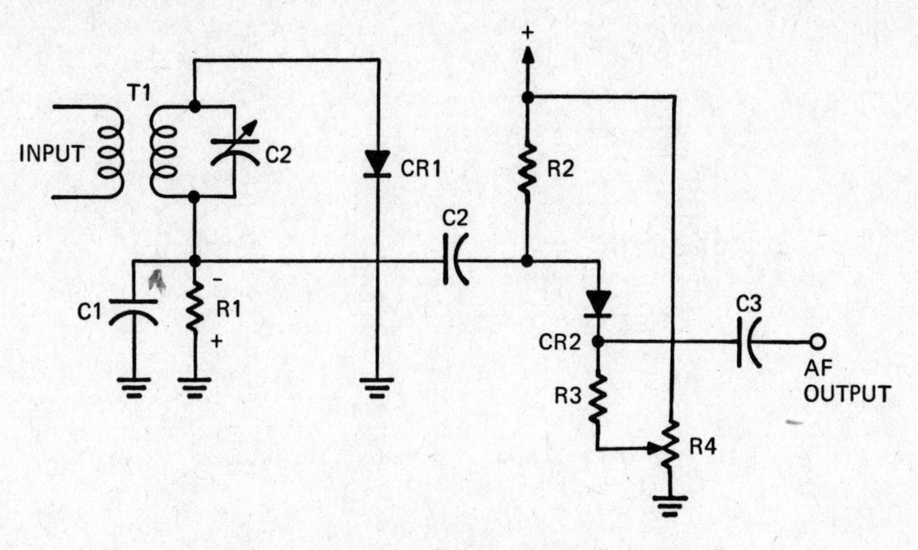

Figure 11-3. Noise limiter circuit employing a diode.

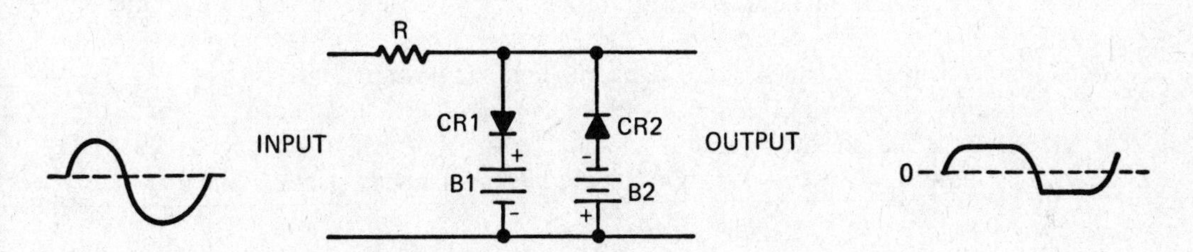

Figure 11-4. Double shunt diode clipper.

If batteries B1 and B2 each supply 1.5 volts DC and both CR1 and CR2 require 0.8 volt of forward bias to conduct, their clipping level will be increased to 2.3 volts. If the input signal is a 10 volt peak-to-peak sine wave, the output signal level will be limited to 5.4 volts peak-to-peak. Flat-topping occurs when the forward bias across either CR1 or CR2 reaches 2.7 volts.

Phase Inverter, Autotransformer

The input signal is fed through C1 to one end of center-tapped autotransformer T1. Because of magnetic induction, T1 acts as a phase inverter. When the control grid of V1 is made positive-going by the AC input signal, the control grid of V2 will be negative-going, and vice versa. When this push-pull power amplifier circuit is used, it is not necessary to use an electronic phase inverter ahead of it.

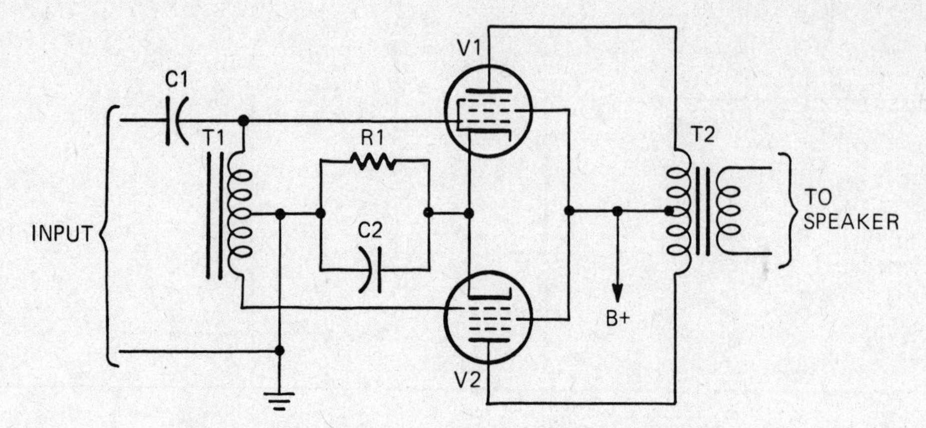

Figure 11-5. Phase inverter circuit, autotransformer.

Phase Inverter, Dual Triode

This phase inverter circuit shown in Figure 11-6 provides both gain and interfacing between a single-ended input source and the input of a balanced push-pull amplifier. Triode V1 functions as a conventional resistance-coupled single-ended amplifier; V2 provides the 180 degrees out of phase signal required for push-pull operation.

When the AC input signal is positive-going V1 plate current rises and the signal at the plate of V1 is less positive (negative going). The signal of output X is negative, as is the signal at point Y (junction of R5 and R6), but its amplitude is smaller. The negative

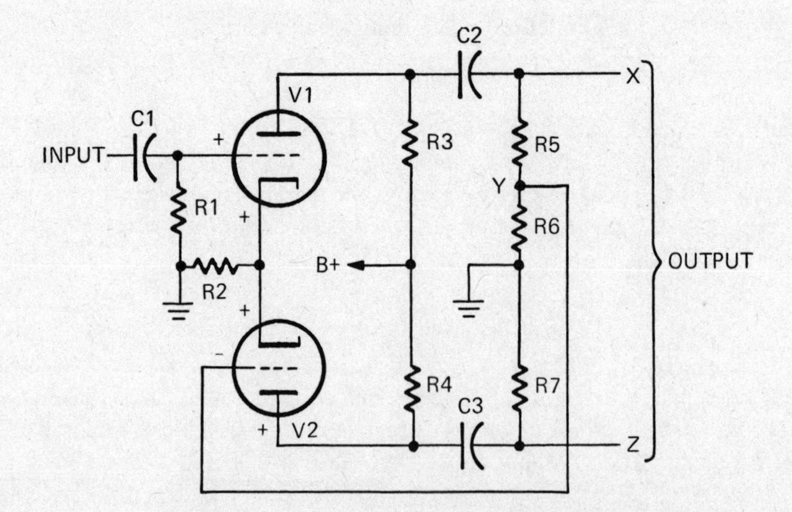

Figure 11-6. Phase inverter circuit, dual triode.

signal voltage at Y is fed to the grid of V2, causing its plate current to fall, and V1 plate voltage becomes more positive. The signal at output Z is positive. When the AC input signal polarity reverses, the opposite takes place: the output signal at X is now positive and at Z it is negative.

Therefore, when X is negative with respect to ground, Z is positive with respect to ground. And, when X is positive, Z is negative. The voltages at X and Z should be of opposite polarity and also of equal amplitude. The amplitude of the signal at Z is determined by the level of the signal at Y and the gain of V2. If V2 provides 10 times voltage amplification, for example, the voltage at Y should be 10 percent of the voltage at X.

Phase Inverter, Philco

To avoid using an extra tube as a phase inverter to obtain push-pull operation, Philco engineers devised the clever circuit shown in Figure 11-7. Tubes V1 and V2 are identical power pentodes or beam power tubes. V1 acts as both a power amplifier and the source of the input signal for V2. When the AC input signal is positive-going, V1 plate current rises and V1 plate voltage becomes less positive (negative-going). At the same time, V1 screen grid current rises as does the voltage drop across R4 which is in series with the screen grid. V1, therefore, has two output loads, half of the primary of output transformer T and R4. The voltage at the screen grid of V1 is negative-going when the voltage at the control grid is positive-going.

The signal developed across R4 is coupled to the control grid of V2 through C2, and is 180 degrees out of phase with the signal at the control grid of V1. Since V2 also inverts

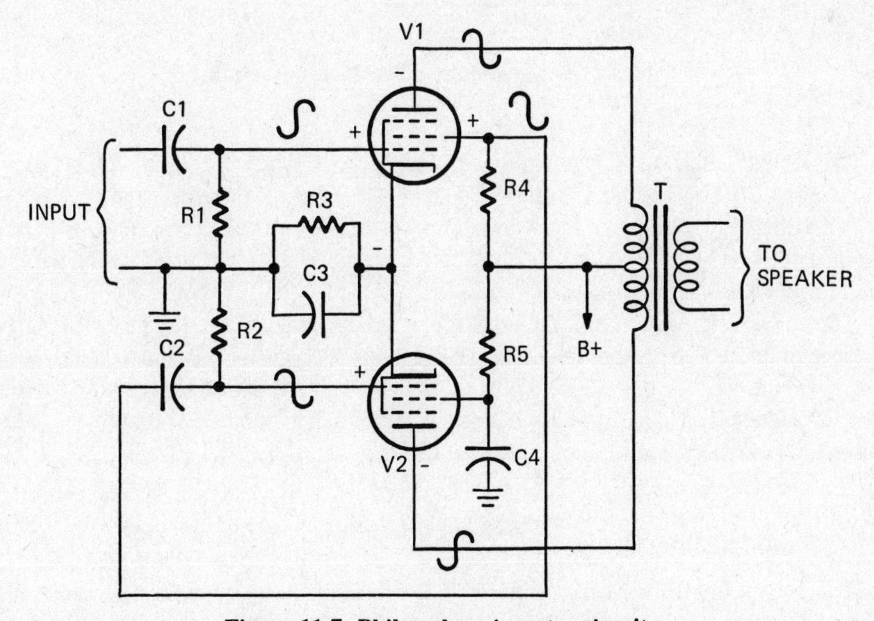

Figure 11-7. Philco phase inverter circuit.

the signal, the signal voltages at the plates of V1 and V2 are also 180 degrees out of phase. Although V2 screen grid current flows through R5, the screen grid voltage is held constant by bypass capacitor C4.

Phase Inverter, Single Triode

The single-tube phase inverter shown in Figure 11-8 utilizes the cathode-follower principle. When the input signal is negative-going, the signal at output X is positive-going because of the 180-degrees phase inversion introduced by the tube. At the same time, the signal at Z is negative-going since it is derived from the cathode of the tube (the signals at the cathode are in phase with the signals at the grid).

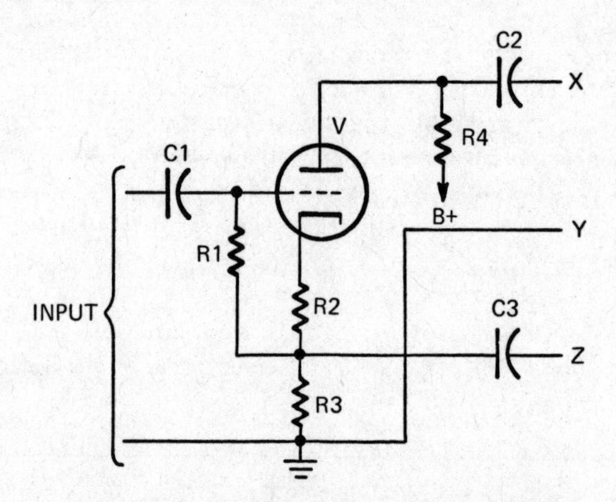

Figure 11-8. Single-tube phase inverter circuit.

The signal voltage at X is positive with respect to Y when the signal voltage at Z is negative, also with respect to Y, and vice versa. Although of opposite polarity, the X-Y and Z-Y voltages should be of equal amplitude. This is achieved when the resistance of plate load resistor R4 is equal to the resistance of cathode load resistor R3. The value of current through R3 is the same as through R4.

The value of R2 determines the level of cathode bias applied to the grid. If R2 has a resistance of 2000 ohms and plate current (through R2, R3, and R4) is 1 mA, there will be a 2-volt drop across R2 and the grid will be 2 volts negative with respect to the cathode. Since R2 is not bypassed by a capacitor, degenerative feedback is introduced which reduces gain but, at the same time, reduces distortion.

Pentode Limiter Circuit

The limiter circuit shown in Figure 11-9 is similar to a conventional IF amplifier stage, except for the grid-leak resistor and capacitor to supply bias to the sharp cutoff

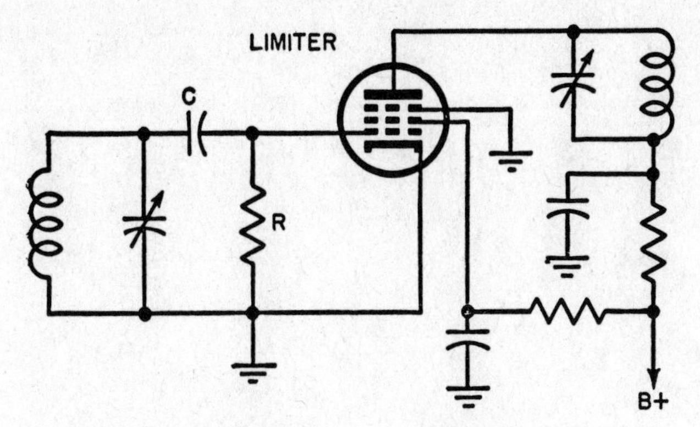

Figure 11-9. Pentode limiter circuit.

pentode. The R-C combination produces a negative DC bias voltage that is almost equal to the positive peak input voltage. When the signal to the grid is positive, the tube is quickly driven into saturation (plate current can rise no higher). When the signal to the grid is negative, the tube is quickly driven into cutoff (no plate current). The short disturbances in the signal caused by noise voltages that are longer than the time constant of the R-C circuit are clipped off. Longer variations caused by fading of the signal will appear in the output. It is common practice to follow the first short-time-constant limiter with one that has a longer time constant to allow for these slower fadings.

Schmitt Trigger

The Schmitt trigger is a device whose state depends upon the amplitude of the input voltage. In the quiescent state, with no input to the collector of Q1, Q1 will be cut off, and Q2 will be conducting. When the input level reaches a sufficiently positive level to turn on Q1, the base of Q2 will fall to the level present at the collector of Q1. As the base of Q2 becomes more negative, Q2 cuts off. When the input at R1 falls below the value needed to keep Q1 in conduction, Q1 will again be cut off and Q2 will again conduct. The network consisting of R3 and C tends to smooth out the signal felt at the base of Q2, and for this reason the Schmitt trigger is useful for waveform restoration, signal level shifting, squaring sinusoidal or non-rectangular inputs, and for DC level detection.

Shunt Diode Clipper

Clipping or limiting either the positive or negative peaks of a signal can be done with a diode connected as shown in Figure 11-11. When input X is positive with respect to Y, diode CR1 does not conduct, and passage of the signal to the output is opposed only by the resistance of current-limiting resistor R. When X is negative with respect to Y, CR1 is forward biased and conducts. It very nearly short-circuits the output signal. Most of the signal voltage is dropped across R which protects CR1 from damage.

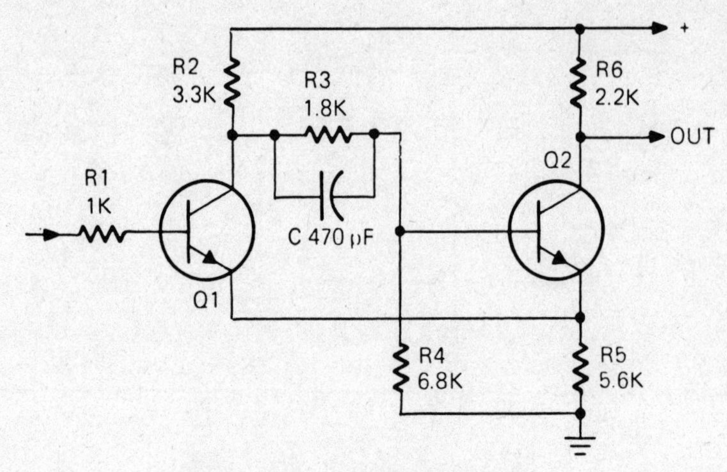

Figure 11-10. The Schmitt trigger.

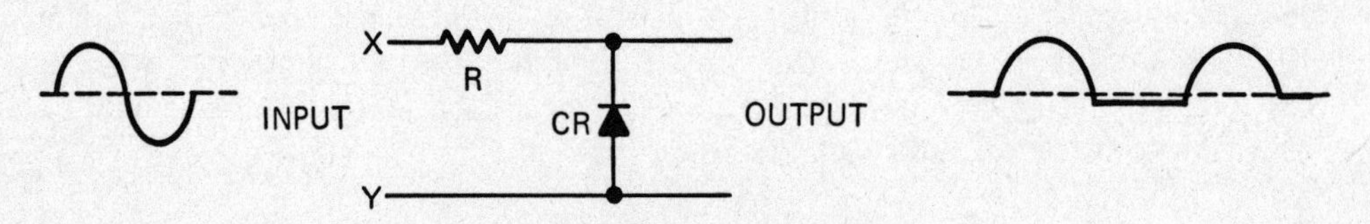

Figure 11-11. Shunt diode clipper.

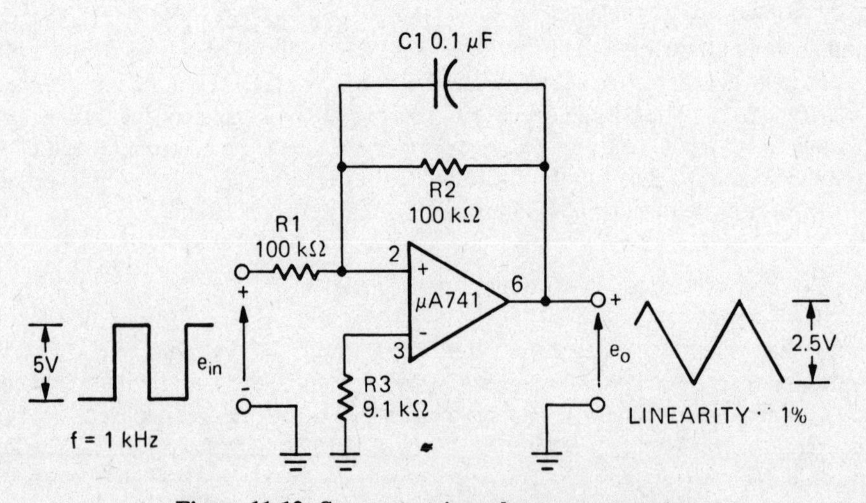

Figure 11-12. Square to triangular wave converter.

Square to Triangular Wave Converter

Figure 11-12 demonstrates the use of a commercially available integrated circuit, the Fairchild mA741, in a configuration that converts a square wave to a triangular wave. The time constant of the C1R2 network in the feedback path of the operational amplifier IC is such that the square wave applied to the input appears as a triangular wave at the output. The amplitude of the output waveform in relation to that of the input is determined by the value of input resistor R1 and by the ratio of R2 to R3.

12

Signal Generation Circuits

There are two basic types of signals: *digital*—it's there or it isn't—and *analog*—it's there but it varies in amplitude or frequency. The digital signal can be modulated by varying its duration, timing, rate of occurrence, or amplitude. The analog signal can be modulated by varying its amplitude, frequency, or phase.

A battery produces a steady-state signal which, if switched or modulated, can be used for transmission of information. So can the AC obtained from a powerline. But, a signal generator is usually some kind of "oscillator" which converts DC into AC signal or a pulse or amplitude modulated signal. The electronic oscillator (accidentally discovered by De Forest and intentionally devised later by Armstrong utilizing De Forest's electron tube) converts DC into AC.

Covered in this chapter are numerous circuits which generate signals. They include various types of oscillators that produce various kinds of waveforms.

Arc Oscillator

In the circuit shown in Figure 12-1, oscillations are produced by an electric arc at the series resonant frequency of L2 and C1. L1, an RF choke coil, prevents the RF energy from getting into the DC power supply. Rheostat R is used for controlling arc current.

Arc Transmitter

Before high power electron tubes were developed, the arc transmitter was quite popular. It generates a CW (continuous wave) signal. Figure 12-2 is a circuit of a simple arc transmitter. DC for the arc is fed in through rheostat R with which arc current can be

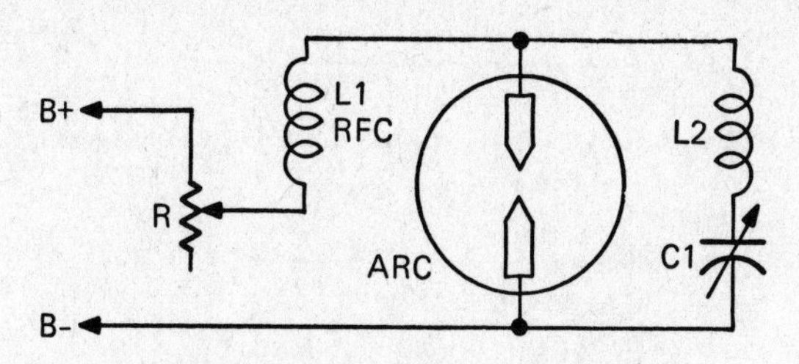

Figure 12-1. Arc oscillator circuit.

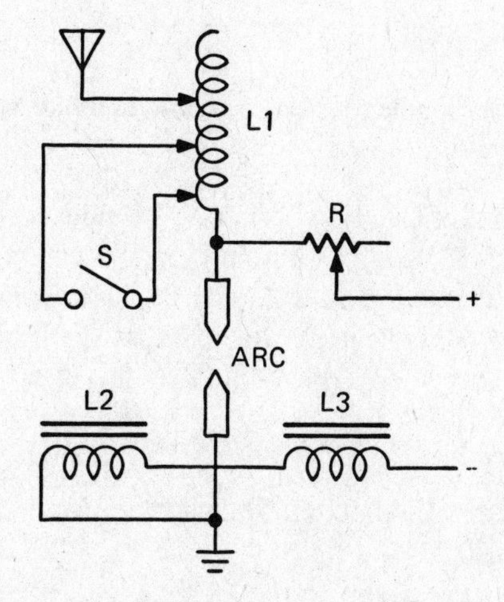

Figure 12-2. Simple arc transmitter circuit.

adjusted and through reactors L2 and L3. The arc is started by momentarily touching its two electrodes together and then separating them. As long as there is arcing between the electrodes, a signal is generated. The frequency of transmission is determined by the tapped inductor, L1. In this circuit, switch S represents a telegraph key. When the key is open, the radiating signal is at one frequency. When the key is closed, the frequency is shifted. This is one of the earliest FSK (frequency shift keyed) transmitters. Since arc transmitters are now obsolete, this circuit is shown here only for historical purposes.

Astable Multivibrator

Figure 12-3 illustrates a simple transistor astable multivibrator. Because of manufacturing techniques and slight component differences, transistors Q1 and Q2 will not conduct at exactly the same rate when power is applied. If it is assumed that Q1 conducts more heavily than Q2, the collector voltage of Q1 will initially keep transistor Q2 cut off

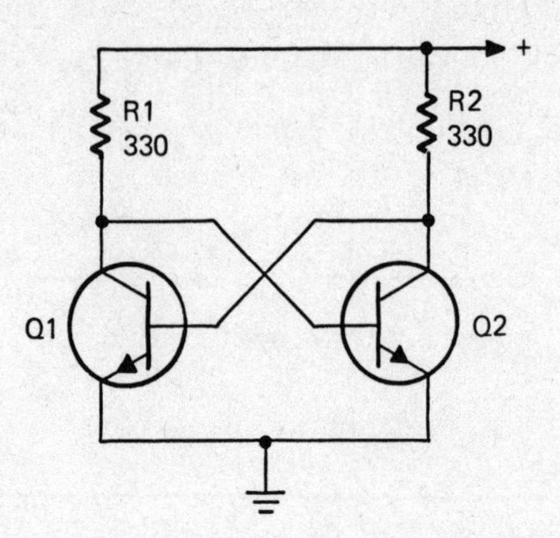

Figure 12-3. Simple transistor astable multivibrator circuit.

since it is applied to the base of Q2. However, as Q1 conducts *more* heavily, the collector voltage will drop, and Q2 will be allowed to turn on. When Q2 comes on, the inital high collector voltage (also applied to the base of Q1) will cut Q1 off. However, as the current through Q2 increases, and the collector voltage drops, it will reach a point where Q1 is again turned on and Q2 is turned off. The output may be taken from the collector of either transistor.

Astable Unijunction Transistor Multivibrator

The astable multivibrator circuit shown in Figure 12-4 employs fewer components than the conventional multivibrator. The circuit functions as follows: In astable operation, capacitor C1 is charged through resistor R2 and diode CR1. During the charging cycle, the

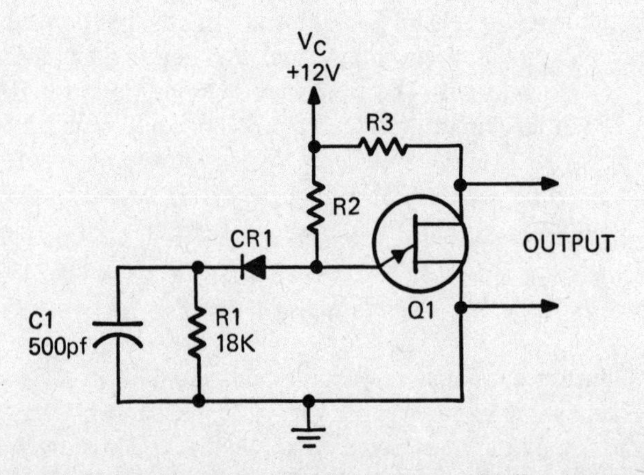

Figure 12-4. Astable unifunction transistor multivibrator circuit.

diode conducts, but the unijunction transistor Q is cut off. When the capacitor voltage is equal to the peak point potential of the unijunction transistor, the transistor conducts and the voltage at the base of the transistor falls very nearly to ground potential. The diode is cut off. Then C1 charges through R1 until the upper plate of C1 is about equal to the junction potential of the unijunction transistor. The diode conducts again. This causes the current through the transistor to decrease and cut off the transitor. Then C1 begins to recharge and the cycle is repeated over and over again. The output waveform is a square wave whose duration is determined by the time constant of C1 and R1, or the reciprocal of the frequency.

Bistable Multivibrator

Two versions of bistable multivibrator circuits are shown in Figure 12-5. In both circuits, two inputs are required to produce one full cycle of output. There are only half as many rising edges from one tube plate. For this reason, the circuit is called a scale-of-two circuit. Since the output will consist of half as many output pulses per second as input trigger pulses, this circuit may be used as a divide by two circuits.

With both grids biased to cut off, a negative trigger pulse applied to one (V1, for example) drives the tube further into cutoff. The resulting rise in plate voltage will result in a voltage rise at the grid of V2. If this rise is greater than the negative trigger pulse at this grid, V2 will conduct with a resultant drop in plate potential, sensed by the grid of V1 to further aid the cutoff of V1. The next pulse will restore the circuit to its original condition. The cathode resistor used in the circuit shown in B is bypassed by a capacitor which absorbs transients caused by switching.

Blocking Oscillator

The circuits of blocking oscillators shown in Figure 12-6 function in a manner similar to an astable multivibrator. One cycle period is much shorter than the following. It may be triggered by a pulse to produce short-duration, high-amplitude pulses. Since it can produce shorter pulses than a multivibrator it is very useful in radar and television applications. The circuits of two types are shown, the free-running in A and the triggered in B. In these circuits, an iron core transformer and a triode tube are used. These oscillators rely upon the cutoff of the triode tube, and the magnetic saturation of the transformer. When the voltage at the grid of the triode rises high enough, plate current flows and an induced voltage is passed to the secondary, causing the grid to draw more current, and plate current to further increase. As plate voltage decreases, the conductance of the tube falls off until the grid voltage turns negative. The regeneration occurs in the opposite direction to cut off the tube.

Bootstrap Sawtooth Generator

The bootstrap generator circuit shown in Figure 12-7 produces a sawtooth waveform when fed a negative-going rectangular input pulse. With no input pulse, triodes V1 and

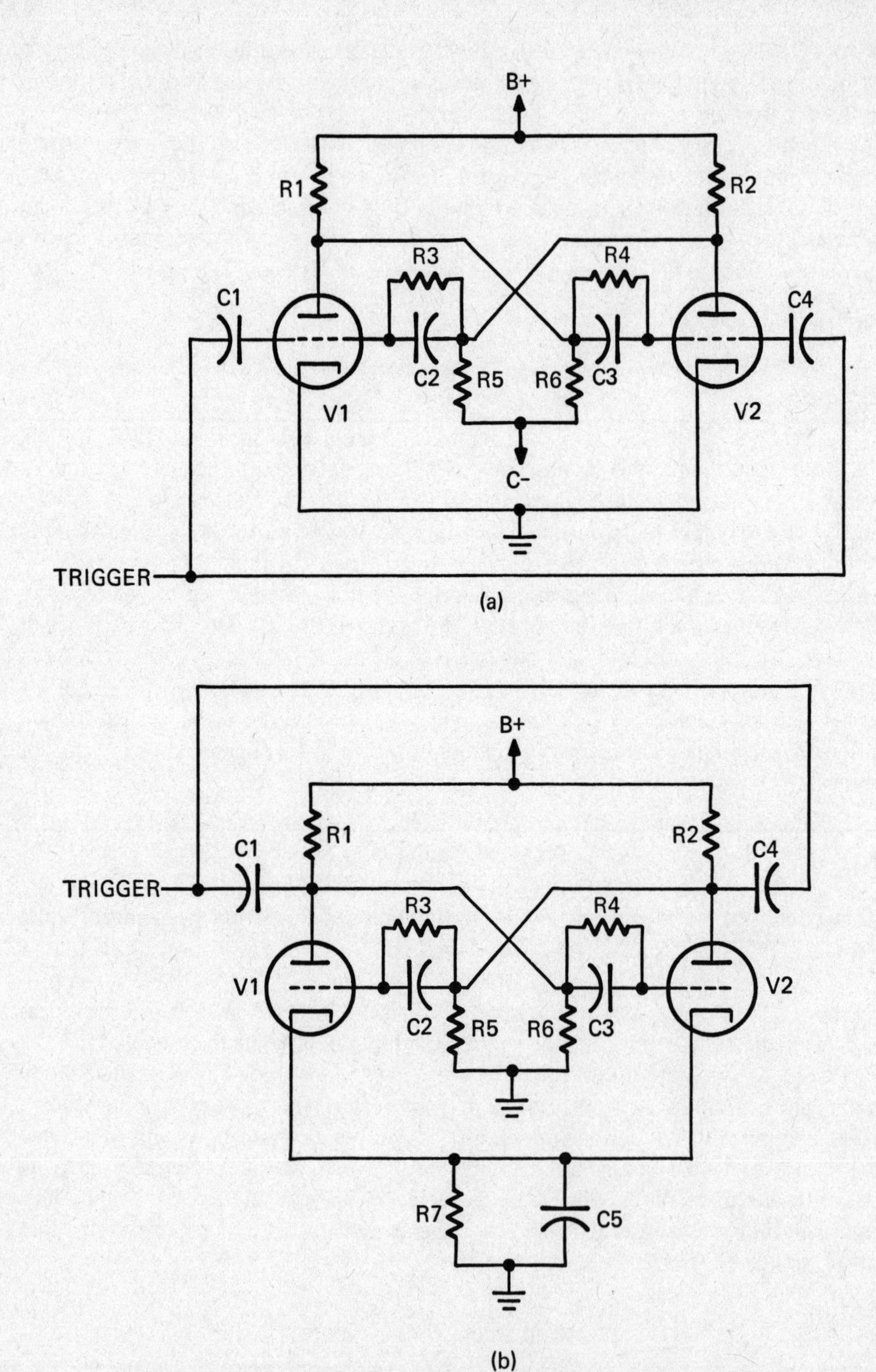

Figure 12-5. Bistable multivibrator circuits.

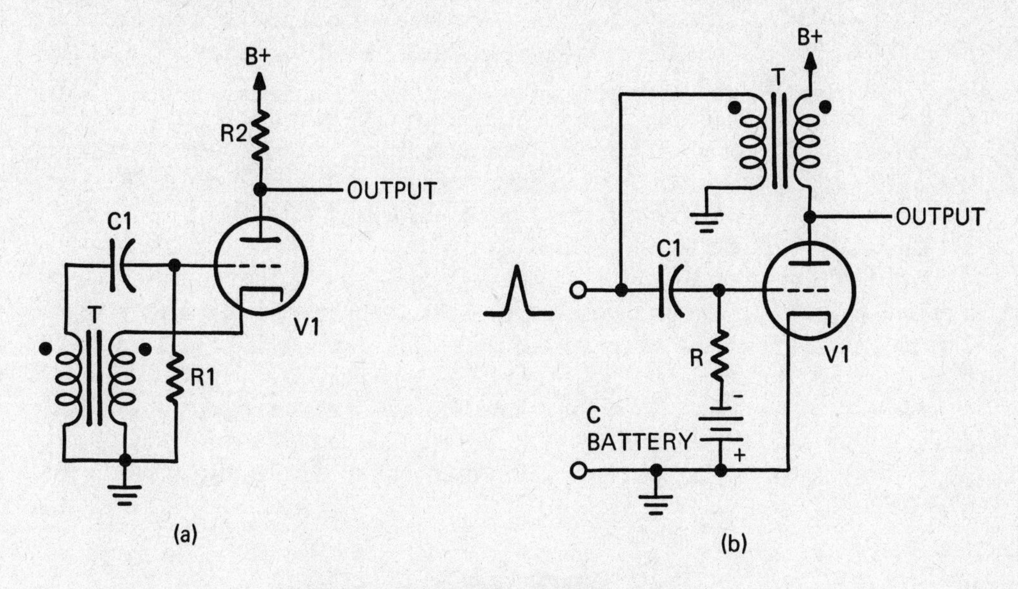

Figure 12-6. Blocking oscillator circuits.

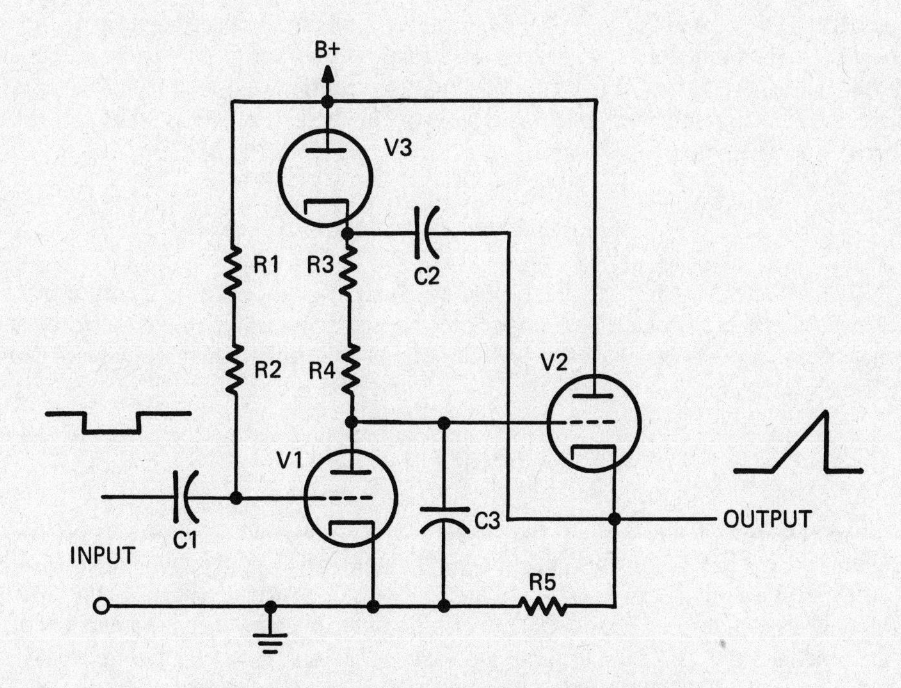

Figure 12-7. Bootstrap sawtooth generator circuit.

V2 and diode V3 conduct and C2 is charged. When the negative gate input pulse reaches the grid of V1, that tube cuts off, and timing capacitor C3 begins charging through R3 and R4. This rising voltage is applied to the grid of V2, which is wired as a cathode follower, and it begins to rise toward the voltage to which C2 was charged during the no-input state. The cathode of V3 becomes more positive because the charge in C3 is added to the charge in C2. The voltage drop across R3 and R4 remains constant as C3 charges. Thus, the current through R3 and R4 remains constant and the voltage across C3 rises linearily, causing a sawtooth output signal to appear across R5.

The bootstrap diode V3 ceases to conduct when its cathode becomes positive with respect to its plate, and effectively disconnects the charging circuit from supply voltage. Thus the voltage of the linear waveform continues to rise slowly until the negative gating pulse is removed from the grid of V1. When V1 again begins conducting, C3 discharges through V1 to its low voltage level. The voltage across R5 is also returned to its low level due to the decreased current through V2. Since the cathode of V3 is now once again slightly less positive than its plate, the diode conducts until another negative impulse is fed to the grid of V1.

Bramco Resonator

The circuit shown in Figure 12-8, the Bramco resonator circuit, uses a vibrating reed resonator to provide an audio output that is extremely stable. When power is applied to the circuit, the vibrating reed, indicated by T on the schematic, swings one way due to the magnetic field from one of the coils. The resonant reed vibrates back and forth and forms a very stable audio oscillator in conjunction with transistors Q1 and Q2. Potentiometer R7 is a level control, transistor Q3 is a common-emitter amplifier, and the output of the resonator is taken from the emitter of transistor Q4. This type of circuit is widely used for selective calling in mobile radio units. (Illustration courtesy of Ledex, Inc.)

Calibration Oscillator

The oscillator shown in Figure 12-9 is a 100kHz crystal-controlled transistor oscillator that can be used for calibration of receivers, etc. Frequency is controlled by the crystal Y and by the values of C1 and C2. Diode D1 insures that the output of the oscillator will be rich in harmonics.

Code-Practice Oscillator

Figure 12-10 illustrates a simple practical transistor oscillator that can be used as a code-practice oscillator by interrupting the supply voltage with a telegraph key. Basically an amplifier, the circuit is caused to oscillate by means of positive feedback to the base of Q1 through capacitor C1. Tone control is by means of R1, which varies the amount of bias to the base of transistor Q1. Changing the value of C1 will also vary the frequency of the oscillation.

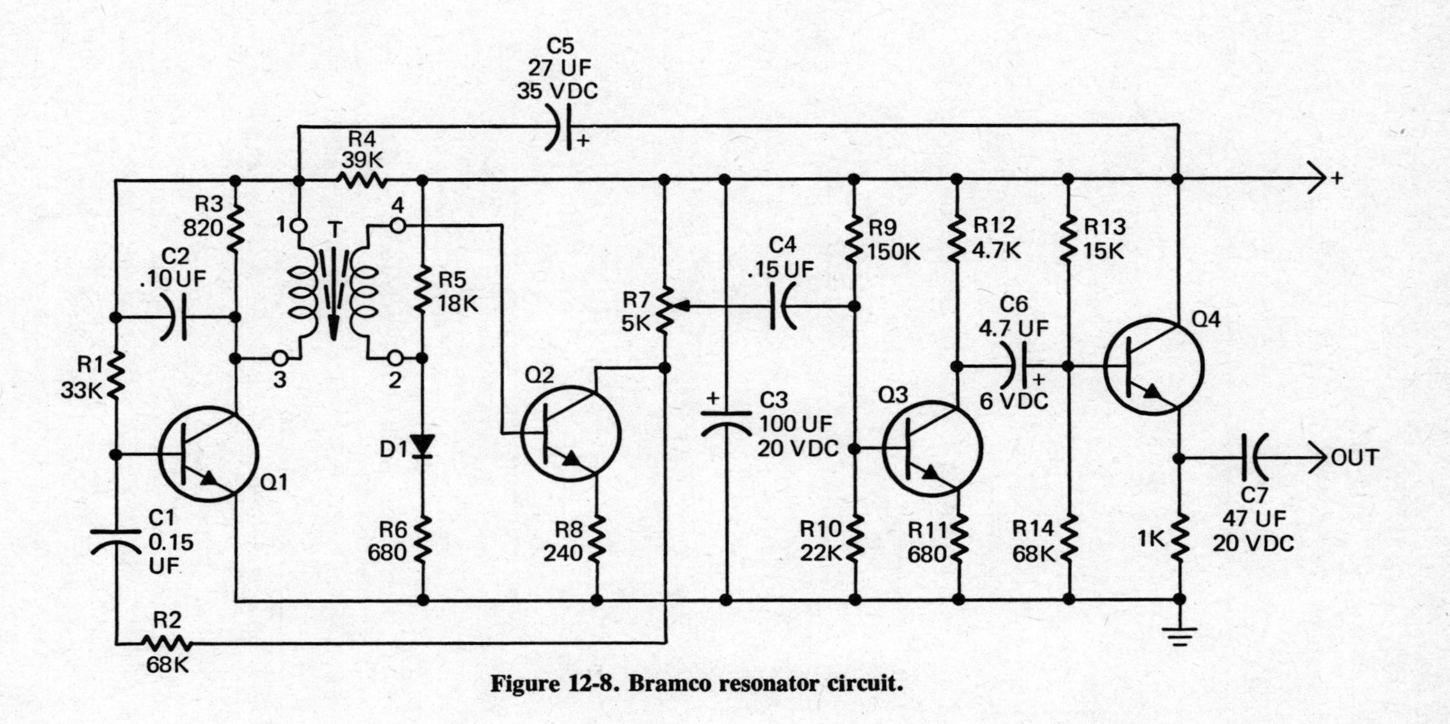

Figure 12-8. Bramco resonator circuit.

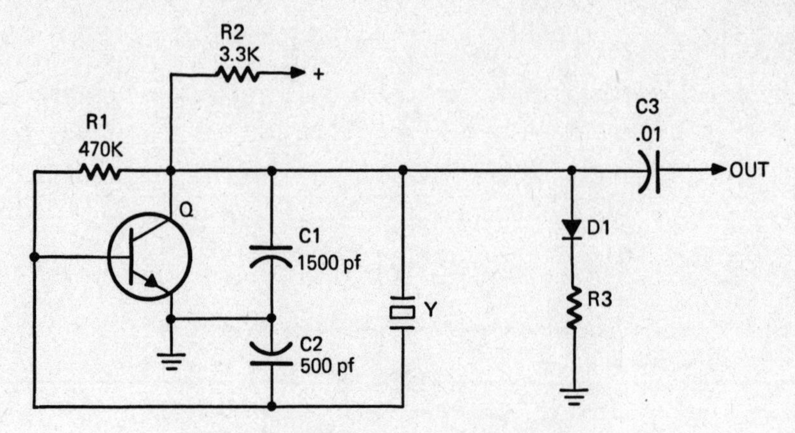

Figure 12-9. Calibration oscillator circuit.

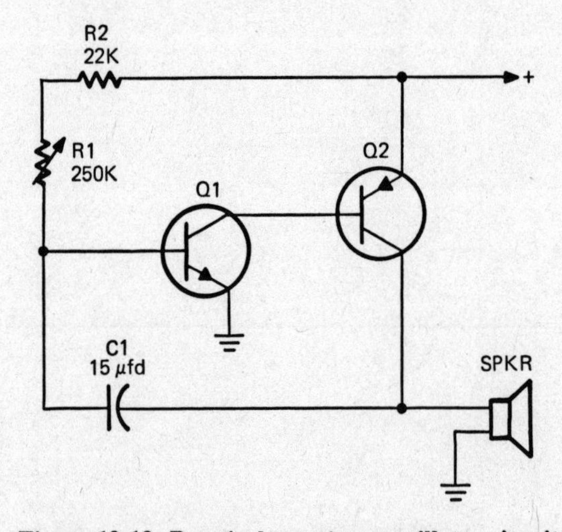

Figure 12-10. Practical transistor oscillator circuit.

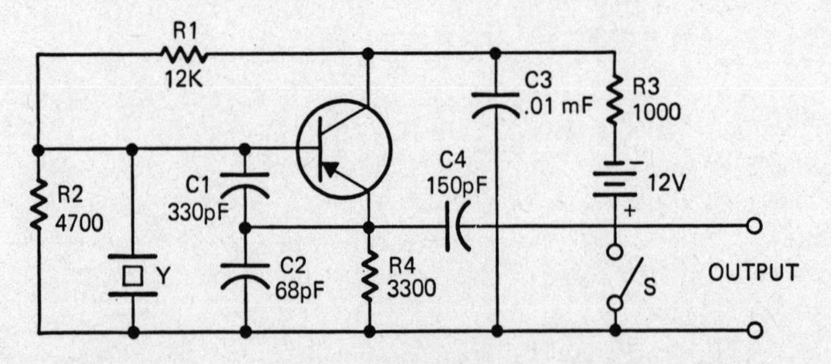

Figure 12-11. Crystal-controlled Colpitts oscillator circuit.

Colpitts Oscillator, Crystal-Controlled

Figure 12-11 illustrates a practical, crystal-controlled Colpitts oscillator circuit suitable for use in transmitters, as a test oscillator, or other applications. Notice that the crystal Y in the tank circuit insures frequency stability.

Colpitts Oscillator, Series Fed

In the Colpitts oscillator circuit shown in Figure 12-12 collector supply voltage is fed through inductor L, which, in conjunction with C1 and C2, determines the frequency of oscillation. The capacitors across the collector tank coil form a voltage divider. A portion of the signal is fed back to the emitter to sustain oscillation. Resistors R1, R2, and R3

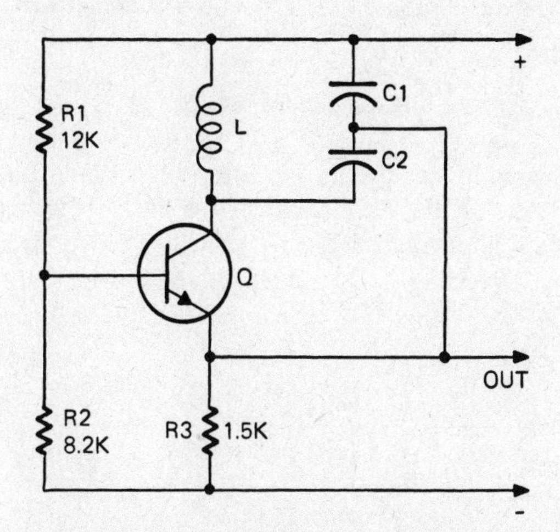

Figure 12-12. Series fed Colpitts oscillator circuit.

form the base bias network for the transistor. For fixed-frequency operation, L usually has an adjustable ferrite core which enables tuning of the oscillator to the required frequency.

Colpitts Oscillator, Shunt Fed

The Colpitts oscillator circuit shown in Figure 12-13 employs a triode tube and can be used for generating either AF or RF energy. The amount of feedback is determined by the capacitance ratio of C1 and C2. The frequency of oscillations is determined by the inductance of L and the net capacitance of C1 and C2. The oscillator can be made tunable by using a variable inductor for L or variable capacitors for C1 and C2. Grid leak bias is developed by C3 and R1. Feedback is routed from the output to the input through C4. Resistor R2 isolates the plate from ground.

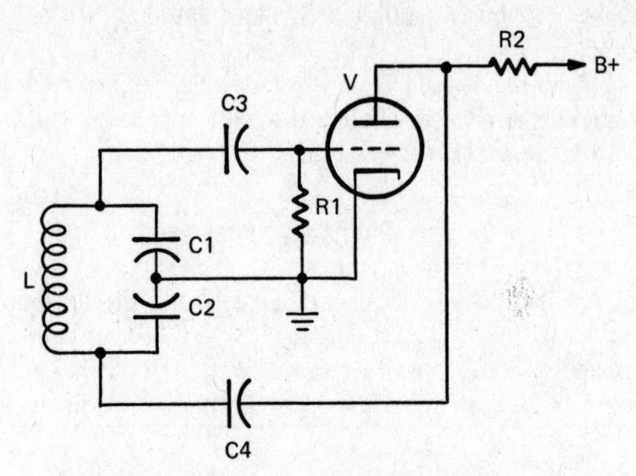

Figure 12-13. Shunt fed Collpitts oscillator circuit.

Crystal-Controlled Exciter

Two triodes are used in the crystal controlled exciter circuit shown in Figure 12-14. Triode section V1A is the oscillator which employs a shunt fed Pierce circuit. Its output is fed through C2 to the grid of triode section V1B which functions as an RF amplifier or frequency multiplier, depending upon the frequency to which its plate tank circuit is tuned.

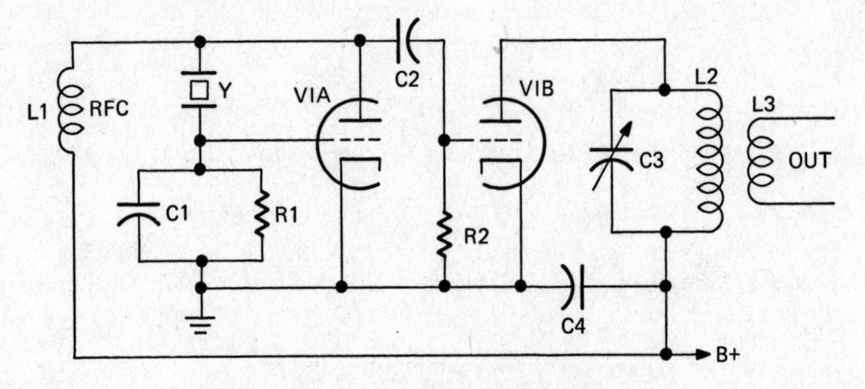

Figure 12-14. Crystal-controlled exciter circuit.

Crystal-Controlled Hartley Oscillator

A simple crystal-controlled Hartley oscillator is shown in Figure 12-15. The crystal Y is in the feedback path from the tank circuit to the base of transistor Q. At all but the crystal frequency (or a crystal overtone frequency) the feedback path is open.

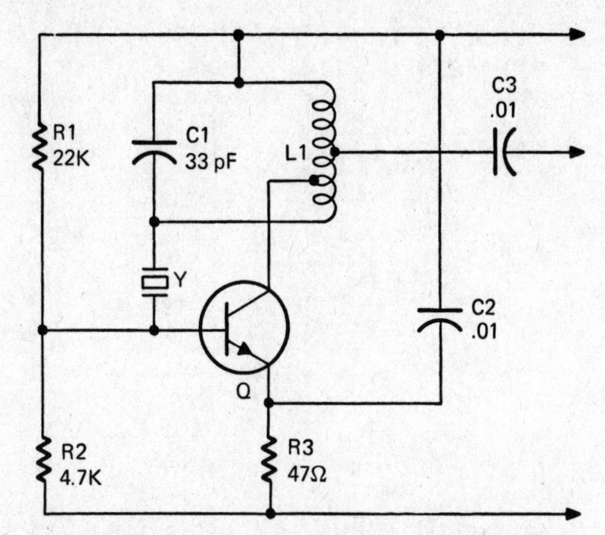

Figure 12-15. Crystal-controlled Hartley oscillator circuit.

Crystal-Controlled Tunnel Diode Oscillator

A third overtone crystal is used in the tunnel diode oscillator circuit shown in Figure 12-16. With 1.5 volts applied, the power output of the oscillator is in the microwatt range. When variable inductor L has a nominal inductance of approximately 0.34 microhenry, and when using an appropriate crystal, the oscillator will operate at a frequency within the 45-50 MHz range. (Illustration courtesy of General Electric Company.)

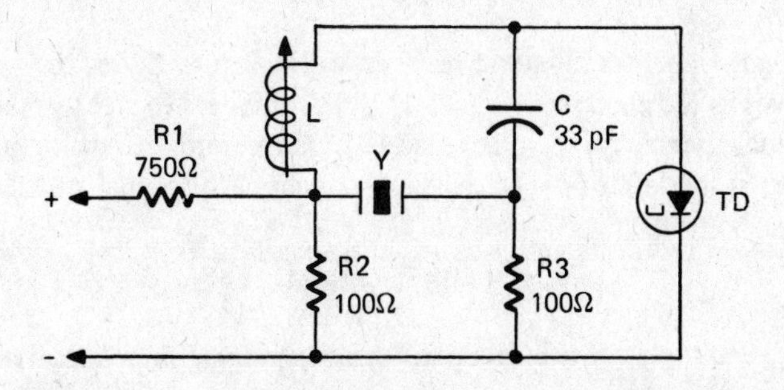

Figure 12-16. Crystal-controlled tunnel diode oscillator circuit.

Crystal-Controlled Unijunction Oscillator

Figure 12-17 shows an oscillator which uses a unijunction transistor. The crystal placed in the feedback path selects the frequency, and C1 gives frequency control over a narrow range above and below the crystal frequency. Addition of diode D1 at the junction

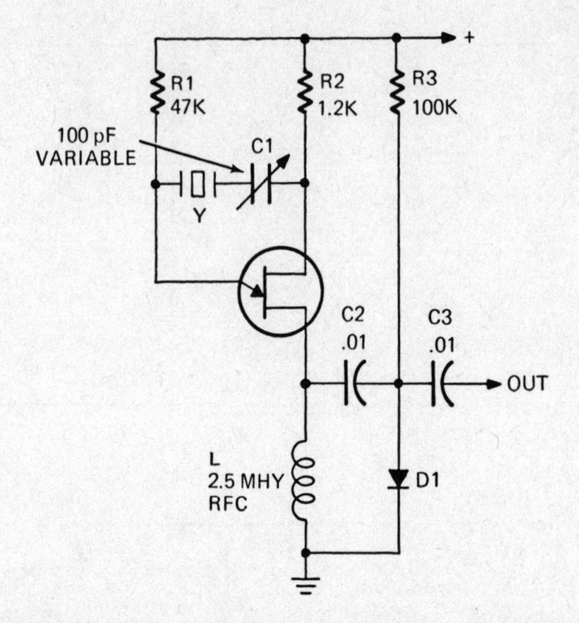

Figure 12-17. Crystal-controlled unijunction oscillator circuit.

of C2 and C3 insures that the output will be rich in harmonics, suitable for use as a calibration oscillator or other application where an output that is rich in harmonics is desired.

Electron-Coupled Oscillator

In the electron-coupled oscillator circuit shown in Figure 12-18, the screen grid of a pentode tube serves as the plate of a Hartley oscillator. Since the screen grid is bypassed to ground at the signal frequency, the plate load is isolated from the oscillator circuit. Plate current, however, is modulated by the electron stream at the signal frequency rate.

Franklin Oscillator

The Franklin oscillator circuit shown in Figure 12-19 employs a two-stage amplifier. In this circuit, each tube shifts phase 180 degrees so that the total phase shift fed back to the grid of the left-hand tube is 360 degrees, thus maintaining oscillation at a frequency selected by the L1/C2 tank. The values of Cb are very small (typically 5-50 Pf) so that both tubes are operated near the Class A region. The tremendous amplification of the two triodes enables operation with very little coupling through the Cb's to the resonant circuit, meaning that the frequency is influenced very little by changes in tube parameters. The Franklin oscillator is very stable, and is often used as a frequency-determining stage, but not as a power oscillator.

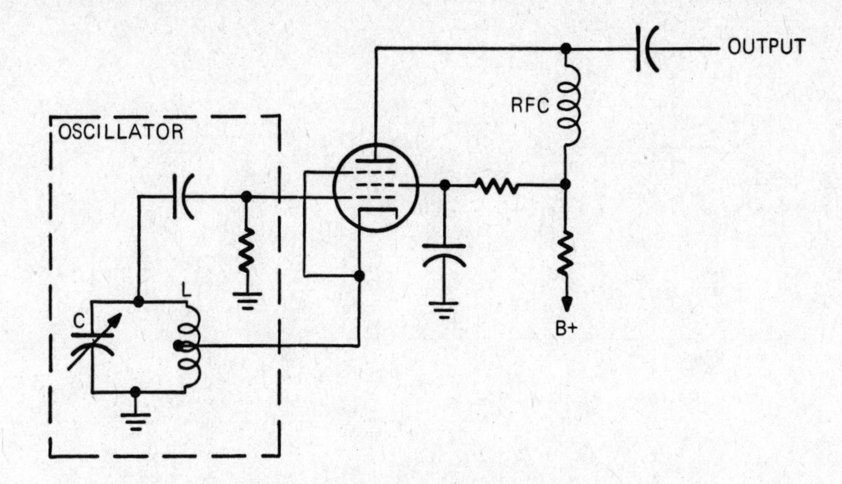

Figure 12-18. Electron-coupled oscillator circuit.

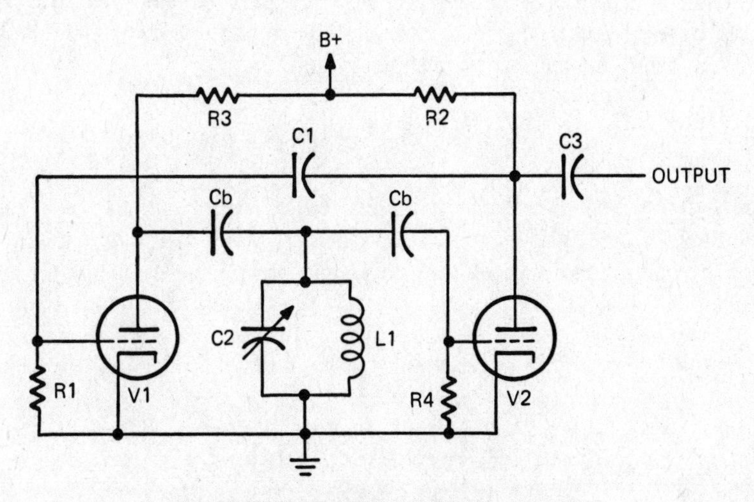

Figure 12-19. Franklin oscillator circuit.

Free-Running PRF Generator

The pulse repetition frequency (PRF) generator is a free-running device which produces short-time-duration, large-amplitude pulses. The transistor in the circuit shown in Figure 12-20 conducts for a short period to produce a pulse and is then cut off for a longer period of time. Transformer T1 provides the necessary regenerative coupling from the collector to the base of the transistor Q1. Capacitor C1 and resistor R1 form the RC circuit which determines the time constant in the base-emitter circuit.

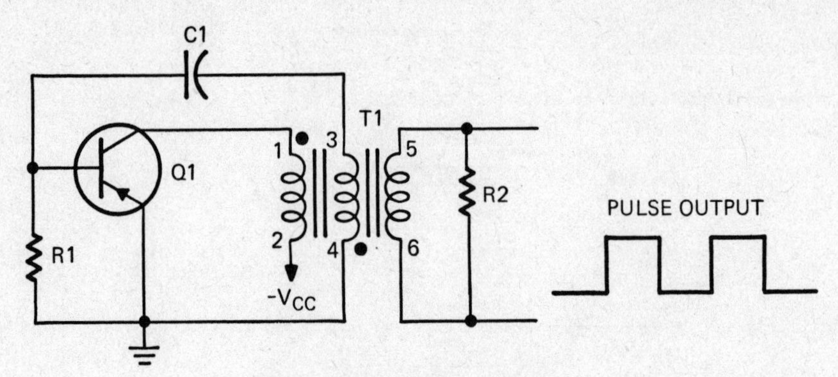

Figure 12-20. PRF generator.

The output pulse width depends primarily upon the inductance of transformer T1, while the output frequency depends mostly upon the time constant of R1 and C1. When power is applied, the transistor saturates quickly, producing the leading edge of the pulse. When the magnetic fields collapse, they produce the trailing edge of the pulse. Q1 is then held cut off by the discharge of C1 through R1 until the transistor again becomes forward-biased, saturates, and begins another cycle.

Harmonic Oscillator

In the harmonic oscillator shown in Figure 12-21 the fundamental frequency is selected by the crystal Y, and the tank in the cathode circuit of the tube is tuned to this frequency. The tank of the plate circuit, consisting of variable capacitor C4 and trans-

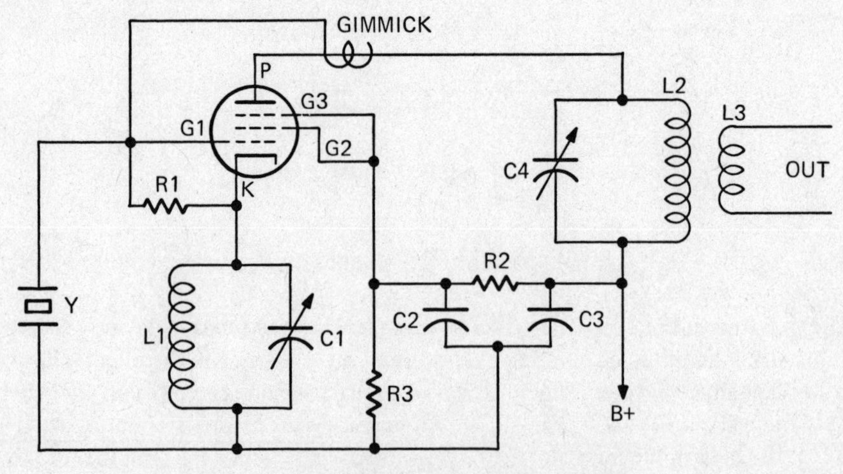

Figure 12-21. Harmonic oscillator circuit.

former primary L2, is tuned to the desired harmonic of the crystal frequency. Feedback to maintain oscillation is supplied by the ''gimmick,'' a small coil of insulated wire wrapped around the lead from the plate of the tube. The air space and insulation between the gimmick and the plate lead form the dielectric, and the two leads between the plates form what is actually a small capacitor.

Hartley Oscillator

A Hartley oscillator circuit using an NPN transistor is shown in Figure 12-22. The collector tank circuit is tuned by inductor L and variable capacitor C. Feedback to sustain oscillations is applied to the emitter from a tap on the inductor L which is used as an autotransformer. A voltage gain in excess of 10 is required to produce and sustain oscillations at the frequency determined by the values of L and C. Resistor R1 is required to prevent the transistor from saturating and become biased to cutoff. (Note: A PNP transistor can be used by reversing the supply voltage polarity.)

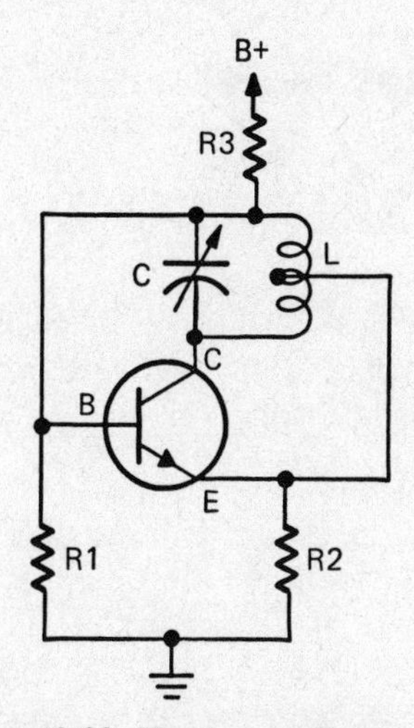

Figure 12-22. Hartley oscillator circuit.

IC Bridged-T Oscillator

In the bridged-T (or twin-T) oscillator circuit shown in Figure 12-33 an IC operational amplifier (op-amp) is used. Because of the feedback from the output to the non-inverted input (+) through the 1000-ohm potentiometer and lamp Il, the circuit could oscillate at any frequency. However, because the bridge-T network is a full network, negative feedback to the inverting input (-) is minimum at its null frequency and the circuit will oscillate at the null frequency. Lamp Il, by increasing resistance with an increase in output level, helps to limit positive feedback and improve stability. (Illustration courtesy of *Ham Radio*.)

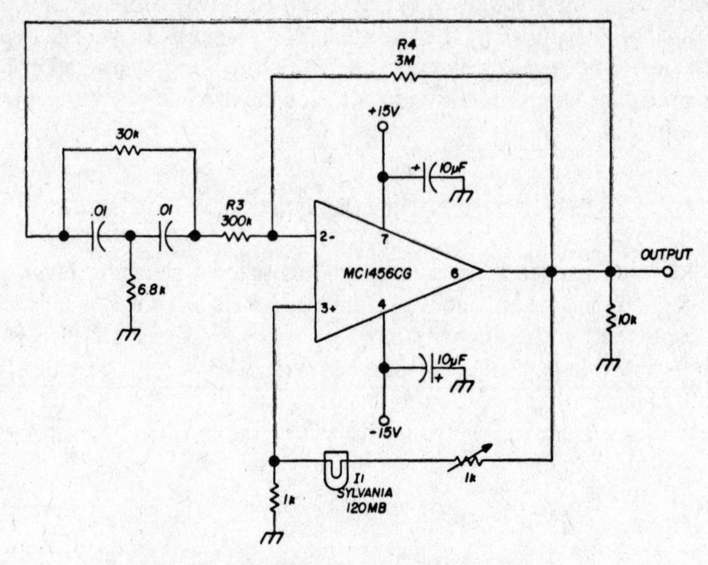

Figure 12-23. Bridged-T oscillator circuit.

IC Function Generator

Figure 12-24 shows how a commercially available integrated circuit can be used in a function generator circuit. The output from IC terminal 4 is a triangular wave, while the output from terminal 3 is a square wave, as shown. By connecting various value

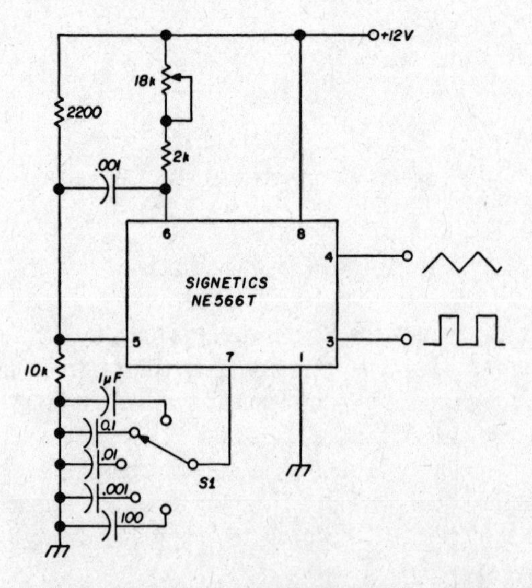

Figure 12-24. IC function generator circuit.

capacitors between terminal 7 and ground as selected by switch Sl, bandswitching is provided. The 18,000-ohm potentiometer is used for fine frequency adjustment. Nearly complete coverage is available from 20 Hz to 1 MHz. (Illustration courtesy of *Ham Radio*.)

IC Phase-Shift Oscillator

The oscillator circuit shown in Figure 12-25 uses an integrated circuit operational amplifier (op-amp) and a phase-shift network to produce oscillations. Resistor R3 determines the input impedance into which the RC network must operate, and the ratio of R3 and R4 determines the closed-loop gain of the amplifier. Each of the three 10,000-ohm resistors and its associated 0.033 uF capacitor provides 60 degrees of phase shift. The total phase shift by the network is 180 degrees. Since the phase-shifted output is applied to the inverting input, the amplifier will oscillate if gain is adequate. (Illustration courtesy of *Ham Radio*.)

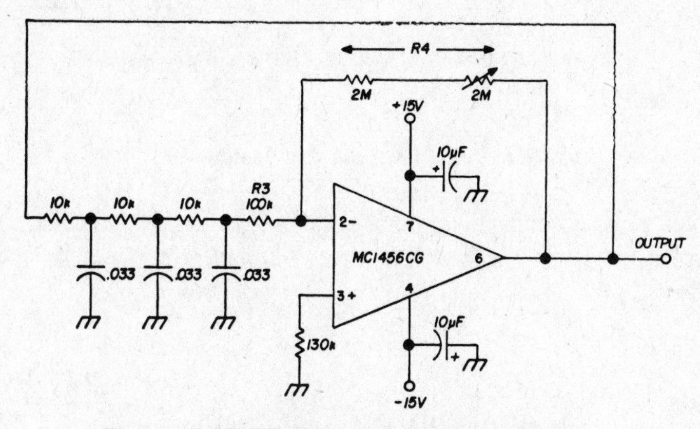

Figure 12-25. IC phase-shift oscillator circuit.

Induction Heating Oscillator

High power electronic generators are used in numerous industrial, scientific, and medical applications. In the circuit shown in Figure 12-26, plate supply voltage for the three power triodes is furnished through transformer T from a three-phase AC powerline. The three tubes are connected in parallel except for their plates, each of which is fed from a separate winding of transformer T. In this Hartley oscillator circuit, the cathodes of the triodes are connected to the tap of the tank coil L1. Grid leak bias is developed by C2 and R.

Unrectified AC is fed to the plates of the tubes. The tubes themselves act as self-rectifiers. For example, V1 operates during the AC half cycle when the voltage from the transformer makes its plate positive. The same is true of V2 and V3. However, since the source is three-phase 60 Hz AC, the generated signal is modulated at a frequency of 180 Hz.

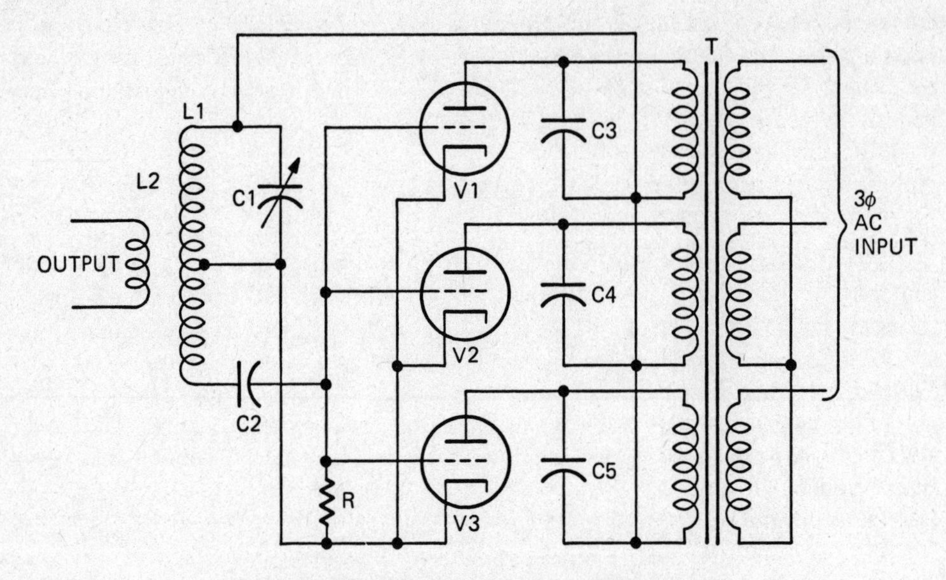

Figure 12-26. Induction heating oscillator circuit.

Inductively Coupled Arc Transmitter

The circuit of an arc transmitter whose output is inductively coupled to the antenna system is shown in Figure 12-27. When an arc is present between the electrodes, RF

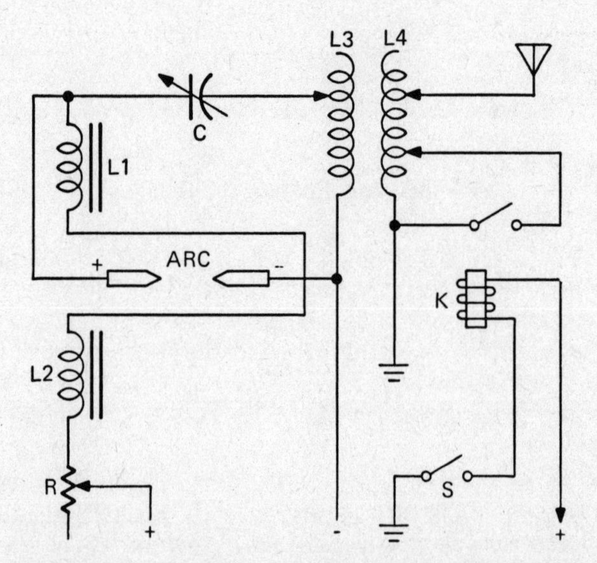

Figure 12-27. Inductively coupled arc transmitter circuit.

energy is generated at the series resonant frequency of variable capacitor C and L3. Since L3 is a tapped coil, the frequency can be varied over a wide range. The DC for the arc is fed through L2 and L1 and also through rheostat R with which arc current is controlled. L1 and L2 prevent the radio frequency energy from getting back into the DC power source.

The oscillating magnetic field around L3 is inductively coupled to L4, the antenna coil. The antenna system is tuned to be resonant at the transmitting frequency by adjusting the top half of L4. Since the arc cannot be keyed on and off to provide a radio telegraph transmission, the power output is raised and lowered by shorting out part of the antenna coil with the contacts of relay K. Switch S represents a telegraph key.

Meecham Bridge-Controlled Crystal Oscillator

The oscillator whose circuit is shown in Figure 12-28 is one of the most stable frequency generators in use. It is used where high precision is required. In the Meecham oscillator, the output of a bridge circuit is amplified and fed to the input of the bridge. This bridge circuit will produce the correct phase shift for oscillation at the series-resonant frequency of the crystal located in one arm of the bridge. Therefore, oscillations can occur only at the crystal frequency. As a rule, the triode tube is operated as a class A amplifier. Resistor R1 is a thermistor that increases its resistance as current increases its temperature. This causes an unbalance in the bridge and reduces the feedback. Frequency stability can be as high as one part in 100,000,000.

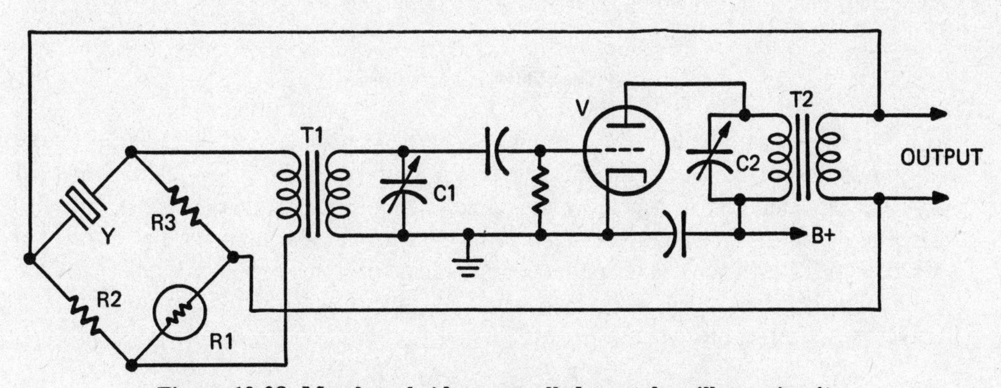

Figure 12-28. Meecham bridge-controlled crystal oscillator circuit.

Meissner Oscillator

Before the development of crystal controlled oscillators, the Meissner oscillator circuit shown in Figure 12-29 was very popular. It employs both a grid tank circuit and a plate tank circuit which are inductively coupled to each other and to the antenna system through the antenna coils shown at the left side of the diagram. C3 and R provide grid leak bias. C4 is a bypass capacitor which grounds the cold end of T2 and C2 at the signal frequency. For radio telegraph transmission, the plate supply voltage or the cathode circuit

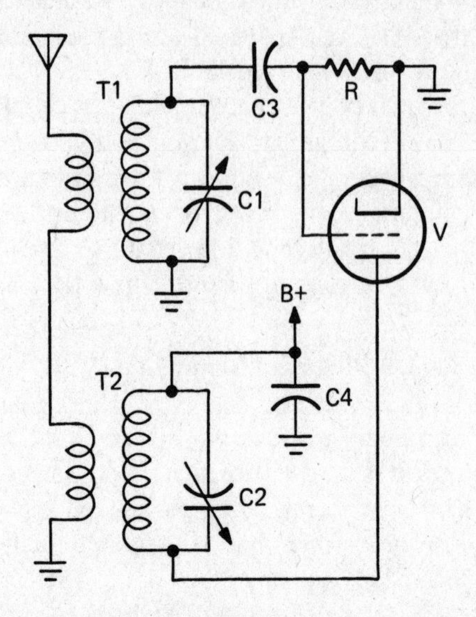

Figure 12-29. Meissner oscillator circuit.

would be opened and closed with a telegraph key. For radio telephone transmission, modulated plate voltage would be fed in at the point marked B+.

Monostable Multivibrator

The monostable multivibrator is called a gate or one-shot multivibrator. In Figure 12-30 two monostable oscillator circuits are shown, presented in two forms, DC coupled in A and AC coupled in B. A positive trigger pulse applied to the grid of triode V2 negates the negative bias, causes V2 to conduct, and the plate voltage to drop. This action causes V1 to cut off. It will stay cut off until the grid voltage rises in a period of time determined by the time constant of Rg1 and C1. Then V1 conducts, and V2 is cut off until another trigger pulse causes the cycle to repeat.

Monostable Transistor Multivibrator

The circuit of a monostable multivibrator shown in Figure 12-31 will assume a preferential state in operation. It has two states, one stable and the other unstable. When capacitor C is charged through resistor R, the circuit will shift into a stable state; then it is driven by a trigger pulse into an unstable state which is determined by the time constant of R-C. With C charged, the transistor Q2 is turned on, and a pulse is passed to the base of transistor Q1 to neutralize the capacitor C. The trigger input changes the charge in C which starts the cycle again. The circuit is called monostable because it requires a trigger to start the sequence.

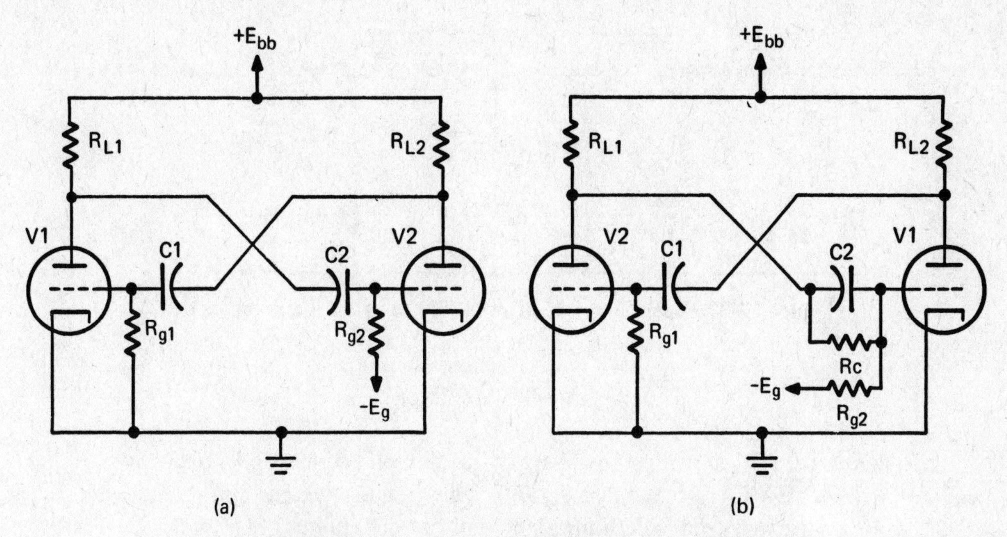

Figure 12-30. Monostable multivibrator circuit.

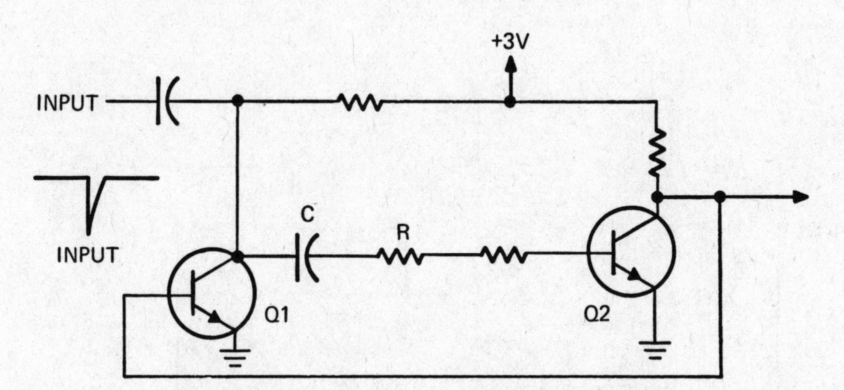

Figure 12-31. Monostable mulivibrator circuit.

Neon Lamp Siren

A pair of neon lamps can be used in the dual relaxation oscillator circuit shown in Figure 12-32 to generate a siren-type signal. The high frequency signal is generated by the relaxation oscillator formed by R3, C2, and neon lamp I2. The low frequency signal is generated by the relaxation oscillator formed by R2, C1, and neon lamp Il. The high frequency signal is modulated by the low frequency signal since current to both oscillators must flow through Rl. The output level of the composite signal is controlled with potentiometer R4.

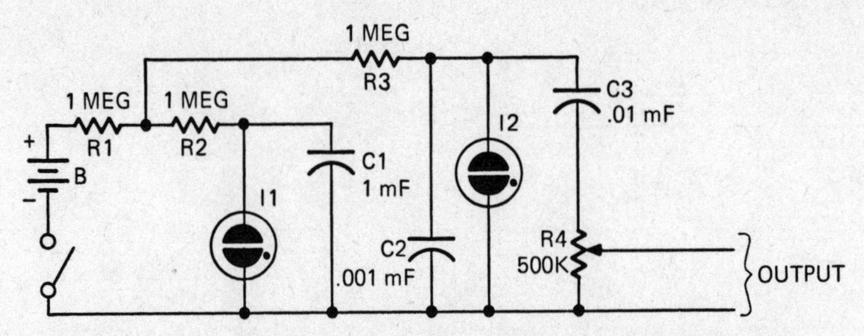

Figure 12-32. Neon lamps used in a dual relation oscillator circuit.

Parallel-T Oscillator

In this circuit, Q1 is made to oscillate by feeding part of the signal back through the RC network C2, C4, R4, R5, R6, and R7. R7 is the frequency controlling means, and C3 is a coupling capacitor used to feed the signal back to the emitter of Q1. Q2 is an emitter follower amplifier that isolates Q1 from the phase shifting network for better stability. Output is taken from the collector of Q2.

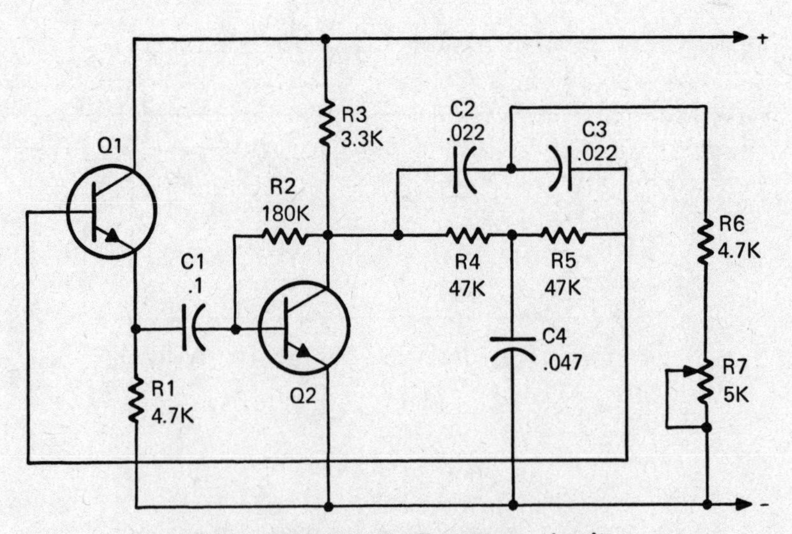

Figure 12-33. Parallel-T oscillator circuit.

Phase Shift Oscillator

The audio oscillator shown in Figure 12-34 has been quite popular because of its inexpensive components and stability. The plate of the high mu triode tube is coupled to

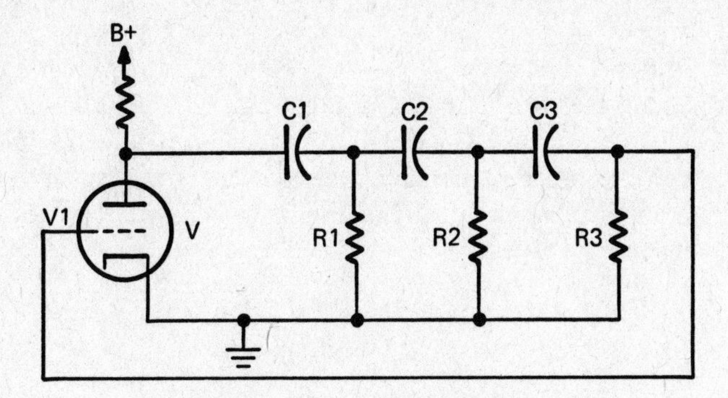

Figure 12-34. Phase shift oscillator circuit.

the grid through the R-C phase shifting ladder network consisting of C1-R1, C2-R2, and C3-R3. Phase shift is introduced by the three capacitors connected in series between the plate and grid of the tube, and the ground return resistors R1, R2, and R3. Oscillations will occur only at the frequency at which the total phase shift produces maximum positive feedback.

Pierce Crystal Oscillator

In the Pierce oscillator circuit shown in Figure 12-35, the crystal Y is the feedback path from the plate to the grid. Oscillation occurs at the series-resonant frequency of the crystal. Grid-leak resistor R must be chosen to make grid bias voltage high enough so that the net voltage across the crystal is low; otherwise the crystal may crack. Unlike most other crystal-controlled oscillators, no tuned plate tank circuit is required and oscillator frequency can be changed by using a different crystal. Plate current is fed through RF choke coil L. The RF voltage developed across L, which is not resonant at the oscillator frequency, is fed out through capacitor C.

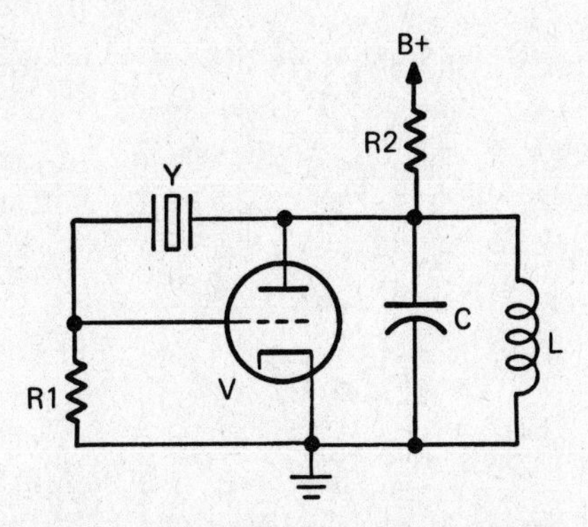

Figure 12-35. Pierce oscillator circuit.

Pierce-Miller Crystal Oscillator

The Pierce-Miller oscillator circuit shown in Figure 12-36 employs a triode tube and has a tank circuit composed of capacitor C and inductor L. The feedback capacitor C1 is actually the grid-plate capacitance of the tube. Since the crystal will vibrate only on the frequency for which it is cut, or a harmonic, the oscillator frequency is very stable if the supply voltages are maintained steady. This oscillator is basically a tuned-plate, tuned-grid oscillator with capacitive feedback.

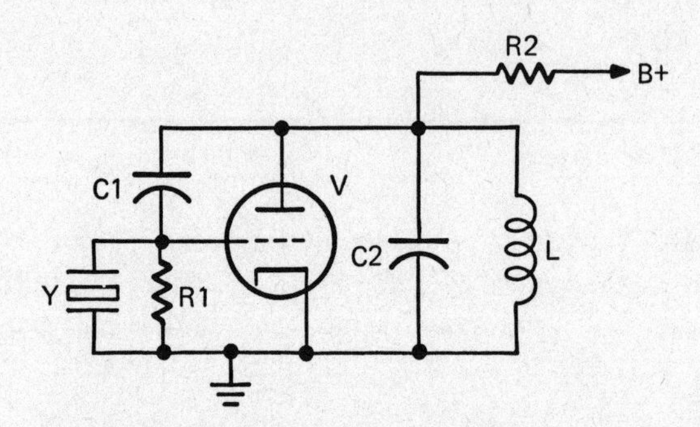

Figure 12-36. Pierce-Miller oscillator circuit.

Quadrature Oscillator

Some modulation schemes and servo systems require both sine and cosine signals from an oscillator, and these complementary signals are available from the circuitry shown in Figure 12-37. By using two operational amplifiers such as the Fairchild uA 741 integrated circuits used here, two outputs are obtained, each with less than 1 percent distortion.

Radiotelegraph Transmitter

Two transistors are used in the low-power CW (radiotelegraph) transmitter whose circuit is shown in Figure 12-38. Transistor Q1 is the tunable oscillator and Q2 is the Class C RF power amplifier. Meter M is a DC milliammeter which indicates Q2 collector current and facilitates tuning of the oscillator tank L1 for maximum drive to Q1 and the tuning of the RF amplifier tank L3 for optimum loading. Both transistors are turned on and off by the telegraph key when transmitting code signals.

RC Oscillator

In low-frequency oscillator circuits where large inductors are clumsy and undesirable, resistors and capacitors can be used. The basics of the low-frequency RC oscillator

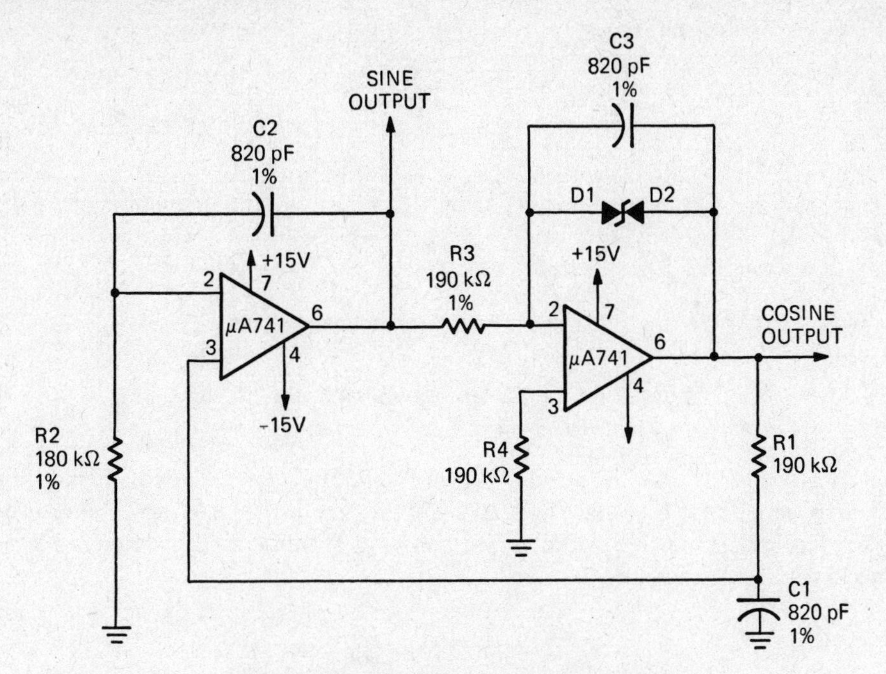

Figure 12-37. Quadrature oscillator circuit.

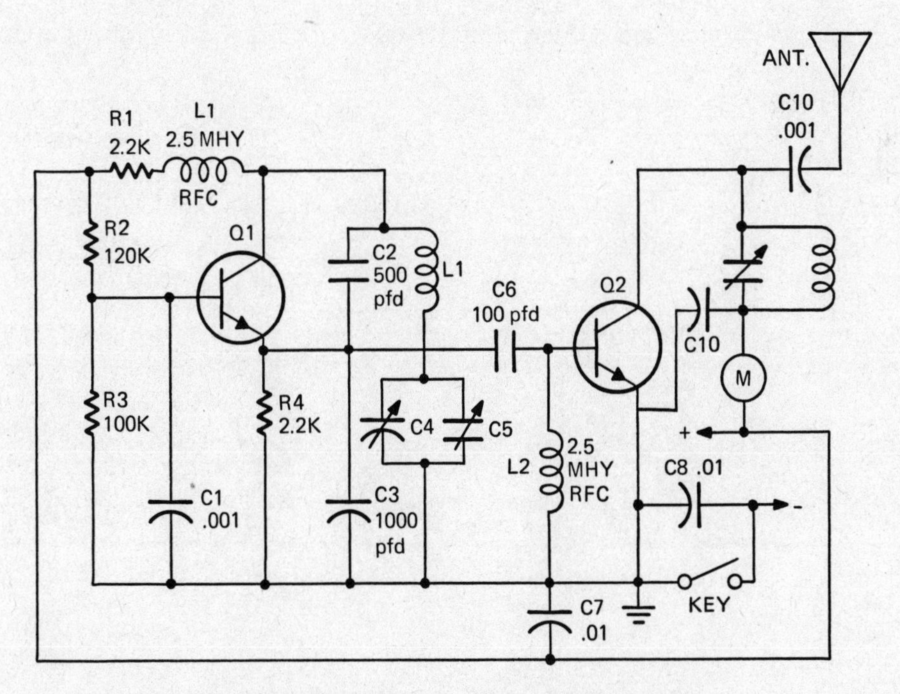

Figure 12-38. Low-power CW transmitter circuit.

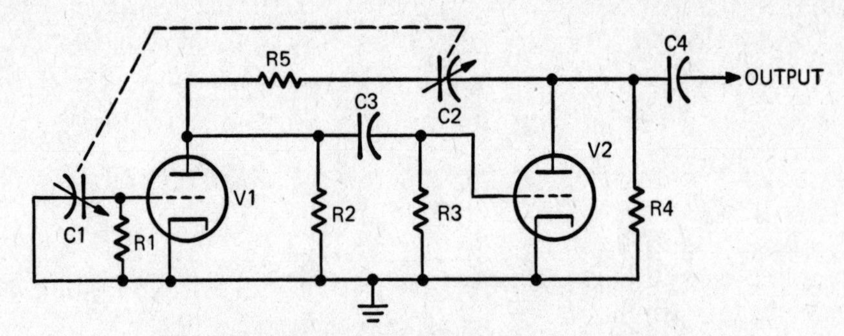

Figure 12-39. Low-frequency RC oscillator circuit.

employing two triode tubes is shown in Figure 12-39. Tuning of this oscillator is accomplished by varying the capacitances of C1 and C2. Time is the reciprocal of frequency. Positive feedback is from the plate of V2 to the grid of V1 through C2 and R5. The output signal is developed across R4.

Saturated Flip-Flop

The unclamped saturated flip-flop circuit shown in Figure 12-40 is an example of an Eccles-Jordan flip-flop. The output voltage levels of this type of flip-flop are practically independent of the transistors and the circuit is simple. The disadvantage is that the turn

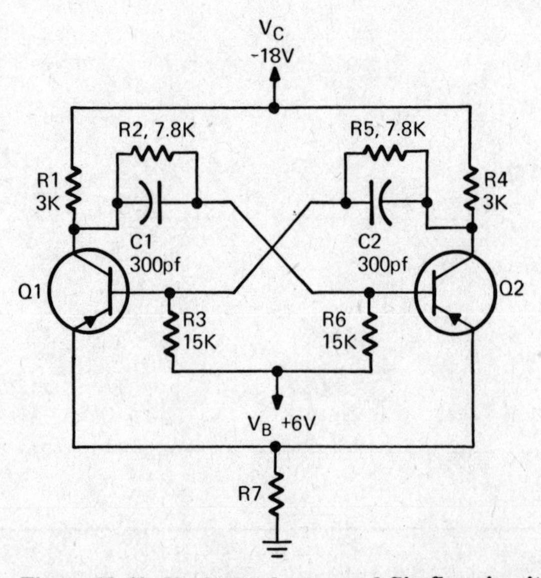

Figure 12-40. Unclamped saturated flip-flop circuit.

off transient is delayed by the storage time. Saturated flip-flops are for low frequency operation (i.e. 200 kHz or lower).

The components are balanced so that R1 is equal to R4, R2 is equal to R5, and R3 equal to R6. The transistors are the same type and as nearly alike as possible. However, if they were exactly alike, the circuit would not function. The dissimilarities of the transistors cause one to conduct before the other when power is turned on. If Q1 conducts first, it will drive Q2 further into cutoff. At this time, the collector of Q1 is near ground potential and C1 is discharging. When the voltage at the base of Q2 starts to go negative, Q2 begins to conduct with a positive-going waveform at the collector which is transferred to the base of Q1 through C2-R5, turning Q1 off and driving Q2 harder into conduction. When C2 has discharged to a slightly negative value, the base of Q1 is again forward biased and the procedure repeats itself. Resistor R7 can be eliminated and the emitters connected directly to ground. However, R7 does improve stability.

Sawtooth Generator

The sawtooth generator illustrated in Figure 12-41 employs a UJT (unijunction transistor) and a capacitor to generate a sawtooth waveform at the output. When power is applied, C_T charges until it reaches a level sufficient to fire the UJT. After the capacitor has discharged quickly through the unijunction, it begins to charge again, creating a sawtooth waveform at the output.

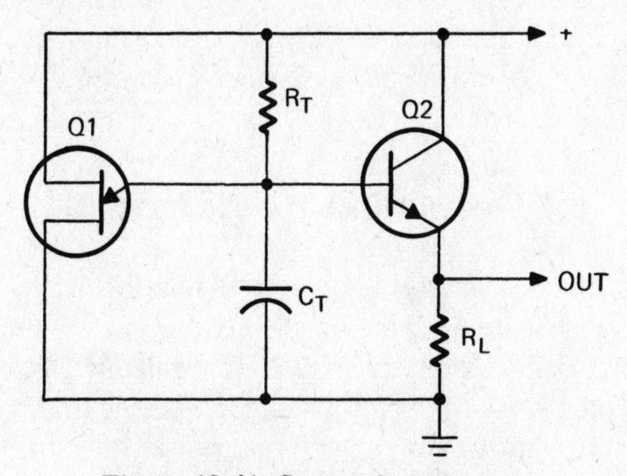

Figure 12-41. Sawtooth generator.

Screen Coupled Phantastron

Figure 12-42 shows one type of trigger circuit which is triggered by positive pulses. The screen coupled phantastron is monostable but it can be astable or bistable. Only tubes such as type 6AS6 are satisfactory for this circuit. The suppressor is coupled to the screen through R2 and C2, and operation is similar to that of a relaxation oscillator. The bias requires that a positive trigger pulse be applied to the suppressor grid to start each cycle.

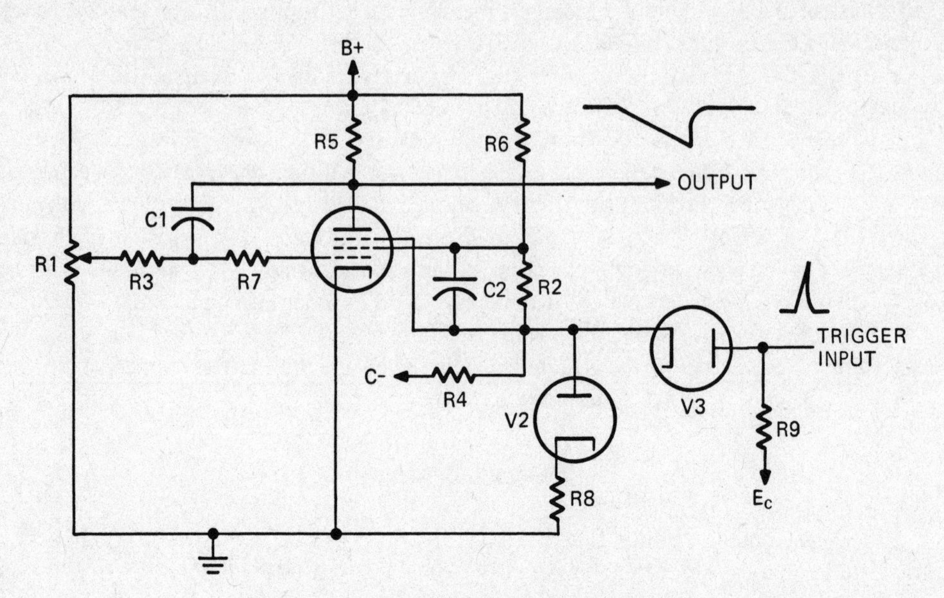

Figure 12-42. Screen coupled phantastron.

Capacitor C1 provides a feedback path from the plate to the control grid. The setting of potentiometer R1 determines the amount of drive to the control grid and the height of the output pulse. V2 and V3 are clamping diodes. V2 ensures that the suppressor cannot go more positive than the cathode when pulses are being generated. V3 decouples the trigger source during the time plate voltage is changing.

Self-Modulated Tunnel Diode Transmitter

The transmitter whose circuit is shown in Figure 12-43 generates an amplitude modulated signal that can be used for remote control purposes. The RF carrier frequency is determined by the crystal, C2, and L2, and the modulating frequency by L1 and C1. The AM signal is coupled to the whip antenna through C3 and loading coil L3. In addition to remote control applications, this transmitter can be used in security alarm applications. The power supply source can be a 1.5 volt flashlight cell. (Illustration courtesy of General Electric Company.)

Self-Rectifying Transmitter

Before it was outlawed by government regulation, many radio telegraph transmitters did not have DC power supplies for furnishing plate voltage. Instead, AC was fed to the plates through a transformer as shown in Figure 12-44. In this circuit, the two parallel triodes are used in a Hartley oscillator circuit. When the AC voltage fed to the plate of V1

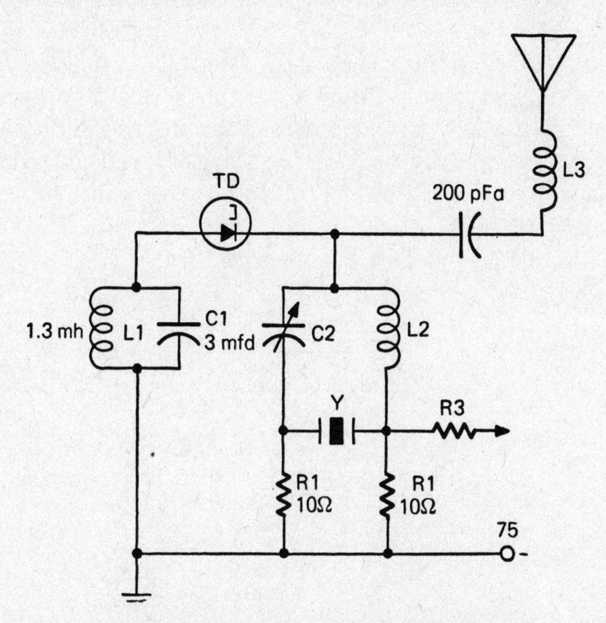

Figure 12-43. Self-modulated tunnel diode transmitter circuit.

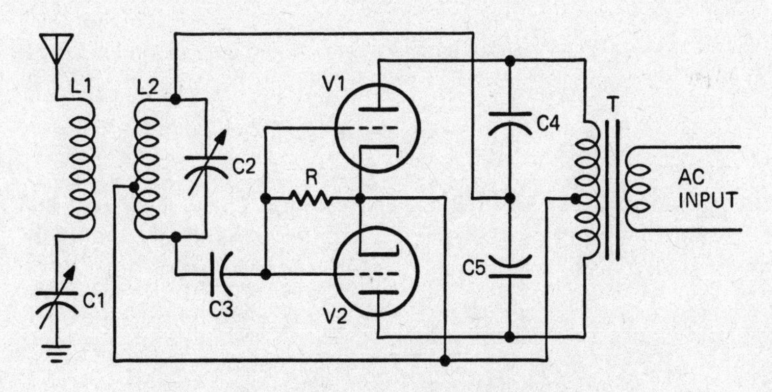

Figure 12-44. Self-rectifying transmitter circuit.

is positive, it operates and V2 does not operate because its plate is negative. During the next half cycle, the plate of V2 is positive and the plate of V1 is negative, and therefore V2 operates. If only one tube were used, the generated RF signal will be modulated at the 60 Hz rate. By using two tubes in this manner, the modulation frequency is 120 Hz since the tubes are part of a full wave rectifier circuit.

The carrier frequency is determined by L2 and C2. The antenna circuit is tuned to resonance with variable capacitor C1.

Series Tuned Emitter, Tuned Base Oscillator

Series resonant LC networks can be used effectively in frequency-stable transistor oscillators. In the circuit shown in Figure 12-45, the oscillator frequency is dependent almost entirely on the LC tank circuit elements and is only slightly affected by transistor properties or oscillator supply voltages. The frequency of oscillation is determined by two series resonant tank circuits L1-C1 and L2-C2. It is necessary that the reactance of C2 be larger than the reactance of L2 to provide a leading reactance. This provides stability without the use of elaborate non-linear components.

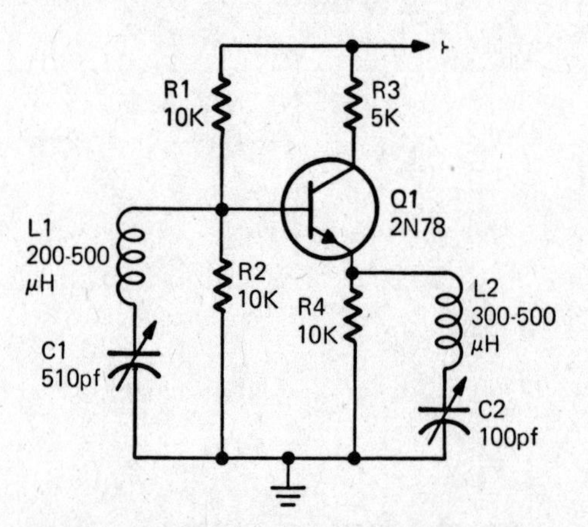

Figure 12-45. Series tuned emitter, tuned base oscillator circuit.

Spark Transmitter

Before the development of high power electron tubes, shipboard and land-based radio telegraph stations employed spark transmitters. As shown in Figure 12-46, the circuitry of a spark transmitter is extremely simple. Transformer T steps up the AC supply

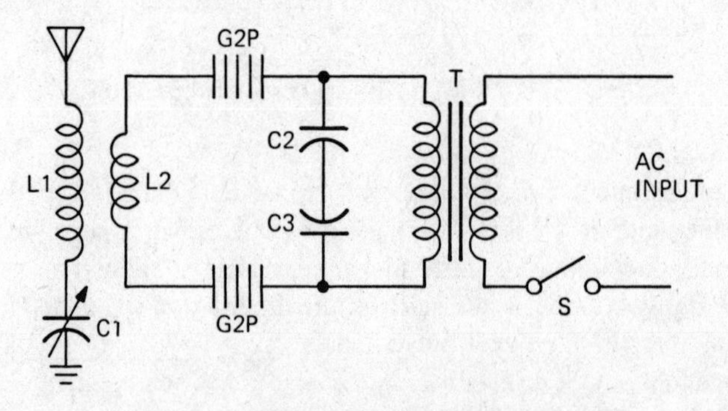

Figure 12-46. Spark transmitter circuit.

voltage to an extremely high voltage—high enough to cause sparks to jump to two spark gaps G2P which are in series with L2. Damped wave oscillations are generated in L2 and are inductively coupled to L1, the antenna coil. The antenna circuit is tuned to be resonant at the transmitter frequency by adjustment of variable capacitor C1.

The use of spark transmitters is now unlawful because they have excessive band occupancy and can cause serious interference to other communications services. In most cases, the AC input voltage was supplied by a 500 Hz alternator which was driven by a DC motor on shipboard or by a 60 Hz AC motor at land-based stations.

Stabilized Transistor Multivibrator

The multivibrator shown in Figure 12-47 is similar to the conventional astable transistor multivibrator, but the addition of a synchronizing pulse through the diode to the base of transistor Q1 stabilizes the frequency of the multivibrator to external circuitry. A

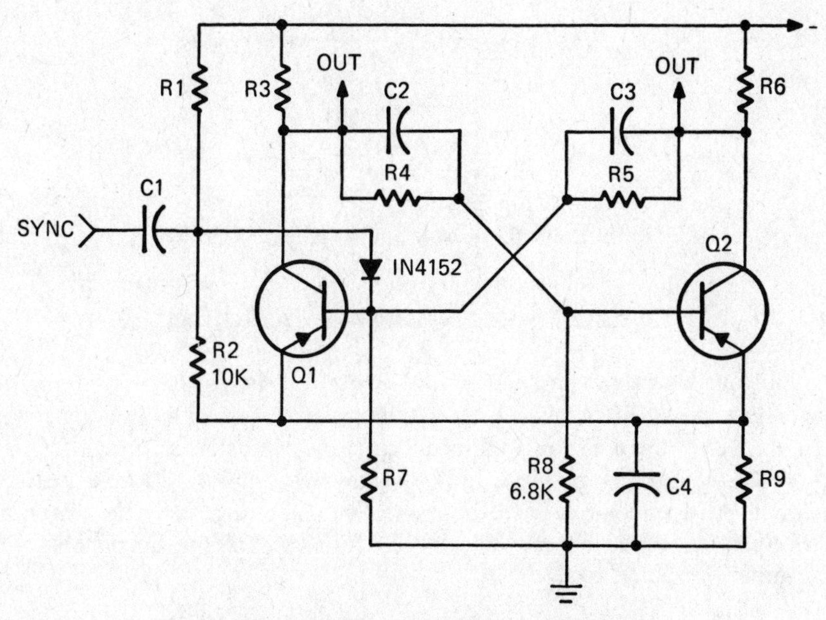

Figure 12-47. Stabilized transistor multivibrator.

positive sync pulse causes Q1 to cut off, raising the collector voltage. Capacitor C2 charges until it reaches a level at which transistor Q2 is cut off, and the resultant high negative potential on the collector of Q2 is felt through C3 at the base of Q1. This forward-biases transistor Q1 until another positive sync pulse is applied, or until C3 charges to a level sufficient to reverse-bias the base of Q1. In practice, the time constant of C3 is selected so that the sync pulse, rather than the charge on C3, will cause transistor Q1 to cut off.

Transistor Audio Oscillator

The audio oscillator shown in Figure 12-48 is a series-fed Colpitts type used for audio frequencies. An unusual feature of this circuit is that C1 and C2 are of equal value. Output resistor R4 will be selected depending upon the load.

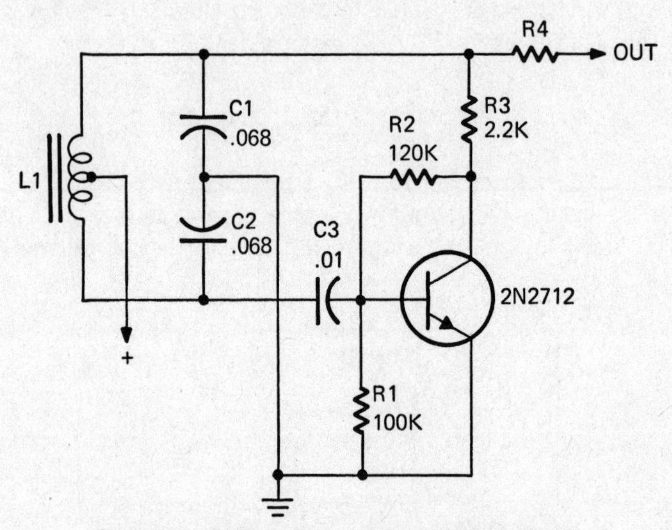

Figure 12-48. Transistor audio oscillator circuit.

Transistor Overtone Crystal-Controlled Oscillator

A third overtone series resonant crystal is used in the oscillator whose circuit is shown in Figure 12-49. RF output is greater than 1 volt RMS across a 2200-ohm load. The feedback path from the collector to the emitter to the transistor is through C1 and the crystal. Since the base is grounded at the signal frequency by C5, the transistor is connected in the common-base configuration. The output signal is taken from a low impedance point on collector tank coil L1. (Illustration courtesy of International Crystal Manufacturing Company, Inc.)

Transistor Phase-Shift Oscillator

Transistor Q, together with its bias network of R1 and R2, forms a simple amplifier, as shown in Figure 12-50. When the phase-shift network of C1, C2, C3, R3, R4, and R5 is added, the circuit becomes a simple phase-shift oscillator. Oscillation will occur at a frequency where there is 360 degree total shift. 180 degrees of this shift is provided by the grounded emitter amplifier Q, and the remaining 180 degrees is furnished by the high pass network. By controlling the value of the RC time constant, R4 provides a means of frequency control.

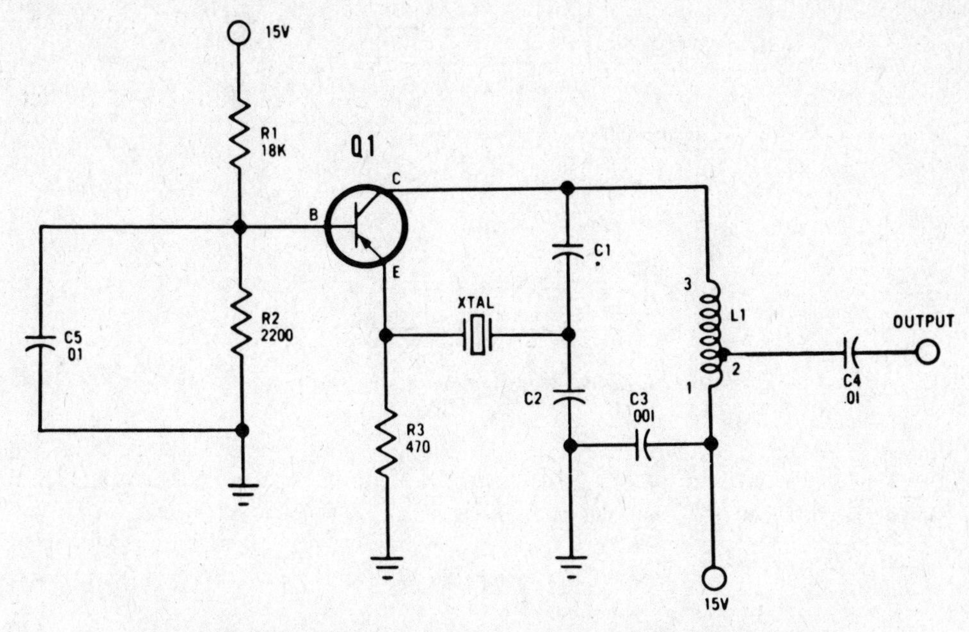

Figure 12-49. Transistor overtone crystal-controlled oscillator circuit.

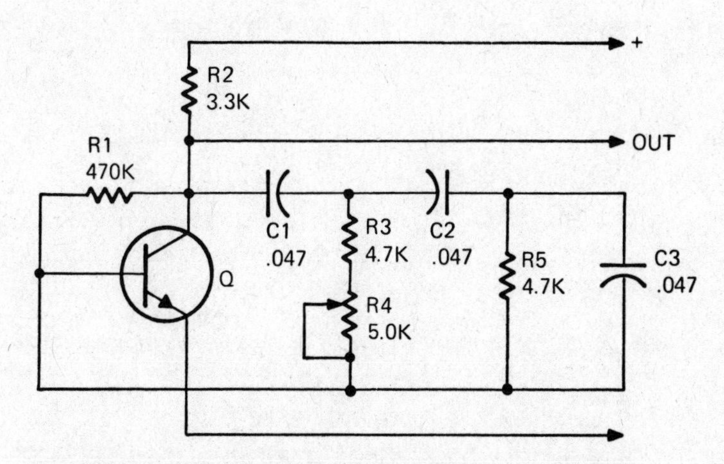

Figure 12-50. Transistor phase-shift oscillator circuit.

Transitron Oscillator

The transitron oscillator, whose circuit is shown in Figure 12-51, is an improvement over the dynatron oscillator. Its operation depends upon the space current returned to the screen by applying a high negative voltage on the suppressor grid. This current is reduced as the suppressor potential rises, and positive feedback results if these two electrodes

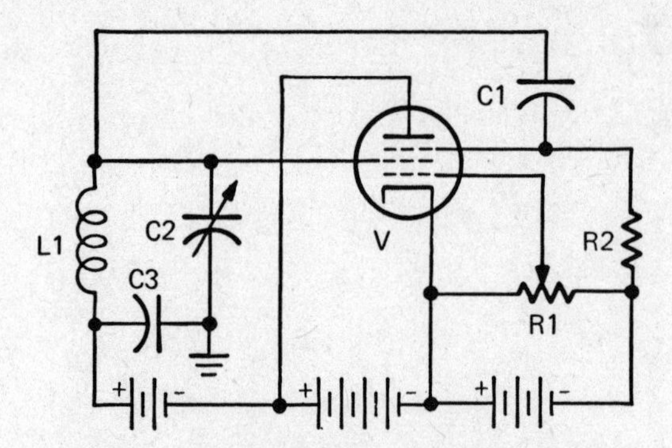

Figure 12-51. Transitron oscillator circuit.

(suppressor and screen grid) are connected through a feedback capacitor (C1). Oscillating frequency is determined by tank circuit L1-C2.

Tuned Grid Oscillator

The triode oscillator circuit shown in Figure 12-52 is regarded as one of the fundamental electronic circuits. It is actually an amplifier whose output is fed back inductively to the input. The oscillation frequency is determined by the values of L2 and C1.

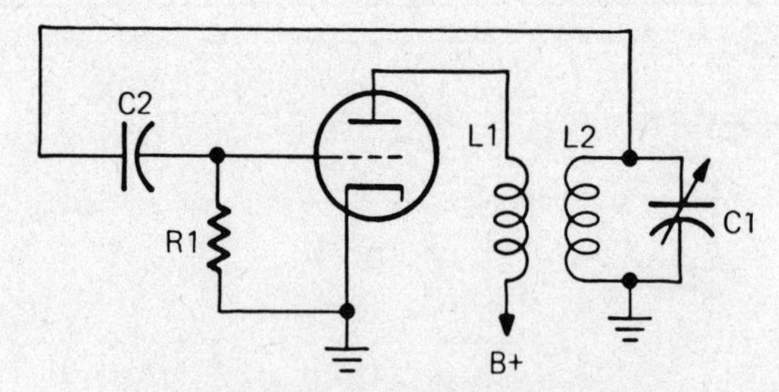

Figure 12-52. Tuned grid oscillator circuit.

Tuned Grid, Tuned Plate Oscillator

The tuned grid, tuned plate oscillator circuit employs a tuned tank circuit at both the grid and plate, as shown schematically in Figure 12-53. Feedback is through the internal grid-plate capacitance of the tube. Oscillation occurs when both tank circuits are tuned to almost the same frequency.

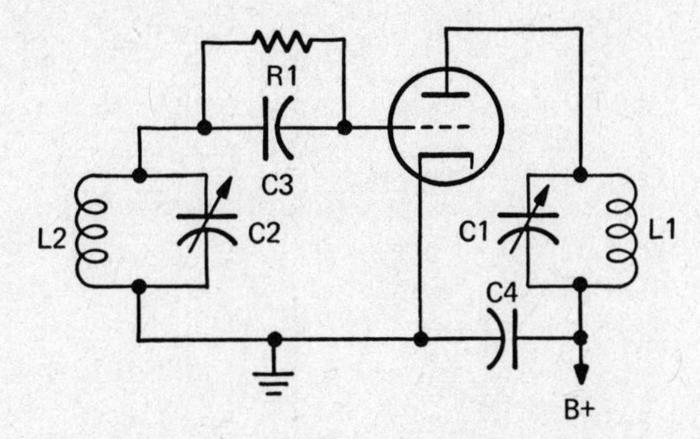

Figure 12-53. Tuned grid, tuned plate oscillator circuit.

Tuned Plate Triode Oscillator.

The oscillator circuit shown in Figure 12-54 employs a resonant plate tank circuit, inductively coupled to a tickler coil in the grid circuit so that positive feedback will cause oscillations to occur.

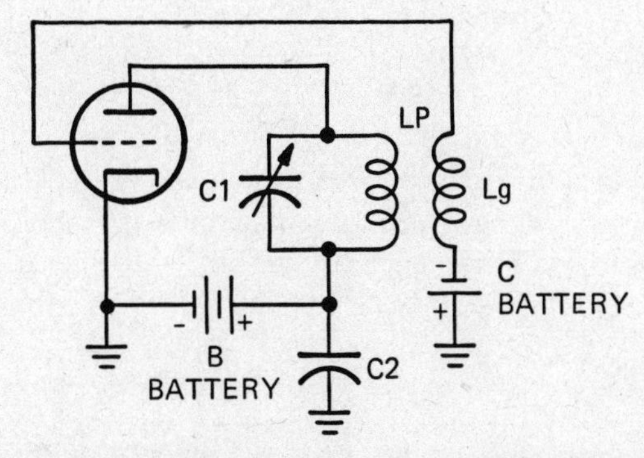

Figure 12-54. Tuned plate triode oscillator circuit.

Tunnel Diode Hybrid Square Wave Generator

A tunnel diode and a bipolar transistor are used in the RC oscillator circuit shown in Figure 12-55. The transistor base is forward-biased through R1 and R2. The frequency of oscillation is determined by the time constant of R2 and capacitor C. Using the values of components indicated in the diagram, the duration of each square wave cycle is 300 microseconds—frequency is approximately 3 kHz. (Illustration courtesy of General Electric Company.)

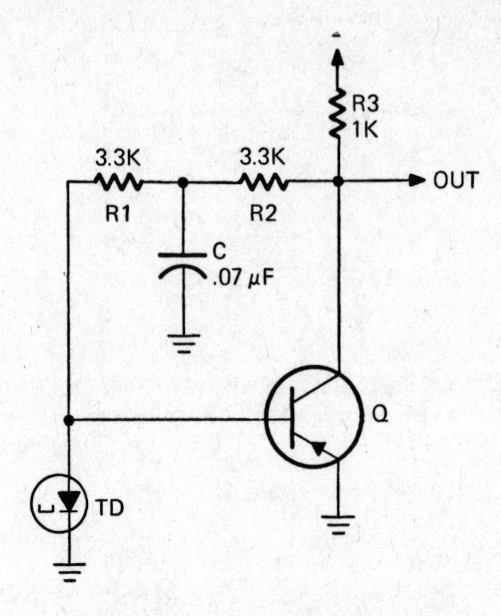

Figure 12-55. RC oscillator circuit.

Tunnel Diode FM Transmitter

A tunnel diode is used as the FM oscillator in the simple FM transmitter circuit shown in Figure 12-56. This circuit is based on an FM wireless microphone circuit developed by General Electric Company. Transmitter carrier frequency is determined by L2 and C2, and antenna loading by L1 and C1. The output of a low impedance microphone is connected to jack J and coupled to the base of the transistor through C5. The

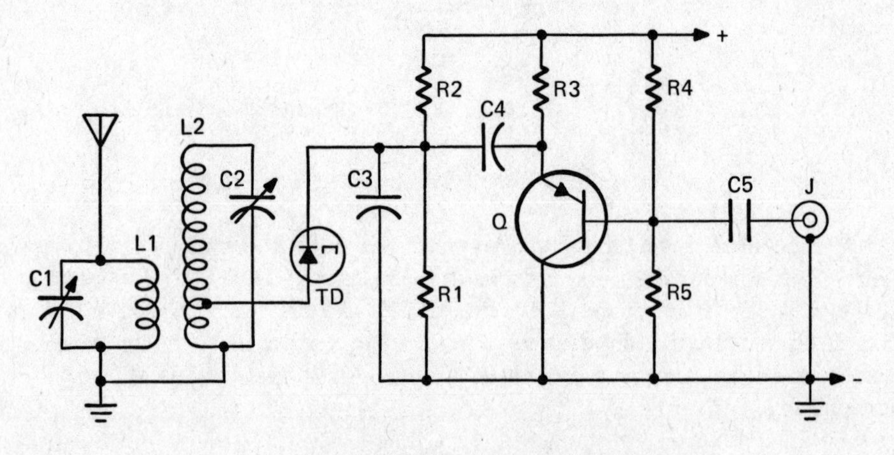

Figure 12-56. Tunnel diode FM transmitter circuit.

amplified audio signal is coupled from the emitter of the transistor through C4 to the tunnel diode. Tunnel diode bias is obtained from the junction of R2 and R1. The power source can be a 1.34 volt mercury cell.

Tunnel Diode Voltage Controlled Oscillator

A voltage variable capacitor (Varicap, Varactor, etc.) is used to control the frequency of the tunnel diode oscillator circuit shown in Figure 12-57. As the input voltage is varied from 0 to 100 volts, the varactor tunes the oscillator frequency through the 12-22 MHz range when using the component values shown in the diagram. A 1.5 volt flashlight cell can be used as the tunnel diode power source and is connected to the + and - points shown at the left side of the diagram. (Illustration courtesy of General Electric Company.)

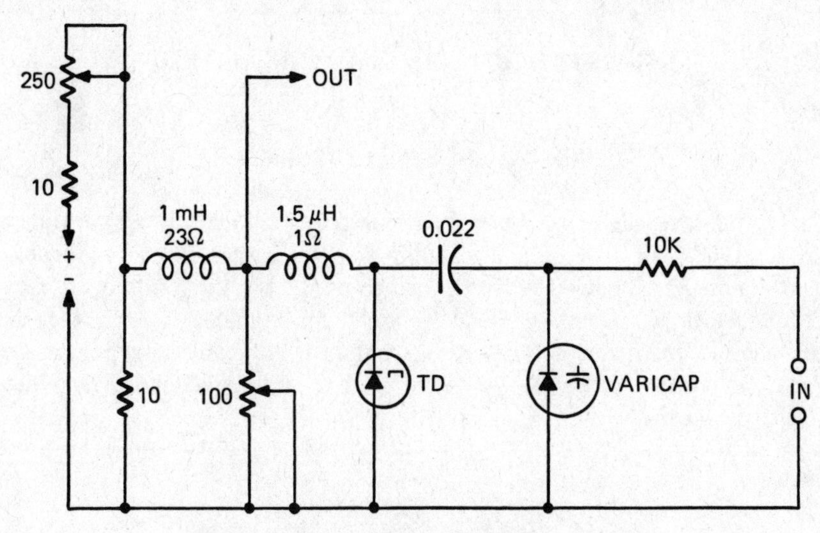

Figure 12-57. Tunnel diode oscillator circuit.

UJT Relaxation Oscillator

Figure 12-58 shows the basic unijunction transistor (UJT) relaxation oscillator circuit. When switch S1 is closed, the UJT is initially reverse-biased. However, as capacitor C1 charges through resistor R1, the positive potential at the emitter of the unijunction transistor Q1 rises until it becomes forward biased. As soon as it is forward-biased, the resistance between B1 and B2 drops, and C1 discharges quickly through R3 and the base-to-emitter (B1 to E) path of the UJT. The waveform at the emitter of the unijunction transistor will be a sawtooth, following the slow charge and rapid discharge of the capacitor C1. A narrow positive pulse appears at B1 because of the high current through through R3 as C1 discharges. A negative pulse appears at B2 as the unijunction transistor "fires" and increases the current through R2.

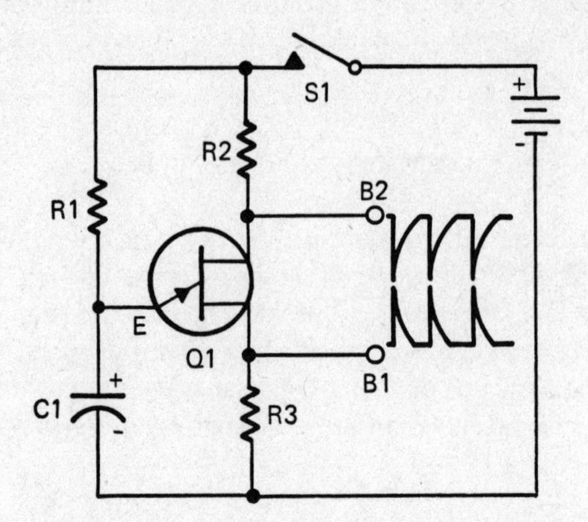

Figure 12-58. UJT relaxation oscillator circuit.

VHF Crystal-Controlled Oscillator

The crystal-controlled oscillator circuit shown in Figure 12-59 will operate at frequencies up to 60 MHz when utilizing a third overtone crystal. Plate load inductor L is adjusted to the required output frequency, which is equal to the fourth harmonic of the crystal's fundamental frequency. (Note: overtone and harmonic differ in that the first overtone is the second harmonic, the second overtone is the third harmonic, and the third overtone is the fourth harmonic.) (Illustration courtesy of International Crystal Manufacturing Company, Inc.)

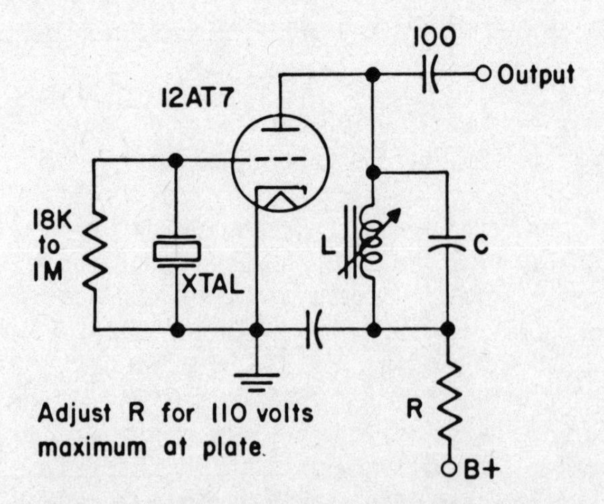

Figure 12-59. Crystal-controlled oscillator circuit.

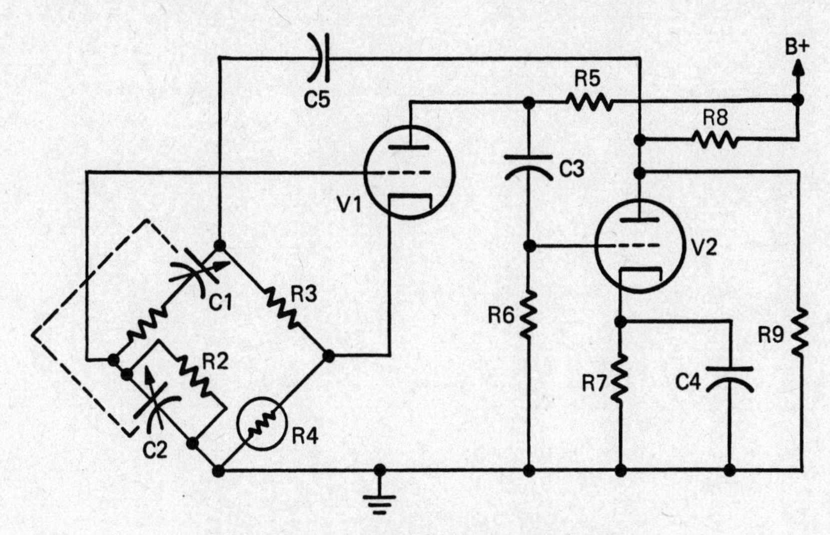

Figure 12-60. Wein bridge oscillator circuit.

Wein Bridge Oscillator

Figure 12-60 shows the circuit of a Wein bridge oscillator which consists of a two stage resistance-coupled amplifier with feedback from the plate of the second tube V2 to the grid of the first tube V1 through the RC bridge. The oscillator functions at a frequency at which the bridge is not quite balanced. If due to some condition a greater flow of current causes thermistor R4 to increase its resistance, the output voltage of the bridge between the cathode and grid of V1 is lowered, and the bridge again becomes balanced. Capacitor C3 provides signal coupling from tube V1 to tube V2.

Index